FLORA OF THE GUIANAS

Edited by

M.J. JANSEN-JACOBS

Supplementary Series
Fascicle 3

THE GUYANAN PLANT COLLECTIONS OF
ROBERT AND RICHARD SCHOMBURGK
(J.A.C. van Dam)

T0132845

2002
Royal Botanic Gardens, Kew

© 2002 The Trustees of The Royal Botanic Gardens, Kew.

First published 2002

ISBN 1 84246 046 3

Designed and typeset at
Media Resources,
Information Services Department,
Royal Botanic Gardens, Kew

Printed in the European Union by
Thanet Press Limited,
Margate, Kent, U.K.

Contents

THE GUYANAN PLANT COLLECTIONS OF ROBERT AND RICHARD SCHOMBURGK

by

JUUL A.C. VAN DAM[1]

1. INTRODUCTION

GENERAL INTRODUCTION

The first plant collections of real importance from Guyana were those of Robert and Richard Schomburgk (Ek, 1990). They explored the interior of British Guiana in the nineteenth century where they discovered many new species and collected many type specimens. Their historical collections are still of great importance. Those who study Guyanan plants will probably come across one of their collections. During their travels several series of plants were collected. Between 1835 and 1839 Robert Schomburgk made three journeys in the interior of British Guiana and made different series of plant collections. Between 1841 and 1844 his brother Richard joined him on his second trip to British Guiana, where they made four journeys and several short excursions. On this second trip, Richard started his own collection series, while Robert continued with his next collection series.

[1] Nationaal Herbarium Nederland, Utrecht University branch, Heidelberglaan 2, 3584 CS Utrecht, The Netherlands. http://www.bio.uu.nl/~herba/

Acknowledgements

The compilation of this book was done in 1999–2000 at the Utrecht University branch of the Nationaal Herbarium Nederland. It was sponsored by the Dutch organisation for Scientific Research (NWO) and supervised by Renske Ek, Ara Görts-van Rijn, and Paul Maas. I would like to thank the Alberta Mennega Stichting for the financial support for electronic equipment and travel expenses and for the additional publication costs for this book. The inclusion of the coloured pages would not have been possible without the grant of the Van Eeden-fonds and the remission of the costs by the Hunt Institute, Pittsburgh and the Natural History Museum, London.

I would like to acknowledge all the help and assistance that I received from many people during the course of preparation of this book. Great appreciation I have for Jan Lindeman, without his efforts I had not been able to publish this book and special thanks to Arian Jacobs-Brouwer for her preliminary work on the Schomburgk brothers. Furthermore, I would like to thank the specialists who gave their information about the Schomburgk collections they had seen during their study of their family for the Flora of the Guianas.

I would like to thank the curators of the herbaria at Berlin and London, for giving me a good hospitality. Especially I would like to mention the assistance of Inge Weinert, she already did a lot of preliminary work in the Pteridophytes herbarium in Berlin and Tini Versteegh for assisting me by photographing the herbarium specimens.

I would like to thank Paul Maas, Ara Görts-van Rijn, and Paul Hiepko for their help with the corrections of the manuscript and Hendrik Rypkema who made the coloured maps. I am very grateful to Peter Rivière for his inspiring information about Robert Schomburgk and for reviewing the manuscript.

The different collection series of the Schomburgk brothers make it rather complicated to find out if the right collector is Robert or Richard, especially when there is only 'Schomburgk' on the label. Those who study their collections will experience difficulty with their enumeration.

This book will list the different series of botanical collections made by Robert and Richard Schomburgk. For many years Dr. J.C. Lindeman from Utrecht had gathered Schomburgk data from a large source of taxonomic literature. For the compilation of the collection lists his data were used, combined with data from herbarium specimens from various herbaria and with information from specialists working on a revision for the Flora of the Guianas. This book also gives a short biography of both brothers and includes information, illustrations, and maps about their travels.

HISTORY OF GUYANA

Before achieving independence from Great Britain in 1966, Guyana is characterised by a varied history. Early Spanish explorers believed that the region was the location of El Dorado, a legendary land of Golden riches. In 1593 a group of 2000 Spanish people searched in vain for the hidden golden land. Between 1595 and 1616, the Englishman Sir Walther Raleigh led three expeditions to the Guyana territory in search for El Dorado. Although Raleigh failed to locate any gold, his efforts resulted in the first mapping of the Guyanese coastline and a description of the country.

The Dutch, at that time known as a trading and maritime nation, arrived in the seventeenth century to establish a colony. They were interested in securing good farmland to grow tropical crops like coffee, tobacco, and cotton. They discovered the great fertility of the soil along the Guyanan coast and many plantations were established. The Dutch succeeded in making the first permanent settlements on the Guyanan coast from where they could operate in the search for gold and trade products like tobacco and dye. The Dutch possessed three separate colonies: Demerara, Essequibo, and Berbice.

During the Napoleonic wars the colonies changed hands several times. In 1781 the British occupied them and a year later the French seized the three settlements. In 1784 the Dutch regained ownership of the colonies again but between 1796 and 1802 Britain had posession of them and from 1803 was the de facto owner. In 1814 European discords concerning the Guianas finally ended. The former three Dutch colonies, Demerara, Essequibo, and Berbice were ceded to Britain under the London convention. They were merged into a single colony, to form British Guiana in 1831. The French secured French Guiana while the Dutch took over the territory of present day Suriname.

Map 1. The expeditions of Robert Schomburgk 1835–1839.

Trip 1: Expedition to the Upper Rupununi River.

Trip 2: Ascent of the Courantyne and Berbice Rivers.

Trip 3: Expedition to the sourcve of the Essequibo River, Mount Roraima and Esmeralda.

4

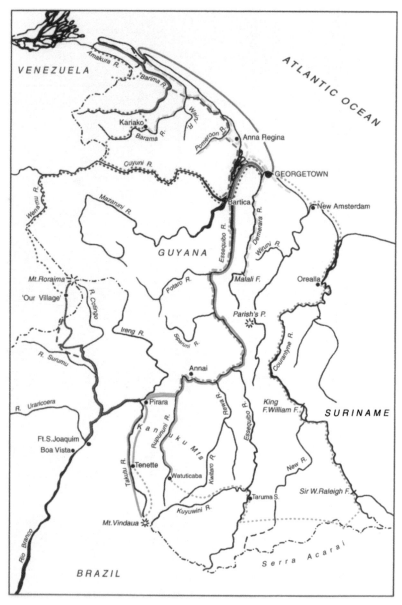

Map 2. The expeditions of Robert and Richard Schomburgk 1841–1844.

        Trip 1: Expedition on the Barima and Cuyuni Rivers.

        Trip 2: Expedition to the source of the Takutu River.

        Trip 3: Expedition to the Roraima Territory.

        Trip 4: Expedition to the Upper Courantyne River.

        Trip 5: Excursions to the Rivers Pomeroon, Moruca, and Demerara.

5

Plate 1. Post at the Cuyuni River.*

*Courtesy of Natural History Museum, London

6

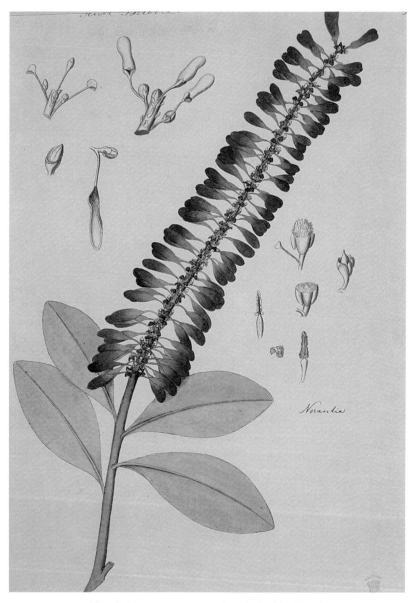

Plate 2. *Norantea guianensis* along the Berbice River.*

7

Plate 3. Arecuna Indians in front of the Mountain Range of Roraima.*

8

Plate 4. *Abolboda macrostachya* var *robustior.**

*Courtesy of Natural History Museum, London

9

Plate 5. Number 391 of Robert's first collection series; *Heteranthera reniformis*
(= *Eichornia diversifolia*).*

10

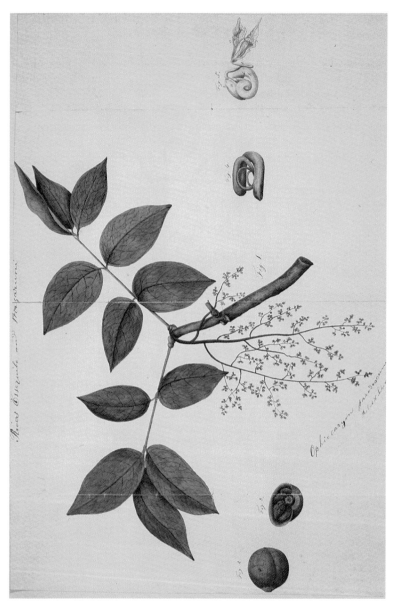

Plate 6. Snake nut tree: *Ophiocaryon paradoxum.* *

THE EXPLORATIONS OF THE BRITISH COLONY BY ROBERT AND RICHARD SCHOMBURGK

The geographical, ethnographical, botanical, and zoological relationship of British Guiana with the rest of the continent was relatively unknown. Robert Schomburgk, under the patronage of the Royal Geographical Society, conducted further surveys. Between 1835 and 1839 he made three separate journeys into the interior as a surveyor and geographer. In 1840 he produced a map of the British colony showing its boundaries. The publication of this map marked the beginning of the border dispute with Guyana's neighbouring countries. The British Government decided to take the matter of the border in hand, and employed Robert Schomburgk to survey the boundaries. The result has been that there is still disagreement over the frontiers of Guyana today.

On Robert's second trip to British Guiana his brother Richard, who was commissioned by the Prussian Government, accompanied him. Together they completed thousands of miles of exploration to reveal the resources and potentials of the British colony. The expeditions contributed to an increasingly available knowledge not only about the flora and fauna but also about the Amerindians and their customs. During their years of travel throughout British Guiana, Robert and Richard came to understand and appreciate the Amerindians, many of whom they employed as guides on their expeditions. Their writings referred to the customs, characteristics, and culture of the different Amerindian tribes and reveal great concern for them. Robert's writings show his constant concern about the fast decreasing tribes in the interior of British Guiana. He was therefore especially eager that any expedition sent to British Guiana should include an artist who could paint members of the different tribal groups and their artefacts. At the time of the expedition illness was causing catastrophic losses among Amerindian tribes. Robert referred to thirteen different tribes; Arawaks, Warraus, Caribs, Accawais, Macusis, Arecunas, Wapisiana, Atorais, Tarumas, Woyavais (Wai-Wais), Maopityans, Pianoghottos, and Drios (Goodall, 1977). Some of these tribes had died out completely by the beginning of the twentieth century. The Schomburgk brothers' detailed descriptions of the Indian tribes and their way of life and Goodall's paintings have left an anthropological record of great importance. Also Richard's careful observations and descriptions of the dress, physical features, living conditions, food customs, languages, and legends of the Amerindians, are now of great value from both an historical and anthropological point of view.

2. ROBERT SCHOMBURGK (1804–1865)

Robert Hermann Schomburgk (figure 1) was born on June 5th 1804 at Freyburg (Freiburg) on the Unstrut River in Thuringia (since 1815 belonging to Prussia), about 50 kilometres West of Leipzig. He was the eldest son of Christiane Juliane Wilhelmine Krippendorf and Johann Friedrich Ludwig Schomburgk. The family consisted of Robert, followed by his sister Linna Theresia (1806), and his brothers Alfred Otto (1809), Moritz Richard (1811) and his youngest brother Ludwig Julius (1819). Robert's father was a Lutheran minister at Freyburg from 1801 to 1820. Robert went to primary school in Freyburg. His family wanted him to enter into business. He was educated therefore in Latin, French, arithmetic, and other sciences which were needed to become a businessman (Schomburgk, R.H., 1859). Robert already showed great interest in nature as a child. While growing plants in his father's garden his affection for orchids started.

At the age of 14 Robert left home to become an apprentice to merchant Krieger in Naumburg. His family moved to Voigtstedt in 1820 where his father continued his work for the Lutheran church. Three years later Robert was sent to work with his uncle Heinrich Schomburgk, a merchant at Leipzig, and lived there until 1828. His uncle encouraged Robert's interest in botany and natural sciences. He gave him the opportunity to be educated in botany. Robert attended lectures and received tuition from professor Schwägrichen, but he never obtained any formal qualification (Rivière, 1998).

In 1828, after the death of his mother to whom he had promised to stay at home, he left for North America to further his commercial career. Shortly after his arrival in New York he entered into business in Richmond, Virginia. After an unsuccessful period as a tobacco producer in Richmond, he left for the island of Saint Thomas in the West Indies in 1829. The somewhat tropical-like vegetation of Richmond already made Robert longing for the warmer regions of the world. At Saint Thomas Robert wanted to extend his business career. A disastrous fire in 1830, however, destroyed most of his belongings and business. This event persuaded him to stop his career as a businessman and to devote himself to botany and natural history.

Robert left for Puerto Rico and travelled around the Virgin Islands, where he visited Anegada. During his visit of this area, which is notorious for shipwrecks and known for both dangerous reefs, and pirate activities, three shipwrecks occurred. The sinking of the Spanish ship called 'Restauradora' with 130 African slaves on board convinced Robert to chart and survey the island in order to prevent future shipwrecks. Between 1830 and 1831 Robert surveyed and mapped the area of the Anegada Passage at his own expense. According to Robert the current maps of Anegada were incorrect, and he decided to make a

13

new one (Schomburgk, R.H., 1859). After a survey of the island he submitted the map and the description of the island to the Royal Geographical Society. This very accurate survey of the island of Anegada was published in two volumes of the Journal of the Geographical Society (Schomburgk, R.H., 1832). His chart and memoranda brought him to the attention of the Admiralty and the Royal Geographical Society. He was asked to conduct further surveys. The secretary of the Royal Geographical Society Alexander Maconochie consulted with John Lindley, Professor of Botany at London University, who suggested

Figure 1. Robert Hermann Schomburgk.*

that Schomburgk should rather go to South America as good collections were already being made in North America (Rivière, 1998). The Royal Geographical Society commissioned him to explore the interior of British Guiana during the years 1835–1839. Robert's interest in British Guiana has been stimulated by the writings of Alexander von Humboldt (1769–1859), who travelled in South America during the early part of the nineteenth century. After reading Humboldt's 'Reise in die Äquinoctial-Gegenden des neuen Continents' Robert realised that Humboldt had travelled to Esmeralda on the Upper Orinoco River, but that the basin of the Essequibo River, Rio Branco, and Courantyne Rivers were still undefined. The Royal Geographical Society of London considered it a matter of importance, to connect Humboldt's observations with those made on the coast of British Guiana, and for this purpose Robert was chosen to lead the explorations in the British colony. The expedition was to have two distinct goals: first, to thoroughly investigate the physical and astronomical geography of British Guiana, and second, to connect his positions with those of Humboldt on the Upper Orinoco River (Rodway, 1889). While, it is not known how much contact existed between Robert Schomburgk and Humboldt before Robert's first expedition to British Guiana, they have certainly collaborated between the two expeditions. Alexander von Humboldt wrote an introduction to Robert's 'Reisen in British-Guiana und am Orinoko während den Jahren 1835–1839' (Payne, 1992).

*Courtesy of Hunt Institute of Botanical Documentation, Carnegie Mellon University, Pittsburgh, Pennsylvania, U.S.A.

Robert Schomburgk's early expeditions in 1835–1839 were notable for tracing the Essequibo River to its source and connecting his survey with that of Humboldt at Esmeralda. The description of British Guiana was published in 1840 (Schomburgk, R.H., 1840a). During these extensive travels in British Guiana, Robert had proved to be a reliable and knowledgeable leader. The Royal Geographical Society in London awarded him their highest award, the gold medal for the year 1839 for his services to geography, zoology, and botany. The King of Prussia decorated him with the order of the Red Eagle (Rodway, 1889; Schomburgk, R.H., 1859).

As a result of his knowledge about the interior of British Guiana he was appointed boundary commissioner by the British Government. The boundary commission worked between 1841 and 1844. Richard, his younger brother was recommended by Alexander von Humboldt to the King of Prussia to join this expedition as both collector and historian. Together they surveyed the boundaries of the colony. Robert pleaded that the boundary of British Guiana should be defined 'for the benefit of the Amerindians' (Schomburgk, R.H., 1841d). Back in Georgetown, after completing thousands of miles of exploration, Robert received a letter from the Geographical Society in Paris. He was rewarded with a silver medal for his researches in British Guiana (Lasègue, 1970). While residing in Georgetown up to the time of his departure Robert took great interest in the formation of the Agricultural and Commercial Society, of which he was elected an honorary member on the 1st of May 1844. He fitted up an observatory where he carried out his astronomical observations and supported the establishment of the Botanical Gardens in Georgetown.

In 1844 Robert returned to Britain. Queen Victoria knighted him for his services in British Guiana in 1845. In October 1845, one year after his brother Richard, Robert became a member of the German Academy of Science (Kaiserlich Leopoldino-Carolinische Deutsche Akademie der Naturforscher); he chose 'Aublet' as his academic epithet (Anonymus, 1860). One year later he left for Barbados as a diplomat for the British Government. After a survey of eleven months on the island Robert constructed a map and wrote the 'history of Barbados'. He also went to Santo Domingo (Dominican Republic) as British Consul until 1857. From 1857 until 1864 his next post was in Bangkok as Consul General. Robert did not loose his interest in botany. Robert's plant collections from the Antilles, Singapore, Malaysia, and Bangkok can still be found at different herbaria all over the world. In 1864 Robert retired to Germany, with his health broken, and died in Berlin on 11 March 1865.

3. RICHARD SCHOMBURGK (1811–1891)

Moritz Richard Schomburgk (figure 2) was born on 5 October 1811 at Freyburg on the Unstrut River in Thuringia (since 1815 belonging to Prussia). He was the fourth child and third son of the Schomburgk family. Richard Schomburgk was educated at a Freyburg primary school and by a private tutor. At the age of 14 Richard began a gardening apprenticeship at Merseburg, about fifty kilometres east of Voigtstedt. This was followed by his military service with the Royal Prussian Guard at Berlin from 1831–1834. Richard worked in the royal gardens of Sanssouci at Potsdam from 1835 until 1840. While Richard was working at Sanssouci, he got experience in growing vines, fruit trees, and vegetables as well as in the care of shrubs and trees and in the cultivation of popular flowers. There is no definite evidence that Richard received any formal training in botany beyond the little that would be given to any apprentice gardener. He was familiar with a variety of landscape design styles in Berlin and thoroughly trained in techniques of propagation and garden management (Payne, 1992).
Some time during his years at the royal gardens of Sanssouci, Richard came to the attention of Alexander von Humboldt. Thanks to the influence of Alexander von Humboldt Richard was encouraged to accompany his brother to British Guiana (Schomburgk, M.R., 1848). When Robert was finally appointed by the British Government to lead the 1840–1844 expedition, Richard was commissioned by the Prussian Government to make collections for the royal museums and the Botanical Gardens in Berlin. From the Museum of Natural History in Berlin Richard received some training in collecting specimens.
Richard and Robert Schomburgk left for British Guiana on October 29th 1840. Not long after Richard arrived in Georgetown he contracted yellow fever. It was considered a fatal disease but miraculously, through the excellent medical attention of Dr. Koch of Nuremberg, he survived (Goodall, 1977). On the 19th of April 1841 the first expedition of the boundary commission set out to survey the Waini, Barima, Amacura, Barama, and Cuyuni Rivers, a trip lasting three and a half months.
In December 1841 the second expedition left Georgetown for the interior for the purpose of defining and exploring the Brazilian frontier (Rodway, 1889). It consisted of Richard, his brother Robert, Edward Goodall the artist, and Mr. Fryer, botanist and Robert's secretary who also acted as medical officer during the expedition. Beside these persons a cook and an Amerindian interpreter joined the expedition. They travelled from Pirara to the source of the Takutu River, navigated the Rio Cotingo to its source near Mount Roraima. Richard collected botanical and zoological specimens and kept written records for the subsequently published account of the expedition since he held the position of historian and botanist for the Prussian Government.

Figure 2. Moritz Richard Schomburgk.*

After an absence of four years Richard returned to Berlin with a large botanical, zoological, and ethnological collection in August 1844. Back in Germany, Richard wrote the three volumes 'Reisen in British Guiana', which were published in 1847–1848. The third volume comprised his Flora of British Guiana that was considered his most important work in the field of taxonomy (Britten, 1891). This volume covered not only the flora but also the fauna of British Guiana. Richard was assisted by J.F. Klotzsch, C.G.D. Nees von Esenbeck, F.G. Bartling, A.H.R. Grisebach, and C.H. Schultz (Schomburgk, M.R., 1876). W.E. Roth made a translation into English of the first two volumes in 1922. Richard had made important collections of both animals and plants. He was member of the Horticultural and Geographical Society in Berlin. The title of Doctor of Philosophy was conferred upon him by the German Academy of Science (Kaiserlich Leopoldino-Carolinische Deutsche Akademie der Naturforscher) in consideration of his achievements in botany and natural history.

In 1849 Richard emigrated to Australia with support of Alexander von Humboldt and Leopold von Buch, because he could not find a position that corresponded with his knowledge and abilities. On the boat 'Princess Louise' he travelled together with his brother Otto to Adelaide, Australia. Both were married by the time. At the age of thirty-eight Richard had married Pauline Kneib. The voyage took five months and they reached Adelaide on the 7th of August 1849. The two families started a farm and vineyard. Richard and Pauline had 5 daughters and a son. Both Otto and Richard were active members of the Gawler Institute in South Australia and about 1860 Richard became curator of the Gawler Museum.

In 1865 Richard applied for the Adelaide Botanic Garden's directorship. On the 14th of September that same year he started and at the first meeting of his Board he outlined his plans for the garden (Lamshed, 1955). Richard gave his approval to a number of improvements, including the heating of the

---

*Courtesy of Hunt Institute of Botanical Documentation, Carnegie Mellon University, Pittsburgh, Pennsylvania, U.S.A.

Orchid house to ameliorate the quality and variety of the orchid collection. He realised the building of the Palm House (open 22 January 1877), the Museum of Economic Botany (open 27 May 1881), and a new Fern House. In January 1868 Richard and the board were planning for a new glasshouse, the Victoria House, which would enable the cultivation of the giant water lily *Victoria regia* (=*V. amazonica*). Richard Schomburgk was probably the most notable importer of plant material to Australia in his days. In 1878 the Adelaide Botanic Garden had one of the world's largest living plant collections. Their catalogue listed 8500 species at that time.

Richard had built a considerable herbarium. He received important collections from all over the world, for example from Paris, Chile, Spain, Kew, and St. Petersburg. In the last few years of his life the herbarium was of special interest to Richard. Until his death, he spent most of his time in the Museum working on his herbarium. Richard suffered from gout, a disorder of the joints causing painful swellings. At the age of seventy-nine he died in office in March 1891.

4. THE EXPEDITIONS OF ROBERT SCHOMBURGK 1835–1839

This chapter describes the expeditions of Robert Schomburgk. Between 1835 and 1839 he made three separate journeys into the interior of British Guiana, which are shown in map 1. His main activities are summarised in the following chapters. At the end of each chapter, dealing with one of the expeditions, a table is given, listing in chronological order the localities visited and some of the main events, and notes about the botanical collections. Localities outside the present-day border of Guyana are indicated with a country code of the relevant country. Within the tables several abbreviations are used; the meanings of which are explaned below. Several localities changed names or are written in different ways. In the tables and general text the alternative names for a locality are cited in brackets.

C.	= Cataract(s)	[BRA]	= Brazil
Cr.	= Creek	[COL]	= Colombia
F.	= Fall(s)	[GUY]	= Guyana
I.	= Island	[SUR]	= Suriname
Mt(s).	= Mountain(s)	[VEN]	= Venezuela
R.	= River, Rio, or Río		

EXPEDITION TO THE UPPER RUPUNUNI RIVER
(21 SEPTEMBER 1835–28 MARCH 1836)

Robert Schomburgk was selected by the Royal Geographical Society to explore the interior of British Guiana. His first visit to the interior was a short trip up the Cuyuni River in August 1835 to recruit paddlers for his canoes. He left Georgetown on September 21st 1835, on the expedition to the Rupununi district in the south of British Guiana. The crew consisted of three Europeans: Lieutenant James Haining, Robert Brotherson of Demerara, and Robert himself. There were also four Negroes, variously employed as attendants. The crew of the canoes consisted of five Negroes, five Caribs and two Accaways, and three Macusis. They first went to the Cuyuni Post (plate 1) at the junction of the Mazaruni (Massaroony) and the Cuyuni Rivers, and on October 1st they left there and started travelling up the Essequibo River until they reached the mouth of the Rupununi River. Robert reported to Georgetown that he had collected about 1500 plant specimens. They continued their journey up the Rupununi River to the Annai River. Robert did not succeed in finding a crew to travel higher up the Essequibo River; therefore he chose to proceed up the Rupununi River. On December 1st 1835 he started this trip and after visiting the Indian village of Pirara, he continued

Table 1. Expedition to the Upper Rupununi River.

Date	Location	Observations, collections, and main events
	Robert Schomburgk	
5 Aug 1835	arrives at Georgetown	
21 Sep 1835	departs from Georgetown reaches mouth of the Essequibo R.; passes Hog I. and Fort I. Ampa; Cuyuni R. and Mazaruni (Massaroony) R.	astronomical observations
25 Sep 1835	Cuyuni Post	collections of plants and birds
1 Oct 1835	ascends the Essequibo R.	
8 Oct 1835	foot of Mt. Arisaru (Arissaro)	
9 Oct 1835	mouth of Potaro R.; Benhori-Bumoko (Benhoori-Boomocoo) I.	
	Waraputa F.	Indian picture writings
15 Oct 1835	Akaiwanna and Takwari (Taquiari) or Kumut (Cumuti) Mts.	new orchid species
16 Oct 1835	halts at a large island Siparuni (Siparoony) R.	*Mikania angulata*
18 Oct 1835	Kurupukari (Ouropocari) F.	
19 Oct 1835	Mt. Makari (Maccari)	
22 Oct 1835	Rappu (Rappoo) F.	
23 Oct 1835	reaches mouth of the Rupununi R.	collection of 1500 plant specimens
24 Oct 1835	ascends the Rupununi R. Mt. Makarapan passes the Rewa (Roiwa or Illiwa) R.	drawings of fishes, birds, and plants
26 Oct 1835	streamlet Curassawaka; tributary of the Rupununi R.	
27 Oct 1835	Annai R. and Annai settlement	transport of collections to Annai
Nov 1835	Annai settlement	collects and preserves specimen
1 Dec 1835	starts trip to the Upper Rupununi R.; offset of the Parima Mts.	
2 Dec 1835	Pirara village near Lake Amuku Kanuku (Conocon) Mts.	
15 Dec 1835	starts journey across savanna	
22 Dec 1835	Cartatan; largest falls of the Rupununi R.	views the source of the Rupununi R.
25/26 Dec 1835	Aripai R.	searches for Urari; *Strychnos toxifera*
28 Dec 1835	starts on his return	
2 Jan 1836	Pirara village excursion to the R. Mahu (Ireng or Maú)	collects birds
15 Jan 1836	mouth of the Annai R.; camp at Curassawaka	
26 Feb 1836	departs from Curassawaka	
27 Feb 1836	junction of the Rupununi R. and the Essequibo R.	
29 Feb 1836	starts journey up the Essequibo R.	
5 Mar 1836	King William's C.	
7 Mar 1836	Sir George Murray's C.; inlet Primos	
8 Mar 1836	returns to camp at Rupununi R.	
9 Mar 1836	travels downstream the Essequibo R.	
11 Mar 1836	Kurupukari F. passes Mucu-Mucu steamlet Itaballi (Etabally) F.	accident with a canoe; part of the collection lost
18 Mar 1836	Bartica	
28 Mar 1836	arrives at Georgetown	

up the river. The low water level, the approaching rainy season and severe malaria (he wrote that he would not have mentioned this last, which would be of little interest to the reader, were it not of importance to his further proceedings) prevented his reaching the source of the river.

He returned to the Lower Rupununi River, and on the way made a trip into the Kanuku (Conocon) Mountains to look for the plant the Indians use to make their 'Ourali', 'Wourali', or 'Urari' poison. At Pirara, Robert gathered information about the preparation and the origin of the poison. After a most fatiguing walk through arid savannas and across mountain streams they found the woody twiner, which the Indians called "Urari". Robert named it *Strychnos toxifera*. This specimen is number 155 of his Guyanan plants. Robert sent a description of the plant and mode of preparation to the Linnaean Society of London (Schomburgk, R.H., 1841a: Schomburgk, M.R., 1879).

On December 28th they started travelling back to Georgetown. They halted at Pirara again from where they made several excursions, including one up the Rio Mahu (Ireng or Maú). On their way back they camped at Curassawaka where they joined a great feast before ascending the Essequibo River until they were stopped by a large cataract which they named King William's cataract, after the British monarch. They had to cope with many cataracts before reaching Georgetown on March 28th 1836. At Itaballi (Etabally) Falls an accident happened to one of the canoes and a large portion of the plant collection was lost, including most of those from the savanna.[2] In map 1 the red line indicates the route of Robert's first expedition into the interior of British Guiana and details about this journey are given in table 1.

ASCENT OF THE RIVERS COURANTYNE AND BERBICE (2 SEPTEMBER 1836–MARCH 1837)

After having explored the Essequibo and Rupununi Rivers it seemed desirable to make a choice of some other great rivers in British Guiana. The Courantyne (Corentyn) River was selected for this purpose. The colonists knew little about its course. Robert wanted to know more about the resources and capabilities of this region.

This expedition started on September 2nd. Members were Mr. Vieth, an ornithologist, Mr. Heraut, a draftsman, and the volunteers Lieutenant Losack, Mr. Cameron, and Mr. Reiss. They left for Plantation Skeldon on the Western bank of the Courantyne River. It was difficult to find enough Indians who could join the expedition. This delayed the trip for some days.

---

[2] Unless indicated otherwise, data were taken from Schomburgk, R.H. 1836. Report of an expedition into the interior of British Guayana, in 1835–1836. J. Roy. Geogr. Soc. 6: 224–284.

Finally, the expedition set off on September 19th 1836. They passed Asirikani or Long island, where a range of clay hills is situated, as deposits of the flood tide. Robert entered the Cabalaba River for some days to reach the Avenavero Falls. On their way they encountered a series of considerable rapids, which blocked their way. In order to skirt the obstacles Robert tried to encourage the Caribs to find a passage over land by which transport the canoes and luggage. Unfortunately, the Caribs were very unwilling to collaborate, and the lack of supplies and quarrels between Robert and his crew forced them to return.

Robert unwillingly returned downstream on October 23rd 1836. At Tomatai most of the Caribs who accompanied him left the crew. When he arrived at Orealla Robert discovered that another group of Caribs was trying to ascend the Courantyne River in order to cross over to the Essequibo River by land. These Caribs worked for Chief Smith, a man who had visited their settlement a few months earlier. This man had asked the Caribs to proceed to a Macusi country, with the intention to trade slaves. The Caribs had the impression that Robert and his crew were bound in the same direction and had supposed that his presence would interfere with their plans. Robert discovered that the Caribs had withheld their knowledge of a path where, by means of a creek, the falls might have been passed. Frustrated by the failed attempt, he decided to ascend the Berbice River, thus hoping to avoid the obstacle of the falls and find more cooperative Indians inhabiting this region. With this in mind he arrived at New Amsterdam early in November 1836.[3]

His crew, except for Lieutenant Losack, was the same as before; the boats crew consisted of Arawaks, Warraus, and three Caribs. On November 25th they left New Amsterdam. During this expedition they had to pass many cataracts. Moreover numerous frightful caimans, poisonous snakes, spiders, scorpions, and prickly palm leaves made it even more troublesome. In his journal to the Geographical Society Robert described the flora of the river banks and mentioned *Norantea guianensis* as one of the greatest ornaments of British Guiana (plate 2).

On Christmas day they spent some time at falls without a name. They called them 'Christmas Cataracts'. Not only did they make little progress because they had to cope with many cataracts, they were also hindered by numerous trees fallen across the stream. January 1st 1837 Robert even feared that he might have to stop his expedition again. Unexpectedly, he found a giant water lily, which is native to the Amazon and surroundings and is known as *Victoria regia*. Robert sent a drawing of the flower to England together with a humble request for permission to dedicate it to the Queen and that it may bear her royal name.

[3] Data were taken from Schomburgk, R.H. 1837a. Diary of an Ascent of the River Corentyn in British Guayana, on October 1836. J. Roy. Geogr. Soc. 7: 285–301.

Table 2. Ascent of the Rivers Courantyne and Berbice.

Date	Location	Observations, collections, and main events
	Robert Schomburgk	
2 Sep 1836	starts second expedition	
9 Sep 1836	halts at Plantation Skeldon	
19 Sep 1836	travels up the Courantyne (Corentyn) R. for 40 miles	
21 Sep 1836	Post Orealla	meteorological observations
25 Sep 1836	passes Asirikani (Long Island)	baseline measurements
4 Oct 1836	Avenavero F. [SUR]	
11 Oct 1836	passes Assipura stream [SUR]	
18 Oct 1836	Wonotobo (Mavari Wonotopo) F. [SUR]	discovers new orchids and some cacti
23 Oct 1836	starts journey downstream	
Nov 1836	arrives at New Amsterdam	
25 Nov 1836	departs from New Amsterdam Berbice R.; Plantation Noitgedacht and Dageraad	
28 Nov 1836	Fort Nassau, Wiruni (Wieroni) R.; arrives at Wikki (Wickie)	celestial observations
	walks across savanna until the brook Etoni	collects savanna plants
4 Dec 1836	departs from Wikki; stops at mouth of Kabilibiri (Kabiribirie) R.	water temperature measurements
11 Dec 1836	Yariki brook	Indian picture writings
13 Dec 1836	Marlissa F.	
15 Dec 1836	Parish's Peak	*Norantea guianensis* and *Marcgravia* species
18 Dec 1836	Itabru F.	
25 Dec 1836	Christmas C.	
1 Jan 1837	travels further up the Berbice R.	discovers *Victoria regia* (=*V. amazonica*)
30 Jan 1837	walks to the Essequibo R.	*Pekea tuberculosa* (=*Caryocar nuciferum*)
30 Jan 1837	meets Caribs of the Courantyne expedition	
31 Jan 1837	Essequibo R.	
1 Feb 1837	returns to camp on the Berbice R.	
9 Feb 1837	halts at Christmas C.	accident with Mr. Reiss
16 Feb 1837	Parish's Peak	variety of ferns
17 Feb 1837	Itabru F.	
28 Feb 1837	mouth of Wiruni R.	
4 Mar 1837	Post Seba near Demerara R.	
7 Mar 1837	makes trip to Ororo-Marali or Great F. in the Demerara R.	
13 Mar 1837	returns to Post Seba and Yakabura (Yucabura) village	
20 Mar 1837	makes excursion to Courantyne R.	collects numerous orchids
31 Mar 1837	arrives at New Amsterdam	

Robert was the first person who observed this species in Guyana, but was not the first one who discovered this water lily. In 1836 E.F. Poeppig, a German explorer, was the first who discovered this water lily in the Amazon. He named the species *Euryale amazonica*. According to the nomenclature rules its official name should be *Victoria amazonica*, however, it is observed that the species is still frequently referred to as *Victoria regia*. (Prance, 1974). Unfortunately the difficulties increased every day, a large part of the crew fell ill and the low riverbanks were flooded which made them unsuitable for building a camp. Some members of the crew became mutinous and fled during the night, taking some provisions and equipment with them. On January 29th they reached the point where the path from the Courantyne River to the Essequibo River crossed the Berbice River and two days later the Caribs of the Courantyne River who had been planning an expedition to the Macusi country arrived at the crossing. Robert himself made the land journey to the Essequibo River but the lack of provisions and the low water level prevented him from crossing over to the Courantyne River. On the 2nd of February 1837 they began their return. Descending the Christmas Cataracts a terrible accident happened; a canoe overturned in the rapids of the Berbice River and Mr. Reiss was drowned.

On his way back, Robert made an excursion to Parish's Peak. After a rest at Peereboom he proposed to descend the Wiruni (Wieroni) River as far as possible. They followed the Wiruni River until they reached Yakabura (Yucabura). From there they continued their journey over land and through several swamps. Finally Robert reached Post Seba on the Demerara River and made an excursion to the Ororo-Marali or Great Fall in that river. On March 15th Robert returned to Wikki (Wickie) whence he also made an excursion over land towards the Courantyne River, which he never reached because he suffered from an attack of rheumatism. After an absence of four months Robert arrived at New Amsterdam on March 31st.[4] In map 1 the blue line indicates the route of Robert's second expedition into the interior of British Guiana and details about this journey are given in table 2.

EXPEDITION TO THE SOURCE OF THE ESSEQUIBO RIVER AND MOUNT RORAIMA (12 SEPTEMBER 1837–9 NOVEMBER 1838)

Robert's next expedition was aimed at exploring the Essequibo River to its sources, and travelling to Esmeralda on the Upper Orinoco River in Venezuela to connect his survey with that of Alexander von Humboldt. After an attack of

[4] Data were taken from Schomburgk, R.H. 1837b. Diary of an Ascent of the River Berbice in British Guayana, in 1836-1837. J. Roy. Geogr. Soc. 7: 302–350.

yellow fever Robert left Georgetown on September 12th to travel up the Essequibo River. He was accompanied by Mr. Vieth an assistant naturalist, Mr. Morrison a draftsman, Mr. Le Breton, who took charge of the supplies, and several Warrau Indians as part of his boat crew. They sailed to Ampa where Mr. Peterson, the coxswain, joined the crew.

On the 1st of October they reached the Takwari (Taquiari) or Kumut (Cumuti) Mountains were they discovered Indian picture writings. Robert made copies of them for further examination. On October 16th they reached the Rupununi River and set up their camp at the mouth of the Rewa (Roiwa or Illiwa) River. Several crew members got fever and the food became scarce; therefore Robert travelled to Annai, a Macusi settlement, to get some cassava bread. After his return they started to follow the Rewa River.

At the end of October they saw Mount Ataraipu at a distance and reached the junction of the Rewa and the Kwitaro (Guidaro or Quitaro) Rivers. On November 3rd they halted at Pukasanta, a settlement inhabited by Carib and Atorai Indians, on the banks of the Kwitaro River. They stayed for three days to gather supplies and to recover. Meanwhile Robert was looking for the flowers of *Bertholletia excelsa*, the Brazil nut. Unfortunately he did not find any flowers, only fruits.

On the 6th of November they set off into the direction of Mount Ataraipu or Devil's Rock, known for its pyramidal shape. The base of this mountain is forested, while the top is bare. They climbed the rock and found numerous plants including several cacti. In the morning of November 10th, they reached the Tene-nuaro brook from where they continued over land towards the Kwitaro River. Meanwhile the canoes were sent back to the Rupununi River. They walked through forests, crossed several streamlets and entered the savanna. On November 17th the Wapisiana Indians living in Watuticaba kindly welcomed them. Most of these Indians had never seen white people before. Their village was placed in the middle of granite boulders, with their own particular flora. Robert found and collected the orchid *Epidendrum bicornutum* (=*Caularthron bicornutum*) and another orchid species, belonging to a new genus, he had first found on the Courantyne River. John Lindley in London named this orchid genus *Schomburgkia* (Jones, 1973). During the following days Robert crossed the savannas visiting various settlements of Taruma Indians on the way. He descended the Kuyuwini River to its junction with the Essequibo River and then started to ascend that latter river towards its sources.

On December 15th they left the Essequibo River and entered the Caneruau River. The next day they marched into the valley of the Caneruau River and viewed Serra Acarai for the first time. The following day they reached the watershed between the basins of the Essequibo and the Amazon Rivers and crossed it to visit a Barokoto village. Unfortunately the rainy season had set in, preventing Robert from continuing his journey. They returned to their canoes in the Caneruau River and started to ascend the Essequibo River. Many fallen trees were obstructing their way, forcing them to walk again. After a three days painful march, they arrived at one of the sources of the Essequibo River, were they hoisted the British flag.

On January 6th 1838 they reached the mouth of the Kuyuwini River and proceeded towards Pirara village where they arrived on March 21st. At Pirara Robert collected plants, birds, and insects, waiting for Peterson to come with new supplies from Georgetown. Meanwhile Thomas Youd, a missionary who wanted to set up a mission at Pirara was welcomed. He told Robert the sad news that Carmichael Smyth, the Governor of British Guiana and Robert's patron had died.

At the end of May Robert made an excursion to the Kanuku Mountains where he collected several orchid species and the shell *Ampullaria guyanensis*. They climbed Mount Ilamikipang not only to enjoy the view over the savanna but also to search for the flowers of the 'Urari' plant. Robert found the plant with fruits and took some 'Urari' bark with him. Robert described the vegetation: containing many undescribed species of Myrtaceae, several species of *Epidendrum*, *Pleurothallis*, *Brassavola*, *Maxillaria*, and *Tillandsia*. Unfortunately, one of Robert's guides had hurt his foot and they had to return to Pirara again. Peterson also arrived at Pirara with new provisions and merchandise. Robert informed the commandant of Fort São Joaquim that he wanted to move from Pirara to Fortaleza in order to determine the position of that place, which had always been considered as the eastern boundary of Brazilian British Guiana. Robert got permission and accompanied by Youd, he left Pirara and travelled to Fort São Joaquim, where Senhor Pedro Ayres welcomed them. Robert started his experiments to extract poison from the 'Urari' bark. He observed the effects of the poison and succeeded in killing two fowls within half an hour (Schomburgk, R.H., 1841a).

On August the 16th they left the Fort and travelled down the Rio Branco. Robert started to explore the mountain range Serra Grande or Carumá (Carauná). Near there he observed a large cactus tree (a species of *Cereus*) and on Mount Carumá he collected three species of *Epidendrum*, *Zygopetalum rostratum* (=*Z. labiosum*), and species of *Oxalis*, *Verbena*, *Mimosa*, *Cassia*, and *Eugenia*.

Before he left on this exploration of Serra Grande Robert witnessed the arrival of a party of Brazilians whom he described as slavers but whose ostensible activity was to recruit paddlers for official canoes. After his trip to the Serra Grande he met the party with their 40 Indian captives, mostly women and children taken from a settlement on the right bank of the Takutu River (Rivière, 1995; 1998). Robert returned to Fort São Joaquim where the Indians were held until they were shipped downstream on August 25th. Robert was deeply angered by this event and informed the president of the

Table 3. Expedition to the source of the Essequibo River and Mount Roraima.

Date	Location	Observations, collections, and main events
	Robert Schomburgk	
12 Sep 1837	departs from Georgetown	
21 Sep 1837	travels up the Essequibo R.; reaches Ampa	
1 Oct 1837	lands at Takwari (Taquiari) or Kumut (Cumuti) Mts.	Indian picture writings
16 Oct 1837	stops at the mouth of Rewa (Roiwa or Illiwa) R.	
31 Oct 1837	junction of the Rewa R. and Kwitaro (Guidaro or Quitaro) R.	
3 Nov 1837	reaches Pukasanta village on bank of the Kwitaro R.	*Mora excelsa* and *Apeiba tibourbou*
6 Nov 1837	pitches camp at Karabiru F.; views Mt. Ataraipu or Devil's Rock	cacti
17 Nov 1837	visits Watuticaba village	*Schomburgkia marginata* and *S. crispa* *Amaryllis belladonna* (=*Hippeastrum elegans*)
25 Nov 1837	travels towards Marudi-Karawaimentau (Carawaimi) Mts.	
26 Nov 1837	halts at a Taruma settlement	
1 Dec 1837	passes tributaries of the Kwitaro R.	
4 Dec 1837	descends the Kuyuwini R. towards junction with the Essequibo R.	Indian picture writings
8 Dec 1837	ascends the Essequibo R.	
9 Dec 1837	passes streamlets of the Quitiva R.; looks at Wanguwai Mts.	Indian sculpture
10 Dec 1837	junction with Kassikaityu R.; halts at Taruma settlement	
15/16 Dec 1837	enters the Caneruau R.; follows the Caneruau valley by foot	
19 Dec 1837	halts at a Barokoto settlement; starts his return	
27 Dec 1837	follows the Essequibo R. on foot	sources of the Essequibo R.
6 Jan 1838	mouth of Kuyuwini R.	
16 Jan 1838	crosses the Marudi-Karawaimentau Mts.	
Jan/Feb 1838	walks over the savanna to the banks of the Rupununi R.	
15 Feb 1838	Rupununi R.	
3 Mar 1838	arrives at Annai village	
16 Mar 1838	departs from Annai village	
21 Mar 1838	arrives at Pirara village	collects plants, birds, and insects Mr. Youd arrives
15 May 1838	Pirara village	
28 May 1838	makes excursion to Kanuku (Conocon) Mts.	
29 May 1838	reaches watershed of the Ireng R. and Rupununi R.	orchids
30 May 1838	crosses the Curassawaka, Rinaute, and Nappi streamlets	
1 June 1838	climbs Mt. Ilamikipang	Urari plant; orchids and Myrtaceae species
4 June 1838	returns to Pirara	constructs a map of the Upper Essequibo R.

		astronomical observations
Jun/Jul/Aug	stays at Fort São Joaquim [BRA]	
16 Aug 1838	departs from Fort São Joaquim; travels towards R. Branco [BRA]	
19 Aug 1838	explores Mt. Carumá (Carauná) or Serra Grande [BRA]	lichens, ferns, grasses, orchids, and bromeliads
22 Aug 1838	returns to Fort São Joaquim [BRA]	invasion of Brazilians
20 Sep 1838	departs from Fort São Joaquim; ascends the R. Takutu in NE direction [BRA]	
22 Sep 1838	passes R. Surumu (Zuruma) [BRA]	
25 Sep 1838	Ireng R., Pirara R., and Takutu R.	view over Pakaraima Mts.
26 Sep 1838	Pirara village	
8 Oct 1838	starts trip to Roraima; crosses savanna near Pirara	*Curatella americana*, Gramineae, and Cyperaceae species
19 Oct 1838	crosses R. Muyang [BRA]	*Hyptis membranacea* (=*Hyptidendron arboreum*)
20 Oct 1838	continues towards sandstone range of Humirida	Orchidaceae, Proteaceae, and Melastomataceae species
25 Oct 1838	follows the R. Cukenam (Kukenaam, Kukenán) and halts at an Arecuna settlement [VEN]	joins a feast
2 Nov 1838	explores the Roraima mountain range [VEN]	*Utricularia humboldtii* and *Heliamphora nutans*

Aborigines Protection Society in London that same day about the barbarous kidnapping. Robert suspected that the inhabitants of the Pirara mission might be the next victims. He stressed the importance of protecting the Indian tribes near the Brazilian frontier. Robert stated that the boundaries of British Guiana should be soon defined by a government survey in order to prevent such events.[5]

On September 20th 1838 the expedition departed from Fort São Joaquim to Pirara. They started the ascent of the Rio Takutu in a North East direction against a strong current. They passed Rio Surumu (Zuruma) or Rio Cotingo (Cotinga), the lattter name used by Arecuna and Macusi Indians. The journey went along the Western bank of the Ireng (Rio Mahu or Maú) River. They arrived at Pirara the 26th of September and were welcomed by Youd, the missionary who helped Robert to obtain new crew members.

On October 8th all arrangements were completed and Robert and his crew started their journey to the Roraima region. They had to wade through the Pirara streamlet with water up to their necks and their luggage on their heads. They entered the Unamara River, which is surrounded by mountains on all sides. Continuing their route towards the sandstone range of Humirida, they crossed several brooks of the Muyang River. According to Robert, the

---

[5] Unless indicated otherwise, data were taken from Schomburgk, R.H. 1841b. Report of the third expedition to the interior of Guayana, comprising the journey to the sources of the Essequibo, to the Carumé Mountains, and to Fort São Joaquim, on the Rio Branco, in 1837–8. J. Roy. Geogr. Soc. 10: 159–190.

vegetation was very interesting. He was gratified to find here a species of *Cyathea*, the first arborescent fern he had seen in the interior of British Guiana. Robert noted that the sandstone region had its own flora, almost every shrub was new to him. Crossing the Yaiwara River they had a nice view over the Roraima mountain range, of which Mount Roraima is the highest (plate 3). At an Arecuna settlement along the Cukenam (Kukenaam, Kukenán) River a party was prepared for them, as messengers had announced their coming. The Indians, dressed in their gayest ornaments, were dancing and repeatedly singing the song of 'Roraima'. On November 2nd they approached Mount Roraima, which was seldom free of clouds. Robert explored the Roraima region and described it very accurately. In one of the swampy savannas near Mount Roraima he found a species of *Abolboda* (plate 4). In map 1 the green line indicates the route of Robert's third expedition into the interior of British Guiana and details about his journey to Mount Roraima are given in table 3.

JOURNEY TO ESMERALDA AND FORT SÃO JOAQUIM (21 NOVEMBER 1838–20 JUNE 1839)

After a stay of 25 days in the Roraima Territory, they set out in the direction of Esmeralda on the Río Orinoco. They marched over the savanna towards the Rio Parimé, when an accident happened. A rattlesnake bit one of the crew members. After a three days rest the injured man had recovered, and Robert decided to continue. On December 6th they finished their march over land and began to follow the Rio Uraricoera (Parima) until they reached the point where the route lay across the mountains to the Río Merewari (Merevari), tributary of the Río Orinoco. The course of the Río Merewari was very winding as it had to force its way through a succession of hilly ranges. They left the Río Merewari, which flows away to the North, and started to follow the Río Cannaracuna. However, the low water level and many boulders forced them to continue their journey over land. The next part of the journey was extremely fatiguing, many times their canoes had to be unloaded and the luggage carried over land. On January 26th they found themselves back in the fluvial system of the Río Merewari. Robert saw Mount Paba where the Río Merewari has its source.
On the last day of January they entered the Río Orinoco system. Robert could not get any support from the Maiongkong Indians in order to find the sources of the Río Orinoco. They were very afraid to accompany him in order to reach the sources of the Río Orinoco they had to pass the Karishanas Indians, who have their territory in the mountains between the Río Orinoco and Río Ocamo and who had killed some Maiongkongs in the past. Robert had to return and set off into the direction of Mount Warima where he

recognised several plant species, which resembled the species he had previously found near Roraima.

They continued their journey towards Cerro Marahuaca (Mount Maravaca). At the foot of this mountain they halted at a Maiongkong settlement, where Robert discovered a new *Arundinaria* (*A. schomburgkii = Arthrostylidium schomburgkii*) species. The Guidaro and Maiongkong Indians, who inhabit this region, use this species of bamboo to make their blowpipes. Just after their departure from this settlement a disastrous accident happened to one of the canoes while passing a fall. It got filled with water and sunk together with its load. Fortunately, the Maiongkong Indians were able to make a new canoe.

On 21st February they entered the Río Orinoco where fresh water dolphins escorted them. Their progress was not without difficulty; several times they ran aground on sand banks. The next day Robert saw a fine savanna extending to the foot of the mountains, which according to Humboldt's description, had to be the savanna of Esmeralda. Robert's objective was realised; his observations, commenced on the coast of British Guiana, were now connected to those of Alexander von Humboldt at Esmeralda.[6]

On February 25th they left Esmeralda, descending the Río Orinoco towards Tamatama. They entered the Río Casiquiare, the natural channel between the Río Orinoco and the Rio Negro. After a 176 miles journey they reached the junction of the Rio Guainia and the Rio Negro. Small villages like Santa Isabel and São Gabriel were visited. They had difficulty in finding the mouth of the Rio Padauiri (Padauari) where Robert had hoped to find Vieth and Le Breton who had travelled down the Rio Branco to meet him. However they had already started their return journey and it was not until the reached Barcelos (Barcellos) or Mariua on March 24th that he met up with them. At the beginning of the 18th century Barcelos had been a populated place, but since the seat of the Government had been removed from there to Manaus only twenty houses were still inhabited. On March 30th they continued together towards Pedrero (Moura) on the Rio Negro. They spent Easter at Pedrero and after three days arrived at the mouth of the Rio Branco. Violent thunderstorms made their progress slow. In the afternoon of April 22nd they reached Fort São Joaquim again. Seven months and two days had elapsed since their departure from the Fort, during which period they had made a circuit of about 2200 miles.

After a short stay at the Fort, Robert wanted to leave for Pirara as soon as possible. What Robert had feared had become reality. When on the evening of the 1st May Robert and his crew reached the village, they found Pirara

---

[6] Unless indicated otherwise, data were taken from Schomburgk, R.H. 1841c. Journey from Fort São Joaquim, on the Rio Branco, to Roraima, and thence by rivers Parima and Merewari to Esmeralda, on the Orinoco, in 1838–9. J. Roy. Geogr. Soc. 10: 191–247.

occupied by Brazilian soldiers. The Brazilian Government had accused Youd of setting up a missionary on Brazilian territory. Youd left Pirara and found another location for his missionary at Urwa on the East bank of the Rupununi River.

In June 1839 Robert and his crew started their return to Georgetown. On June 17th the people from Bartica gave them a warm welcome. Robert, soon after his arrival in Georgetown, wrote a letter to Henry Light, Governor of British Guiana, where he mentioned the occupation of Pirara by Brazilian soldiers and stressed the urgent need to fix the boundary between Brazil and the British colony. In his view the best defence the Indians could have against the Brazilians was to live in British territory. Due to Robert's efforts the Governor of British Guiana and the British Government in London were convinced that the boundaries of the colony had to be fixed. In April 1840 Robert was commissioned to lead an expedition to survey the boundaries[7]. In map 1 the green line indicates the route of Robert's third expedition into the interior of British Guiana and details about his journey from Mount Roraima to Esmeralda and Fort São Joaquim are given in table 4.

Table 4. Journey to Esmeralda and Fort São Joaquim.

Date	Location	Observations, collections, and main events
	Robert Schomburgk	
21 Nov 1838	descends Mt. Tariparu [BRA/VEN-border]	
24 Nov 1838	halts at the foot of Mt. Marua [BRA]	one of the crew members bitten by a rattle snake
27 Nov 1838	marches over the savanna to R. Parimé [BRA]	
7 Dec 1838	reaches the mouth of the R. Uraricoera (Parima) [BRA]	
17 Dec 1838	passes the mouth of R. Uruwé [BRA]	
25 Dec 1838	near Serra Pacaraima [BRA/VEN-border]	*Elizabetha* species
7 Jan 1839	enters the R. Merewari (Merevari) [VEN]	
22 Jan 1839	junction of the Avenima streamlet and R. Merewari [VEN]	mountain range part of Serra Mey (Sierra Mai)
26 Jan 1839	stops at bank of R. Birima [BRA/VEN-border]	observes Mt. Paba as source of R. Merewari
31 Jan 1839	enters the R. Orinoco system [VEN]	
1 Feb 1839	arrives at a Maiongkong settlement [VEN]	
2 Feb 1839	Mt. Warima [VEN]	Bromeliaceae, Orchidaceae, and Commelinaceae species

---

[7] Data were taken from Schomburgk, R.H. 1841d. Journey from Esmeralda, on the Orinoco, to San Carlos and Moura on the Rio Negro and thence by Fort São Joaquim to Demerara, in the spring of 1839. J. Roy. Geogr. Soc. 10: 248–267.

Date	Event	Notes
7 Feb 1839	Cerro Marahuaca (Mt. Maravaca) [VEN]	cock-of-the-rock (*Rupicola elegans*)
10 Feb 1839	enters the basin of R. Padamo (Paramu or Maquiritari) [VEN]	*Lycopodium* species
11 Feb 1839	halts at a Maiongkongs settlement [VEN]	discovers *Arundinaria schomburgkii* (=*Arthrostylidium schomburgkii*)
20 Feb 1839	passes last cataract of R. Padamo [VEN]	fresh water dolphins and many fishes
21 Feb 1839	enters the R. Orinoco [VEN]	astronomical observations
22 Feb 1839	arrives at Esmeralda village [VEN]	data of Robert and A. von Humboldt are connected
25 Feb 1839	departs from Esmeralda; descends the R. Orinoco until Tamatama [VEN]	
26 Feb 1839	R. Casiquiare; passes Pamoni settlement [VEN]	Indian picture writing
3 Mar 1839	reaches junction with R. Negro [COL/VEN-border]	
10 Mar 1839	passes Santa Barbara and arrives in São Gabriel [BRA]	crosses the equator
12 Mar 1839	large rapids in R. Negro [BRA]	Indian picture writings
15 Mar 1839	reaches Santa Isabel [BRA]	
19 Mar 1839	reaches the entrance of R. Padauiri (Padauari) [BRA]	
21 Mar 1839	crosses the R. Negro and reaches Bararoa (Tomar) [BRA]	two palm species and a turtle species *Chelys fimbriata*
24 Mar 1839	arrives at Barcelos (Barcellos) or Mariua on the R. Negro [BRA]	
27 Mar 1839	departs from Barcelos [BRA]	
29 Mar 1839	mouth of the R. Branco (Parima or Urariquira) [BRA]	
30 Mar 1839	passes Ilha da Pedra	Indian picture writings
30 Mar 1839	stays at Pedrero (Moura) on the R. Negro [BRA]	
2 Apr 1839	returns to the mouth of the R. Branco [BRA]	
5 Apr 1839	Santa Maria on the E bank of R. Branco [BRA]	
19 Apr 1839	passes R. Mucajai (Mocajahi) [BRA]	
22 Apr 1839	reaches Fort São Joaquim [BRA]	
27 Apr 1839	ascends the R. Takutu towards Pirara	
1 May 1839	Pirara village	occupation by Brazilian soldiers
11 Jun 1839	reaches junction with Essequibo R.	
17 Jun 1839	Bartica	
20 Jun 1839	arrives at Georgetown	

5. THE EXPEDITIONS OF ROBERT AND RICHARD SCHOMBURGK 1841-1844

This chapter describes the expeditions of Robert together with Richard Schomburgk. Between 1841 and 1844 they made four different journeys into the interior of British Guiana and at the end Richard made three additional excursions. Most of the time the Schomburgk brothers travelled together but occasionally they separated. In map 2 the various colours indicate their different journeys. An unbroken line displays the route of Robert together with Richard, whereas a dotted line displays the route of Robert alone and the broken line displays the route Richard took. At the end of each chapter, dealing with one of the expeditions, a table is given, listing in chronological order the localities visited, and some of the main events, and notes on the botanical collections. In the tables it is indicated whether Robert travelled with his brother Richard or seperately. The meaning of the abbreviation is explained at the beginning of chapter four. Several localities changed names or are written in different ways. In the tables and general text the alternative names for a locality are cited in brackets.

EXPEDITION TO THE BARIMA AND CUYUNI RIVERS
(19 APRIL 1841-30 JULY 1841)

Robert's urgent recommendation to Governor Light that the boundary between Brazil, Venezuela, and the British colony should be fixed, resulted in his employment by the British Government as the leader of the Boundary expedition. Robert and Richard Schomburgk left Germany for British Guiana on the 29th of October 1840. They arrived in Georgetown in January 1841 and spent a long time there. Richard had an attack of yellow fever and had to be fully recovered before leaving. On April 18th, the boundary survey finally got under way. The crew consisted of Robert and Richard, accompanied by Mr. King, superintendent, Mr. Eichlin, artist, Mr. Glascott, assistant-surveyor, Mr. Peterson, first coxswain, and Mr. Cornelinsen, second coxswain. They left for the Barima River region and the frontier with Venezuela. Their travel began along the coast. In the afternoon of April the 21st they arrived at the mouth of the Waini River, where they camped on a sand bank. Robert assisted by Glascott determined the geographical position of the entrance of the Waini River. They ascended the Waini River to the Mora Creek, which connects the Waini River with the Barima River. The Warrau Indians who inhabit these rivers called it Morawan River (Schomburgk, R.H., 1843a).
They visited Cumaka, a settlement of Warrau Indians on the bank of the Aruka River. Some of the crew members had to recover from serious illnesses. During their stay at Cumaka, many of Richard's collections were

destroyed by mould, since the short rainy season had set in. Together with six Warrau Indians, Robert and Richard travelled to the mouth of the Barima River to conduct a navigational survey of this river. They returned to Cumaka village on May 20th.

Many members of the expedition suffered from dysentery and Richard's feet were inflamed due to jiggers. They were not able to join Robert on his trip to the Amakura and Aruka Rivers. Therefore only Eichlin and Robert travelled to the Amakura River and halted at Assecura, a settlement of Warrau and Arawak Indians. The small size of the river and the cataracts prevented them from travelling all the way up the Amakura River. They had to return to Cumaka village, where they arrived on the 10th of June.[8]

On the 14th of June both brothers left Cumaka to continue their survey of the Barima River. June 19th they entered Curawava (Caruawa), a Barima River tributary. They halted at Manari settlement. Robert described this area as a productive one where crops like sugarcane, corn, and bananas were growing well. After a short stay at Manari Richard, King, Hancock, Stöckle, and twelve crew members started their return to Georgetown, taking two loaded canoes with them. Robert, Glascott, Peterson, and Eichlin continued their journey up the Barima River on which they had to pass numerous rapids. The Uropocari Fall was the largest they had to face. Many fallen trees obstructed the Barima River causing them to continue over land. They made their way through bushes and swamps, following the left bank of the Barima River. June 30th they passed a river without a native name. They decided to call it Rocky River. Their supplies were now completely exhausted and they returned to Manari. From Manari Peterson and Glascott returned to Georgetown by boat, while Robert and Eichlin travelled to the Cuyuni River. They descended that river until they reached the junction of the three Rivers Essequibo, Mazaruni, and Cuyuni on July 27th. That same day they received a warm welcome from the other members of the expedition at Bartica Grove. The next morning they all left for Georgetown where they arrived on July 29th.

Robert and Richard had spent three and a half months exploring the Rivers Waini, Barima, Amakura, Barama, and Cuyuni, and during this period they had travelled over 700 miles. Robert acquired an accurate knowledge of the course of these rivers and discovered that their actual courses differed from those presented in extant maps. He was now able to construct a corrected map with the right course of the rivers.[9] He took particular care to base his reports from actual exploration and information obtained from the Indians,

---

[8] Unless indicated otherwise, data were taken from Schomburgk, R.H. 1842a. Expedition to the lower parts of the Barima and the Guiana Rivers in British Guiana. J. Roy. Geogr. Soc. 12: 169–178.

[9] Data were taken from Schomburgk, R.H. 1842b. Excursion up the Barima and Cuyuni Rivers, in British Guiana in 1841. J. Roy. Geogr. Soc. 12: 178–196.

as well as from the evidence of Dutch remains at the Barima River and on the Cuyuni River. He was able to ascertain the limits of Dutch possessions and the zone from which all trace of Spanish influence was absent. After his first boundary survey Robert submitted to Governor Light his plans for surveying the Brazilian boundary. In map 2 the purple line indicates the route of Robert's and Richard's expedition to the Barima and Cuyuni Rivers and details about this journey are given in table 5.

Table 5. Expedition to the Barima and Cuyuni Rivers.

Date	Location	Observations, collections, and main events
	Robert and Richard Schomburgk	
19 Apr 1841	depart from Georgetown	
21 Apr 1841	travel along the coast and reach the mouth of the Waini R.	many water birds; Robert measures geographical positions
27 Apr 1841	leave for Cumaka village on the banks of the Aruka R.	many of Richard's collections destroyed by mould
10 May 1841	travel to the mouth of the Barima R. [VEN]	navigational survey of the river
20 May 1841	return to Cumaka village	
27 May 1841	Robert leaves for the Amakura R. [VEN]	
10 Jun 1841	Robert returns to Cumaka village	
17 Jun 1841	continue survey of the Barima R.	
18 Jun 1841	reach mouth of the Kaituma R.	
19 Jun 1841	reach Curawava (Caruawa) R. and halt at Manari village	
22 Jun 1841	Richard leaves for Georgetown	
	Robert Schomburgk	
24 Jun 1841	travels further down the Barima R.	
28 Jun 1841	passes Uropocari F.	
29 Jun 1841	Rocky R.; starts his return to Manari village	
8 July 1841	marches to the Barama R.	
11 July 1841	reaches Kariako (Cariacu) village starts journey on the Barama R.	
15 July 1841	reaches path to the Cuyuni R.	
19 July 1841	Haiowa village near the Cuyuni R.	
22 July 1841	starts journey downstream the Cuyuni R.	
27 July 1841	reaches mission at Bartica Grove	meets the other members of the expedition
	Robert and Richard Schomburgk	
28 July 1841	leave for Georgetown; visit Fort Zeelandia	
29 July 1841	arrive at Georgetown	many collections damaged

EXPEDITION TO THE SOURCE OF THE TAKUTU RIVER
(23 DECEMBER 1841–22 MAY 1842)

In December 1841 the members of the boundary expedition left Georgetown to explore and define the Brazilian frontier. Beside Robert and Richard, the following people joined this trip: Fryer, assistant surveyor, Goodall, artist, Sororeng, interpreter, a number of Macusi Indians as carriers and guides, and nine men for the canoes.

On 23rd of December they left Georgetown and travelled up the Essequibo River until Ampa where they spent Christmas. They continued towards Bartica Grove to pick up the missionary Youd, who was to be reinstated in his mission at Pirara by military force. After six weeks they approached Pirara Landing, and when they entered Pirara village, they found that only three Brazilians remained there. Soon after their arrival, the detachment of British troops also arrived (Goodall, 1977). Governor Light had given strict instructions to take Pirara, reinstate the missionary, protect the Indians from the Brazilians, assert British rights to the territory and afford facilities for settlement by British subjects. In the meantime the Brazilian and British Governments had agreed that Pirara should be declared neutral territory until the boundary questions were settled (Rivière, 1995).

After a two months stay the members of the boundary expedition finally left Pirara. Robert intended first to explore the Rio Cotingo (Cotinga) in order to reach the Pakaraima Mountains. However, the dry season had started which made it impossible to ascend that river. So Robert decided to explore the Takutu River instead. On March 24th they started to walk over the savanna to the junction of the Pirara and Ireng Rivers. The first boundary markers were erected at the confluence of the Takutu and Ireng Rivers. 'In the name of Her Majesty Queen Victoria', Robert declared the right bank of the River Takutu to form the South Western boundary of the British colony. A number of other markers were set up at various points along the whole course of the river. Robert noted that the object of the Brazilian slave raid in 1838, had been a settlement on the right bank of the Takutu River. The markers would help protect the Indians by identifying them as Her Majesty's subjects (Rivière, 1995). The Brazilians, however, gave consent to Robert's proceedings but warned him that the markers he put up would be regard as being of only scientific value and not valid boundary markers.

At the junction of the Ireng and Takutu Rivers one of the crew members, named Petry, while shooting birds, accidentally shot himself. He seriously injured his shoulder and was carried back to Pirara by Fryer. Meanwhile Robert verified the distances between the Kanuku and the Pakaraima Mountains. The expedition continued its march to the junction of the Takutu and Ireng Rivers, from where they started to ascend the Takutu River. This river winds considerably where it approaches the Kanuku Mountains. At

Table 6. Expedition to the source of the Takutu River.

Date	Location	Observations, collections, and main events
	Robert and Richard Schomburgk	
23 Dec 1841	depart from Georgetown	
27 Dec 1841	travel up the Essequibo R.	
30 Dec 1841	pass Itaballi F. on the foot of the Ahoro F.	
5 Jan 1842	reach Mt. Arisaru (Arissaro)	
6 Jan 1842	pass the Potaro R.	
7 Jan 1842	Waraputa F.	hieroglyphic rock writings
Feb 1842	mouth of the Rupununi R.	
12 Feb 1842	ascend Rupununi R. to Yupukarri (Wai-ipukar) Inlet	British soldiers also arrive at Pirara
Feb 1842	Richard makes an excursion to Kanuku Mts.	bell bird, cock of the rocks, and *Strychnos toxifera*
24 Mar 1842	depart from Pirara; walk to junction of Pirara R. and Ireng R.	accident with a shotgun; two *Eugenia* species
2 Apr 1842	walk to junction of Ireng R. and Takutu R.	
5 Apr 1842	march along the Takutu R.	astronomical measurements
14 Apr 1842	Mt. Taquiara or Mariwette (Maloca do Manda) [BRA]	Robert discovers a new species of *Genipa*
15 Apr 1842	pass Scabunk F.	Indians hurt by stingrays
19 Apr 1842	arrive at Tenette village	trigonomical survey
23 Apr 1842	depart from Tenette; Mt. Auruparu [BRA]	seeds of *Elizabetha coccinea*
25 Apr 1842	Mt. Wurucoka; pass streamlet Curati [BRA]	
27 Apr 1842	Mt. Tuarutu and Tuarutu village; R. Manatiwau [BRA]	crew member got lost
3 May 1842	Mt. Vindaua and Maripa village	*Bertholletia excelsa*
6 May 1842	reach the Takutu R. and follow it upstream	calculation of the course of the Takutu R. from its source
7 May 1842	start their return to Maripa village	astronomical observations
12 May 1842	depart from Maripa village; start journey over savanna	
14 May 1842	cross Caurua R. and reach Mt. Pinighette	
16 May 1842	return to Tenette village	
20 May 1842	travel around West side of Kanuku Mts. towards Pirara	
22 May 1842	arrive at Pirara village	

some places the water was very shallow. They had to unload the canoes repeatedly and push them over land or march with the canoes over sand banks. The enormous heat on the sand banks, burning faces, blisters on hand and feet, and bites of sand flies made this trip very fatiguing. Moreover, two Indians from the crew got injured from stepping on stingrays.

When they arrived at the Wapisiana settlement Tenette on April the 19th, Robert received the news that Petry was in a reasonable condition. On April the 23rd the expedition continued towards Mount Wurucoka. They continued towards Mount Tuarutu and a village named after this mountain. One of the crew members got lost. After a long search Robert found him in an exhausted condition. On May 6th they reached the Takutu River again. The Indians of Tuarutu told Robert that the sources of the Takutu River were at Mount Vindaua. However, Robert discovered that this river received only its first tributary from Mount Vindaua, and had its source further to the South East. On May 13th Robert started his journey back over the savanna between the Takutu and Rupununi. Richard suffered from severe attacks of fever and decided to return directly to the Cursato Mountains immediately. On the 11th of May Robert reached the highest ground between the Takutu and Rupununi Rivers. He observed the granite Tambaro Hill, where he had been in March 1838. He was now able to connect his present survey with the former. It was impossible for them to return to Pirara by boat due to the shallow state of the Takutu River, and instead they crossed the savanna and passed the Kanuku Mountains on its Western angle. They arrived at Pirara on May the 21st and met Fryer with the wounded Petry. They then stayed at Pirara until the 10th of September.[10] In map 2 the green line indicates the route of Robert's and Richard's expedition to the source of the Takutu River and details about this journey are given in table 6.

EXPEDITION TO THE RORAIMA TERRITORY
(10 SEPTEMBER 1842–5 DECEMBER 1842)

On September 10th 1842, after a long stay in Pirara, the members of the boundary expedition proceeded to navigate the Rio Cotingo (Cotinga) to its sources near Mount Roraima. Richard, accompanied by Fryer and Goodall, marched over the flooded savanna towards the junction of the rivers Nappi and Pirara, while Robert tried to reach the mouth of the Pirara River on horseback. Robert did not succeed as his horse got stuck in the mud and threw him. Robert then found a canoe and was reunited with Richard and his party when he found them trying to cross the Nappi River by boat. They

---

[10] Unless indicated otherwise, data were taken from Schomburgk, R.H. 1843b. Visit to the sources of the Takutu, in British Guiana, in the year 1842. J. Roy. Geogr. Soc. 13: 18–74.

entered the Ireng River and followed its course until they reached the junction with the Takutu River. On September 16th they visited Fort São Joaquim for astronomical measurements. The expedition set out again and made its way up the Rio Cotingo on September 24th. On his way, Richard found many new plant species like *Schultesia benthamiana* and *Schultesia subcrenata*.

This expedition was an extremely difficult one. They got exhausted because they had to struggle against a very strong current and rushing cataracts. Richard and Goodall succumbed to bouts of fever. Fortunately, they had a pleasant interlude at the Macusi village Torong Yauwise where in full ceremonial dress of feathers and paint, war-clubs, bows and arrows the Macusis, Wapisianas, and Arecunas greeted them. Here they rearranged their luggage and continued over land in a North-West direction. They had to climb many mountains, some with very steep ridges. On their way they found flowers they had never seen before. Among these they found their first representatives of the genera *Roupala* and *Ternstroemia*. They continued along the Western spur of the Humirida Range. Richard climbed the mountains, consisting of numerous sandstone ridges. He found *Epidendrum* species, cacti, *Tillandsia*, and the gigantic *Elisabetha regia* in bloom. He also made many collections of ferns and mosses.

On October 24th they climbed Mount Humirida, where every shrub or bush was new to Richard. He had more success in finding flowering specimens than his brother had had on his trip in 1838. On the 25th of October they crossed the Cuino River and got finally a view of Mount Roraima. The journey went on to the Río Kukenam (Kukenaam, Kukenán). In the village Barapang they borrowed canoes to cross the river. On October 28th they built their base camp on the Kukenam River and baptized it 'Our Village'.

Numerous Indians had joined the expedition which now consisted of 100 people. On the 17th of November they continued towards Mount Roraima. After having crossed the Mure River a terrible accident happened. A poisonous snake had bitten an Indian woman, named Kate. Fryer scarified the wound and the Indians tried to suck out the venom. Unfortunately all efforts proved to be useless. She was taken back to the village where she died 63 hours later.

Sadly they continued their journey and started to climb Mount Roraima. Richard observed the different vegetation zones. The vegetation became more and more interesting and Richard kept enumerating plants. He collected new species like *Leiothamnus elisabethae* (=*Symbolanthus elisabethae*) and *Encholirium augustae* (=*Connellia augustae*). Unfortunately many collections got lost during their way down. Drying the plants was complicated because of the moist atmosphere. Goodall was unable to sketch Roraima as his fingers were stiff from the cold and the paper wet with the damp (Goodall, 1977).

After the exploration of the Roraima region Robert decided to split the expedition. Richard and Fryer returned to Our Village to keep the plant collections dry, while Robert and Goodall went higher up to the source of the Rio Cotingo and Cuyuni River to complete the survey of the North-Western frontier.

On December 4th Richard left for Pirara. He and Fryer visited Fort São Joaquim on their return journey and spent the night of 21 December there. Richard provided a full description of the fort and the condition of the soldiers stationed there. Richard decided to return to Pirara and wait for his brother whose return was not expected until April.

On December 5th, Robert and Goodall left for Georgetown via the Cuyuni River. In the basin of the Kamarang (Carimang or Kamurán) River, Robert discovered a new *Calliandra* species, which he dedicated to Sir William Hooker. Moreover, he found two species each belonging to a new genus namely *Lightia guianensis* (=*Euphronia guianensis*) named after Governor Light and *Gongylolepis benthamiana*, named after George Bentham (Schomburgk, R.H., 1847).[11] In map 2 the orange line indicates the route of Robert's and Richard's expedition to the Roraima Territory and details about this journey are given in table 7.

Table 7. Expedition to the Roraima Territory.

Date	Location	Observations, collections, and main events
	Robert and Richard Schomburgk	
10 Sep 1842	cross the Nappi R.; travel to the mouth of the Pirara R.	
13 Sep 1842	follow the Ireng R. [GUY/BRA-border] camp just beyond the junction with the R. Virua [BRA]	
14 Sep 1842	follow R. Takutu [BRA] stay at a camp at the junction with the R. Cotingo (Cotinga) [BRA]	
16 Sep 1842	visit Fort São Joaquim [BRA]	astronomical measurements
22 Sep 1842	travel up the R. Cotingo [BRA]	*Genipa americana* and *Inga* species
1 Oct 1842	reach junction of the R. Surumu (Zuruma) and the R. Cotingo [BRA]	Richard collects many Gentianaceae species
3 Oct 1842	follow the R. Cotingo [BRA] reach Mt. Piriuai (Piriwai) and Mt. Maikangpali [BRA]	
6 Oct 1842	end of the journey over the R. Cotingo [BRA] stay at Torong Yauwise settlement [BRA]	

[11] Data were taken from Schomburgk, M.R. 1848. Reisen in British-Guiana in den Jahren 1840–1844. Pages 111–279.

19 Oct 1842	depart from Torong Yauwise settlement [BRA]	
21 Oct 1842	cross the R. Tupuring [BRA]	*Roupala* and *Ternstroemia* species
24 Oct 1842	climb the Humirida Mts.; cross R. Suapi (Zuappi) [BRA]	Richard collects many ferns and mosses
26 Oct 1842	R. Cukenam (Kukenaam or Kukenán) [BRA/VEN-border]	rock writings
28 Oct 1842	'Our Village' near Canaupang settlement [VEN]	
17 Nov 1842	start journey to Mt. Roraima [VEN]	woman is bitten by a poisonous snake
18 Nov 1842	climb Mt. Roraima [VEN]	collect many ferns and new plant species
21 Nov 1842	start to decline Mt. Roraima [VEN]	*Leiothamnus elisabethae* (=*Symbolanthus elisabethae*) and *Encholirium augustae* (=*Connellia augustae*)
	Richard Schomburgk	
22 Nov 1842	returns to 'Our Village' [VEN]	
4 Dec 1842	leaves for Pirara	
10 Dec 1842	stops at the foot of the Humirida Mts. [BRA]	
17 Dec 1842	landing place of the R. Cotingo [BRA]	
18 Dec 1842	junction of the R. Surumu and R. Cotingo [BRA]	
21 Dec 1842	follows the R. Takutu and reaches Fort São Joaquim [BRA]	
28 Dec 1842	arrives at Pirara village	
	Robert Schomburgk	
22 Nov 1842	travels higher up to the source of the R. Cotingo [BRA]	
24 Nov 1842	arrives at 'Our Village' [VEN]	
5 Dec 1842	departs from 'Our Village' [VEN]	
8 Dec 1842	reaches junction of the R. Cukenam and the R. Yuruani	collects many new plant species
12 Dec 1842	basin of the Kamarang (Carimang or Kamurán) R.	*Calliandra hookeriana* (=*C. rigida*), *Lightia guianensis* (=*Euphronia guianensis*), and *Gongylolepis benthamiana*
14 Dec 1842	crosses the Cutzi R.; reaches Arekuna settlement	
19 Dec 1842	follows the Kamarang R.	
3 Jan 1843	junction of the Wenamu R. and the Cuyuni R. [VEN/GUY-border]	
7 Jan 1843	follows the Cuyuni R. [VEN/GUY-border]	
10 Jan 1843	mouth of Acarabisi R. [VEN/GUY-border]	
18 Jan 1843	enters Bartica Grove	
Jan/Feb 1843	stays at Georgetown	

EXPEDITION TO THE UPPER COURANTYNE RIVER
(15 FEBRUARY 1843–12 OCTOBER 1843)

On February 15th 1843 Robert departed from Georgetown to meet his
brother and the rest of the crew at Pirara village. In the meantime both
Brazilians and Venezuelans had requested that the boundary markers must
be removed (Rivière, 1995). Governor Light had tried to avoid this by stating
that the markers where only placed for scientific purpose, but without
success. Accordingly, after his return to Pirara, Robert sent Fryer to Fort São
Joaquim with orders to efface the boundary markers on the Rio Mahu and
Rio Takutu.
Robert and Richard started to proceed towards Watuticaba (Mowenow ?).
They collected some fresh provisions at a Wapisiana settlement and
ascended the Rupununi River. The banks were covered with new species like
Combretum aurantiacum (=*C. fruticosum*) and *Petrea macrostachya*. When
they reached Watuticaba, Robert and Richard continued separately. Richard
decided to save his valuable collections and returned to Georgetown. On
his way back he passed Pirara village for the last time. The village looked
desolated and was deserted by all residents. The abandonment of Pirara by
the Indians began with Youd's departure and had been furthered by the
military withdrawal. Youd had left Pirara in June 1842 for a visit to
England and died at sea in August 1842. Richard arrived in Georgetown in
June 1843.
Robert decided to complete his commision with a survey of the boundary with
Dutch Guiana, over which there was little disagreement. He and the artist
Goodall marched over the savanna towards the Kwitaro River. On June 18th
they continued their journey in two small boats and six bark canoes and entered
the Kuyuwini River. After a three days journey they came to the Essequibo
River where they halted at various Taruma settlements. When they reached the
mouth of the small river Urana they started to travel over land again. They had
to walk over numerous hills and through swampy ground until they reached the
Darura River. They continued by boat on the Rio Cafuini (Caphiwuin). Many
falls hindered their way, even causing the loss of two canoes.
On August 16th they ascended the Rio Anamu (Wanamu) for several days. They
halted at a Pianoghotto village and then marched towards a Drios settlement,
situated on the Kutari River, in the headwaters of the Courantyne River. They
passed many cataracts while descending the Courantyne River and had to
unload the canoes over and over again. They passed the place where Robert had
not been able to continue when he ascended the Courantyne River in 1836.
They met with great hardships, navigation was very difficult and the canoes
cracked by continuous collisions with rocks. Several Indians suffered from
fever and the amount of provisions was not sufficient. Fortunately, they arrived
at Tomatai settlement where the starving crew got food. After an absence of

eight months Robert and Goodall arrived safely in Georgetown on October 12th. The objectives of the expedition had been fully realised. During his last journey Robert visited several settlements of Tarumas, Atorais, Caribs, Warraus, Taurais, Wapisianas, Drios, Amaripas, Maopityans (Mawakas or Frog Indians), and Pianoghottos. Robert described their customs and appearance in his papers to the Royal Geographical Society, purchased some of their curiosities and brought some of their weapons and household utensils. Goodall made portrait sketches of the different tribes and prints of those sketches have been published by the British Museum (Goodall, 1977). Robert spent a long time in Georgetown, completing the calculations and the maps. On May 19th 1844 he and Goodall took the mailboat for England where they arrived on June 25th.[12] In map 2 the blue line indicates the route of Robert's expedition to the Upper Courantyne Rivers and details about this journey are given in table 8.

After Robert had finished his boundary surveys he was appointed British consul first in Santo Domingo in the Domican Republic and then in Bangkok. Negotiations about the boudaries between British Guiana and both Venezuela and Brazil dragged on for a long time. The line of the former was finally settled in 1899 as a result of USA intervention, and the latter in 1904 as a result of arbitration by the King of Italy. (Rivière, 1995).

Table 8. Expedition to the Upper Courantyne River.

Date	Location	Observations, collections, and main events
	Robert Schomburgk	
15 Feb 1843	departs from Georgetown	
3 Mar 1843	follows the Essequibo R.	observes a comet
24 Mar 1843	arrives at Yupukarri (Wai-ipukari) Inlet	
	Robert and Richard Schomburgk	
30 Apr 1843	depart from Pirara village	
4 May 1843	ascend the Rupununi R.	
8 May 1843	Aripai R.	*Combretum aurantiacum* and *Petrea macrostachya*
9 May 1843	Kanuku Mts.	visit to old camp from journey in 1838
11 May 1843	reach Paruauku Portage pass mouth of Wichabai (Witzapai) R. and Araquai R.	Richard collects several savanna plants
12/14 May 1843	Fryer's C.; Waruwau or Awarra R.	meteorological observations
15 May 1843	reach path to Watuticaba (Mowenow ?) village	
16 May 1843	arrive at Watuticaba village near Mt. Tamboro	

[12] Unless indicated otherwise, data were taken from Schomburgk, R.H. 1845. Journal of an expedition from Pirara to the Upper Corentyne, and from thence to Demerara. J. Roy. Geogr. Soc. 15: 1–103.

Richard Schomburgk

21 May 1843	departs from Watuticaba with Fryer	
24 May 1843	Aripai settlement	*Attalea speciosa*
25 May 1843	returns to Pirara village	
11 Jun 1843	departs from Pirara village; Haiowa F. passes Rappu R.; stays at Waraputa mission	
19 Jun 1843	arrives in Georgetown	

Robert Schomburgk

3 Jun 1843	departs from Watuticaba	many portraits of the people from Watuticaba
5 Jun 1843	reaches foot of the Marudi-Karawaimentau (Carawaimi) Mts.	
6 Jun 1843	follows the Kwitaro (Guidaro or Quitaro) R.	
8 Jun 1843	reaches Kuyuwini R.; halts at a Taruma settlement	meteorological observations
18 Jun 1843	follows the Kuyuwini R.	
21 Jun 1843	follows the Essequibo R.	
27 Jun 1843	reaches a next Taruma settlement	feast
12 July 1843	crosses the Onoro R.	
13 July 1843	Mt. Zibingaatzacko and Darura R.; arrives at a Maopityans village	describes the Maopityans
19 July 1843	enters R. Cafuini (Caphiwuin) [BRA]	*Carapa guianensis, Tachigali pubiflora,* and *Clusia insignis*
29 July 1843	junction of the R. Anamu (Wanamu) and R. Cafuini [BRA]	Indian picture writing; two canoes got lost
3 Aug 1843	Zibi C. [BRA]	one canoe tumbles over; a hunting dog drowns
5 Aug 1843	reaches deserted Pianoghotto settlement [BRA]	Maopityans plunder village; meteorological observations
16 Aug 1843	descends the R. Anamu and R. Iriau [BRA]	
22 Aug 1843	Aramatau R.; Pianoghotto settlement	*Bertholletia excelsa*; astronomical observations
1 Sep 1843	crosses the Kutari (Koetari) R. to next settlement	visits some Drios Indians
5/6 Sep 1843	descends the Kutari R. [GUY/SUR-border]	*Vantanea guianensis*
11 Sep 1843	entrance of the Courantyne (Corentyn) R. [GUY/SUR-border]	
12 Sep 1843	Sir Walther Raleigh's C. [GUY/SUR-border]	Goodall makes a sketch of the cataract
18 Sep 1843	Falls and Rapids of the Thousand Islands [GUY/SUR-border]	*Elizabetha coccinea*
22 Sep 1843	Frederick William C. [SUR]	Robert names this rapid after the King of Prussia
24 Sep 1843	reaches path from Courantyne R. to Essequibo R.	
27 Sep 1843	Great Cataracts in Courantyne R. [GUY/SUR-border]	barometrical observations
1 Oct 1843	arrives at Tomatai settlement [SUR]	
9 Oct 1843	passes New Amsterdam	
12 Oct 1843	arrives in Georgetown	

JOURNEYS TO THE POMEROON, MORUCA,
AND DEMERARA RIVERS
(AUGUST 1843–4 JUNE 1844)

After his return to the coast, Richard made some excursions while his brother was exploring the Courantyne River. He wanted to replace some of the lost or damaged specimens. For this purpose he investigated the fauna and flora along the Pomeroon, Moruca (Morocco or Maruka), and Demerara Rivers. In map 2 the yellow broken line indicates the route of Richard's excursions to the Rivers Pomeroon, Moruca, and Demerara.

Richard, accompanied by Stöckle, started his journey to the sources of the Pomeroon River in August 1843. He wanted to re-collect some of the living orchids, as his first collection had been lost during transportation to Berlin. Richard and Stöckle travelled to Anna Regina, where they collected plants and obtained extra provisions. Richard also observed the variety of fishes in the canals of the plantation at Anna Regina. Together with several Arawak Indians they continued towards Tapakuma Lake. They made botanical excursions and found species like *Cypripedium palmifolium* (=*Selenipedium palmifolium*), and the new species *Tovomita schomburgkiana* (=*T. schomburgkii*) and *T. macrophylla* (=*T. obovata*).

After 8 days they continued towards the Pomeroon River and the mouth of the Arapiacro River. They halted at a missionary house called 'Pomeaco House', and proceeded towards the Carib settlement Kuamuta. This village is named after a bamboo species the Caribs called 'Kuamuta'. On September 9th Richard left the village and continued his trip on the Pomeroon River. He paid a visit to Arraia, another Carib settlement. At this last settlement along the Pomeroon River, he learned how the Caribs used *Clibadium asperum* (=*C. surinamensis*) as their fish poison. Along the banks of the Pomeroon River Richard collected several species of *Vismia* like *V. guianensis*, *V. sessilifolia*, *V. cayennensis*, and *V. latifolia*. Strong currents, stormy weather, and fallen trees prevented them from reaching the source of the Pomeroon River at Mount Imataca. Richard returned to the Sururu River where he found a flowering species of *Strychnos toxifera*. Details about Richard's excursion to the Pomeroon River are given in table 9.

After a short stay at 'Caledonia' estate, Richard left the Pomeroon River. His brother had still not arrived in Georgetown, which made Richard decide to continue his journey towards the Moruca River. He and Stöckle ascended the Moruca River until the mouth of the small Kuamuto streamlet where they halted at a settlement of the same name. On October 7th Richard and three Warrau Indians departed from Kuamuto and continued towards the Barabara River. Richard was fortunate to find many orchids. Trees were covered with flowering species of *Maxillaria*, *Oncidium*, *Pleurothallis*, and *Zygopetalum*. On October 14th they followed the Waini River, inhabited by Warrau and

Table 9. Journey to the Pomeroon River.

Date	Location	Observations, collections, and main events
	Robert Schomburgk	
Aug 1843	travels to Anna Regina	
16 Aug 1843	Tapakuma Lake	botanical excursion into forest
24 Aug 1843	Tapakuma R.	
25 Aug 1843	reaches mouth of Arapiacro R.	
29 Aug 1843	follows the Pomeroon R. until Kuamuta settlement	new species and genera
9 Sep 1843	departs from Kuamuta settlement and follows the Pomeroon R.	
10 Sep 1843	reaches mouth of Sururu R. and Makaiku R. travels further up the Pomeroon R. travels to 'Caledonia' estate reaches mouth of Wakapau R. and Pomeroon R.	*Mora, Lecythis,* and *Sloanea* species *Vismia* species *Siphonia schomburgkii* (=*Hevea pauciflora*)

Akawai Indians, and entered the Barama River. Again Richard was surprised by a *Strychnos* species. It turned out to be a new species. He called it *Strychnos mitscherlichii*, named after the famous chemist E. Mitscherlich. They ascended the Barama River until they reached Kariako (Cariacu) where Robert had been in July 1841. After a very short stay, he returned to the mouth of the Waini River. Along the coast Richard recognised different sea fishes and birds. Unfortunately, the rainy season set in and, suffering from severe attacks of fever, he was not able to travel to the mouth of the Río Orinoco. Instead he decided to return. On his way back, he was delighted when he received the news that his brother had arrived safely in Georgetown. After an absence of four months Richard and Stöckle arrived in Georgetown as well.[13] Details about Richard's excursion to the Moruca River are given in table 10.

On April 1st Richard and several Arawak Indians started an excursion to the Demerara River. They halted at a Dutch establishment for a few days and looked for flowers of the Greenheart (*Nectandra rodiei= Chlorocardium rodiei*). Luckily, Richard found a flowering specimen. This species is named after Dr. Rodie, who discovered the medicinal qualities of the bark. As usual Richard described the vegetation and fauna extensively. He mentioned the occurrence of *Alexandra imperatricis* (=*Alexa imperatricis*), discovered by his brother on his trip to the Cuyuni River. Richard followed the Demerara River until he reached Post Seba. From this point it is possible to take a path towards the Berbice River. Robert had walked this track already in 1836. At Ororo-Marali or Great Fall the journey up the Demerara River ended.

---

[13] Data were taken from Schomburgk, M.R. 1848. Reisen in British-Guiana in den Jahren 1840-1844. Pages 409–464.

Table 10. Journey to the Moruca River.

Date	Location	Observations, collections, and main events
	Robert Schomburgk	
	enters Moruca (Morocco or Maruka) R.	
	reaches mouth of Itabo R. and arrives at Kuamuto settlement	
4 Oct 1843	departs from Kuamuto settlement and enters the Barabara R.	many orchids
14 Oct 1843	follows the Waini R. upstream and enters the Barama R.	*Strychnos mitscherlichii*
17 Oct 1843	reaches Kariako (Cariacu) village	
21 Oct 1843	travels to the mouth of the Waini R.	
	returns towards the Ascota R.	
	reaches Marukao mission and Morocco settlement	
	reaches the mouth of the Pomeroon R. and 'House Pomeaco'	
	Anna Regina	
	Georgetown	

For the last time Richard returned to Georgetown. He did not accompany the other members of the boundary expedition to Europe because he had to take large numbers of specimens, including living palms, orchids, and animals with him. They left in May 1844, and Richard took a merchant ship to Europe on June 4th.[14] Details about Richard's excursion to the Demerara River are given in table 11.

Table 11. Journey to the Demerara River.

Date	Location	Observations, collections, and main events
	Robert Schomburgk	
1 Apr 1844	departs from Georgetown and travels to the Demerara R.	
	reaches Stanley's town and the mouth of Hubabu R.	flowers of *Nectandra rodiei* (=*Chlorocardium rodiei*)
	passes the mouth of Madewini R.	
	arrives at sandbank near Golden Hill	
	passes Post Seba and Seba Hills	
	reaches end of journey at Ororo-Marali or Great F.	
	makes short excursions to the East and West coast	
4 Jun 1844	returns to Europe	

---

[14] Data were taken from Schomburgk, M.R. 1848. Reisen in British-Guiana in den Jahren 1840–1844. Pages 482–512.

6. THE COLLECTIONS OF ROBERT AND RICHARD SCHOMBURGK

In 1839 Robert returned to Europe with a remarkable collection. All geographical information was to be considered the property of the Royal Geographical Society (Rodway, 1889). Many of his valuable maps can still be found at the Society's headquarters in London or at the State Library in Berlin.

Robert presented more than 60 specimens of birds to the British Museum of Natural History in London. His collections of fishes caught in the rivers of the interior British Guiana were preserved in spirit. There is a two-volume book on the fishes of Guiana, including drawings of different species (Schomburgk, R.H., 1860). Preserved animals and skulls were given to the Linnaean Society or to the Zoological Society. An ethnological collection, consisting of weapons and household utensils of the Indians were also sent to London and are now held by the Museum of Mankind. After his first trip Robert even took three Indians back to London. The Indians were introduced to the Aborigines Protection Society and were observed with great interest by the visitors to the Guyana exhibition in London (Rodway, 1889). The Royal College of Surgeons was presented with some skulls and skeleton of Indians.

BOTANICAL COLLECTION OF ROBERT (R.H.) SCHOMBURGK

One hundred and six wood specimens were given to the Model room of the Admiralty (Schomburgk, R.H., 1845a). Seeds were collected for the Horticultural Society, and a considerable number of living orchids were given to the Royal Botanic Gardens at Kew. Some fine botanical illustrations from the expedition, drawn by Robert, are now in the Natural History Museum in London. Six copies of those original watercolour drawings are included in the present book. Some of the drawings contain information about the species, collection number, and locality written by Robert Schomburgk (plate 5). Although there were many losses from water damage, botanical specimens were subsequently sent to London. His collections consisted of thousands of dried plant specimens, an amount of dried fruits, and several other botanical objects preserved in spirit. Dried plants were sent to G. Bentham in London. Both at the Herbarium of the British Museum (BM) in London and in Kew (K) a complete set of Robert's plants, collected during his travels from 1835–1839 and 1841–1843, can still be examined. Duplicates were distributed to various herbaria like: B, BR, CGE, DS, F, FI, G, G-DC, HAL, L, LIVU, LZ (destroyed), MANCH, NY, OXF, P, PH, TCD, U, UPS, US, and W (Stafleu & Cowan, 1967–1988). These acronyms for herbaria are according the Index Herbariorum (Holmgren et al., 1990).

Robert himself did not have the time to study his collections adequately and elaborate the results but there were others to do so. People such as Lindley, Nees von Esenbeck, and W.J. Hooker were ready to investigate the specimens. Many Neotropical plant collections of Robert Schomburgk were studied by Bentham, who also published the plant names of Robert's collections in the Journal of Botany. By tracing Bentham's handwriting it is known that he studied the plant collections of Robert not only in the British Museum in London but also in the Webb herbarium in Florence (Steinberg, 1977). Descriptions of plant species were published in both scientific periodicals of that time and newspapers. One example is Robert's description of *Ophiocaryon paradoxum* or the Snake-nut tree, where he also mentioned the assistance of Bentham. This species derived its name from the inside of the nut, which resembles a coiled up snake (Schomburgk, R.H., 1840b; 1845b). Robert gave a full description of the Snake-nut tree and plate 6 shows a flowering branch of the species and detailed drawings of the nut. The most remarkable species he collected during his first trip was the giant water lily *Victoria regia* (=*V. amazonica*). He brought back seeds and living plants of it. Although other European travellers had reported it, Robert Schomburgk was the first to make a detailed description, which he sent to Britain.

BOTANICAL COLLECTION OF
RICHARD (M.R.) SCHOMBURGK

Richard Schomburgk was commissioned by the Prussian Government to accompany his brother and to collect specimens of flora and fauna of the British Guiana forest for the Royal Museum and the Botanical Gardens in Berlin. On his return trip to Europe salt-water damage, climatic changes, and other factors led to the loss of specimens. Richard lost 80 species of living orchids and only 60 of 200 palms were alive when they reached Berlin. The exact number of dried specimens taken back to Berlin by Richard is not easy to ascertain. Moreover a large part of his collection had been destroyed, due to a fire during the Second World War in Berlin. Fortunately a large amount of his fern collection can still be examined at the herbarium in Berlin (B). Duplicates were distributed to various herbaria like: BM, BR, C, CGE, F, G, GOET, K, MO, L, NY, OXF, P, UPS, US, W (Stafleu & Cowan, 1967–1988).

People such as Klotzsch, Nees von Esenbeck, Bartling, Grisebach, and Schultz 'Bipontinus' assisted Richard by compiling a synopsis of the flora of British Guiana, published in the third volume of his 'Reisen in British-Guiana' (Schomburgk, M.R., 1876). This synopsis contains a general description of the flora of British Guiana and descriptions of the characteristic flora of its four different regions: coast, primitive forest, sandstone formation, and savanna.

BOTANICAL COLLECTION SERIES OF ROBERT AND RICHARD SCHOMBURGK

Mainly due to their different collection series, the enumeration of the Schomburgk collections is rather complex. Robert for example made four series of botanical collections. This section will explicate the different collection series.

Robert's first collection series (Rob. ser. 1)
Robert Schomburgk collected his first series between the years 1835 and 1839 and has to be cited as Rob. ser. 1. Figure 3 shows an example of the plant label of Robert's first collection series. Data about his first collection series are listed in index 1 in the next chapter. Most records of Robert's first collection series were documented.

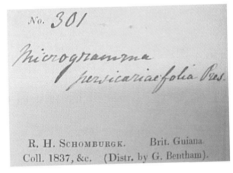

Figure 3. An example of a plant label of Robert Schomburgk's first collection series.

Robert's additional collection series (Rob. add. ser. 1)
Between the years 1835 and 1839 Robert also collected small sets of plants, only distributed to one or two herbaria, better known as his 'single plants' collection. This 'single plants' collection is enumerated separately and can be cited as Rob. add. ser. 1. The collections belonging to this additional series can be recognised by the extra 'S', which is added at the end of each collection number (figure 4). Sometimes the 'S' could be misinterpreted as the number '5'. Data about his additional collection series are listed in index 2 in the next chapter.

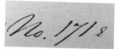

Figure 4. An example of a fragment of a plant label of Robert Schomburgk's additional collection series with the additional 'S' at the end of the collection number.

Robert's second collection series (Rob. ser. 2)

Robert collected his next series between the years 1841 and 1844 and can be cited as Rob. ser. 2. The labels of Robert's collections from his second series have generally two numbers. The first number indicates the number of Robert's second collection series while the second number refers to the collection number of his brother Richard (figure 5). Data about his second collection series are listed in index 3 in the next chapter. Most records of Robert's second collection series were documentented.

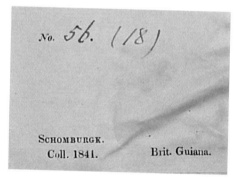

Figure 5. An example of a plant label of Robert Schomburgk's second collection series.

Robert's Roraima collection series

Apart from the above mentioned collection series Robert also collected a small set of 'Roraima' plants in the period between December 1842 and September 1843. Most likely this is also a small set of single plants, because on occasion in his field notes he mentioned that he only found one specimen. One part of this so-called 'Roraima' series, he collected near Mount Roraima after he and his brother separated. This same 'Roraima' series ends with collections from his expedition to the Courantyne River, also without his brother Richard. Data about his Roraima collection series are listed in index 4 in the next chapter.

Richard's collection series (Rich.)
Many of Richard's collections, to be cited as Rich., have corresponding Rob. ser. 2 numbers. Still there are many Rich. collection numbers, which do not have a corresponding Rob. number. Many Pteridophytes for example were collected only by Richard Schomburgk and not by his brother (figure 6). Data about Richard's collection series are listed in index 5 in the next chapter. Unfortunately, many records of Richard's collection series are missing. Presumably many specimens were lost during transportation to Europe. About the corresponding numbers there is a lot of confusion, for wheras both numbers refer to the same species, this does not mean that they were collected at the same time by both brothers. When examining their collection it often turns out that collections with corresponding numbers were collected at different localities.

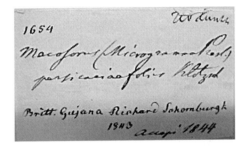

Figure 6. An example of a plant label of Richard Schomburgk's collection series.

Sine numero collections of Robert and Richard Schomburgk
In addition to the different collection series there are also many Schomburgk collections without a number, which makes it even more difficult to identify the real collector. Data about those collections are listed in index 6 in the next chapter.

7. INDEXES OF PLANT COLLECTIONS

This chapter lists all data found about the Schomburgk collections. An important part of the information about the plant collections has been gathered from Bentham's contributions to the flora of Guiana published in the London Journal of Botany, contributions towards the flora of South America published in Hooker's Journal of Botany, and the enumeration of plants collected by Robert and Richard Schomburgk published in the Annals of National History (Bentham, 1839ab; 1840; 1841; 1842abc; 1843; 1844; 1845ab; 1846ab). Further data were taken from the Flora of the Guianas, Flora Neotropica monographs, and from the contributions of B. Maguire and collaborators to the Botany of the Guayana Highlands. In this way most of the Schomburgk collections were recorded. Field notes made by Robert about his first and additional collections series, typed from the manuscript lists preserved at Kew, were used to verify the collections. Furthermore, detailed writings of both Robert and Richard Schomburgk about their journeys were used to check the species they found. For the compilation of the collection lists data from the various sources were combined with additional data from herbarium specimens from various herbaria and information from specialists working on a revision for the Flora of the Guianas.

Unfortunately, there is still a part of the data missing, probably due to losses during the expeditions or on their way back to Europe. Moreover, it was impossible to check all the herbarium specimens for Schomburgk collections at B, BM, and U. Any additions or corrections brought to my attention will be most welcome.

The lists of Schomburgk collections contain information about the original and current plant names, dates, localities, and herbaria including the information whether it is a type specimen or not. In many cases the original names changed after (re)identification. It seemed desirable to publish both plant names. Remarkable is the high amount of 'nomina nuda', mainly introduced by Klotzsch.

No evidence has been discovered that the Schomburgk brothers gave numbers to their gatherings in the sequence in which they collected. The evidence from Robert's letters to Bentham and his field notes points rather to the probability of his having numbered them at convenient stages of the journey when he had the time to rest and prepare the despatch of his boxes of specimens (Sandwith, 1936). Presumably, the Schomburgk brothers sorted their combined collections, and then divided them according to some prearranged system, so that each as far as possible retained a part of each gathering. They probably often included in a single gathering all the specimens supposedly belonging to one species, even though these had been collected at different times and places (Maguire et al., 1969). This could explain the amount of mixed collections of closely resembling species.

One has to keep in mind that the year of a collection sometimes does not correspond with the exact year when one or both of the Schomburgk brothers were at a certain locality. Packages of plant collections were sent to London or Berlin from where they were further distributed to other herbaria all over the world. On arrival at another herbarium, specimens sometimes got labelled with the year of acceptance that differs from the collection date (figure 6).

Noticeable were the numerous herbarium specimens from Paris with number 33. This number is not the collection number, but probably refers to the package number of that collection.

The Guyanan plant collection of Robert and Richard also includes specimens collected outside the present-day borders of Guyana. Some of Schomburgk's herbarium specimens with 'British Guiana' on their label were actually gathered in Venezuela, Brazil, or Suriname. Robert collected for example many plants of his first series in North Brazil on the Rio Negro and the Rio Branco. Some plants of his first and second series could actually have originated from Suriname because he ascended the Courantyne River in 1837 and descended it in 1843 and visited some villages on the Surinamese side. The Schomburgk brothers explored the Roraima region in 1838 and 1842 on the South Western slopes, which belong to the Venezuelan part of the mountain (Prance, 1971; Steyermark, 1981).

Within the lists of collections several abbreviations and symbols are used. The meaning of the abbreviations and symbols are explained below. Names between brackets refer to older or alternative spellings. Three-lettered acronyms, introduced by Weber in 1982, are used as a code for family names (Weber, 1982; Snow & Holton, 2000). Three-lettered acronyms which are used in the index are listed in table 12.

ht	= holotype	n.n.	= nomen nudum
it	= isotype	s.n.	= sine numero
ilt	= isolectotype	?	= unknown
lt	= lectotype	*	= extra collection with the same number
nt	= neotype	†	= destroyed
pt	= paratype	F.	= Fall(s)
st	= syntype	I.	= Island
tc	= type collection	Mt(s).	= Mountain(s)
frag	= fragment	R.	= River, Río or Rio

54

Table 12. Three-lettered acronyms for the families of vascular plants.

ACA	Acanthaceae	CTH	Cyatheaceae	LCS	Lacistemataceae	PLP	Polypodiaceae
ADI	Adiantaceae	CUC	Cucurbitaceae	LCY	Lecythidaceae	POA	Poaceae
ALI	Alismataceae	CUN	Cunoniaceae	LIL	Liliaceae	PON	Pontederiaceae
AMA	Amaranthaceae	CYC	Cyclanthaceae	LNT	Lentibulariaceae	PRT	Proteaceae
ANA	Anacardiaceae	CYP	Cyperaceae	LOG	Loganiaceae	PSL	Psilotaceae
ANN	Annonaceae	CYR	Cyrillaceae	LOM	Lomariopsidaceae	PTR	Pteridaceae
APO	Apocynaceae	DAV	Davalliaceae	LOR	Loranthaceae	QII	Quiinaceae
AQF	Aquifoliaceae	DCH	Dichapetalaceae	LYC	Lycopodiaceae	RHM	Rhamnaceae
ARA	Araceae	DLL	Dilleniaceae	LYG	Lygodiaceae	RHZ	Rhizophoraceae
ARE	Arecaceae	DRO	Droseraceae	LYT	Lythraceae	ROS	Rosaceae
ARS	Aristolochiaceae	DRY	Dryopteridaceae	MEL	Meliaceae	RPT	Rapateaceae
ASC	Asclepiadaceae	DSC	Dioscoreaceae	MIM	Mimosaceae	RUB	Rubiaceae
ASL	Aspleniaceae	DST	Dennstaedtiaceae	MLP	Malpighiaceae	RUT	Rutaceae
AST	Asteraceae	EBN	Ebenaceae	MLS	Melastomataceae	SAB	Sabiaceae
BIG	Bignoniaceae	ELC	Elaeocarpaceae	MLV	Malvaceae	SAP	Sapindaceae
BLE	Blechnaceae	EPR	Euphroniaceae	MNM	Monimiaceae	SAR	Sarraceniaceae
BML	Bromeliaceae	ERI	Ericaceae	MNS	Menispermaceae	SCR	Scrophulariaceae
BMN	Burmanniaceae	ERO	Eriocaulaceae	MNY	Menyanthaceae	SCZ	Schizaeaceae
BNN	Bonnetiaceae	ERX	Erythroxylaceae	MOR	Moraceae	SML	Smilacaceae
BNP	Balanophoraceae	EUP	Euphorbiaceae	MRC	Marcgraviaceae	SMR	Simaroubaceae
BOR	Boraginaceae	FAB	Fabaceae	MRN	Marantaceae	SOL	Solanaceae
BRS	Burseraceae	FLC	Flacourtiaceae	MRS	Myrsinaceae	SPT	Sapotaceae
CAB	Cabombaceae	GEN	Gentianaceae	MRT	Myrtaceae	STR	Sterculiaceae
CAC	Cactaceae	GLC	Gleicheniaceae	MTT	Marattiaceae	STY	Styracaceae
CAM	Campanulaceae	GMM	Grammitidaceae	MTX	Metaxyaceae	SYM	Symplocaceae
CAN	Cannaceae	GNE	Gnetaceae	MYS	Myristicaceae	TEA	Theaceae
CCR	Caryocaraceae	GSN	Gesneriaceae	NYC	Nyctaginaceae	TEC	Tectariaceae
CEC	Cecropiaceae	GUT	Guttiferae	NYM	Nymphaceae	TEO	Theophrastaceae
CEL	Celastraceae	HAE	Haemodoraceae	OCH	Ochnaceae	THL	Thelypteridaceae
CHB	Chrysobalanaceae	HLC	Heliconiaceae	OLC	Olacaceae	TIL	Tiliaceae
CLE	Clethraceae	HMP	Hymenophyllaceae	OLN	Oleandraceae	TNR	Turneraceae
CLR	Chloranthaceae	HPC	Hippocrateaceae	ONA	Onagraceae	TRG	Trigoniaceae
CLU	Clusiaceae	HUG	Hugoniaceae	ORC	Orchidaceae	ULM	Ulmaceae
CMB	Combretaceae	HUM	Humiriaceae	OXL	Oxalidaceae	URT	Urticaceae
CMM	Commelinaceae	HYD	Hydrophyllaceae	PAS	Passifloraceae	VIO	Violaceae
CNN	Connaraceae	ICC	Icacinaceae	PDS	Podostemaceae	VLL	Velloziaceae
CNV	Convolvulaceae	IRI	Iridaceae	PED	Pedaliaceae	VOC	Vochysiaceae
COT	Costaceae	IXO	Ixonanthaceae	PGL	Polygalaceae	VRB	Verbenaceae
CPP	Capparaceae	KRM	Krameriaceae	PHT	Phytolaccaceae	VTT	Vittariaceae
CPR	Caprifoliaceae	LAM	Lamiaceae	PIP	Piperaceae	XYR	Xyridaceae
CSL	Caesalpiniaceae	LAU	Lauraceae	PLG	Polygonaceae		

INDEX 1. ROBERT SCHOMBURGK'S FIRST COLLECTION SERIES (1835–1839).

Nr.	Year	Fam.	Original name	Current name	Herbarium	Location
1	1835	PIP	Artanthe adunca (L.) Miq.	Piper aduncum L.	BM K	Mazaruni R.
2	1836	MRT	Campomanesia glabra Benth.	Calycolpus goetheanus (Mart. ex. DC.) O. Berg	ht: K; it: BM F K US	Essequibo R.
3	1836	APO	Tabernaemontana heterophylla Vahl	Tabernaemontana heterophylla Vahl	BM	Essequibo R.
4		SOL	Solanum radula Vahl	Solanum asperum Rich.		
5		MLS	Clidemia elegans D. Don	Clidemia hirta (L.) D. Don var. elegans (Aubl.) Griseb.		Essequibo R.
6		RUB	Gardenia mussaenda L.f.	Randia mussaenda DC.		
7		CHB	Hirtella paniculata Sw.	Hirtella paniculata Sw.	BM BR F K LE NY P US	Essequibo R.
8		MLS	Clidemia miconioides Benth.	Miconia ibaguensis (Bonpl.) Triana		
8*		MLS	Miconia rubiginosa (Bonpl.) DC.	Miconia rubiginosa (Bonpl.) DC.		
9		OCH	Sauvagesia erecta L.	Sauvagesia erecta L.		Essequibo R.
10		CSL	Vouapa bifolia Aubl.	Macrolobium bifolium (Aubl.) Pers.		Essequibo R.
11	1836	RUB	Diodia articulata DC.	Borreria hyssopifolia (Roem. & Schult.) Bacigalupo & E.L. Cabral	BM	Essequibo R.
11*		RUB	Diodia articulata DC.	Borreria ocymifolia (Roem. & Schult.) Bacigalupo & E.L. Cabral	BM	Essequibo R.
12	1836	SPT	Sideroxylon cuspidatum A. DC.	Pouteria cuspidata (A. DC.) Baehni subsp. cuspidata	BM	Essequibo R.
13	1836	CSL	Campsiandra comosa Benth.	Campsiandra comosa Benth. var. comosa	st: BM K	Essequibo R.
14		LOG	Spigelia schomburgkiana Benth.	Spigelia flemmingiana Cham. & Schltdl.	tc: U	Essequibo R.
15	1836	RUB	Sipanea dichotoma Kunth	Sipanea pratensis Aubl. var. dichotoma (Kunth) Steyerm.	BM U	Essequibo R.
16		MLS	Spennera dysophylla Benth.	Aciotis annua (DC.) Triana		
17	1836	RUB	Oldenlandia herbacea (L.) DC.	Oldenlandia lancifolia (Schum.) DC.	BM	
18	1836	MTX	Amphidesmium blechnoides (Sw.) Klotzsch	Metaxya rostrata (Kunth) C. Presl	B BM K	Souhara on the Essequibo
19		FAB	Nicholsonia cayennensis DC.	Desmodium barbatum (L.) Benth. & Oerst.		Souhara on the Essequibo
20		LOG	Spigelia humilis Benth.	Spigelia humilis Benth.	BM K	Aripara on Essequibo; savanna Upper Rupununi
21		CSL	Cassia diphylla Lam.	Chamaecrista diphylla (L.) Greene		
22		EUP	Phyllanthus guianensis Klotzsch	Phyllanthus carolinensis Walter subsp. guianensis (Klotzsch) G.L. Webster	st: K	Essequibo R.
23		CHB	Hirtella americana L.	Hirtella racemosa Lam. var. racemosa	BM BR F GH K NY US	Essequibo R.
24	1836	OXL	Oxalis schomburgkiana Prog. var. leiocarpa Prog.	Oxalis frutescens L.	lt: K; ilt K US	Essequibo R. and Rupununi R.
25		MIM	Inga disticha Benth.	Inga disticha Benth.	ht: K; it: BM G K NY P U US	Essequibo R.

Index 1. Robert Schomburgk's first collection series (1835–1839) – continued

No.	Year	Fam.	Original name / current name	Types	Locality
26		RUB	Psychotria nervosa Benth. — Psychotria lupulina Benth. subsp. lupulina	tc: BM G U US	Essequibo R.
27	1836	RUB	Psychotria inundata Benth. — Psychotria capitata Ruiz & Pav. subsp. inundata (Benth.) Steyerm.	tc: BM G US	Essequibo R.
28		CHB	Moquilea comosa Benth. — Couepia comosa Benth.	ht: K; it: BM BR CGE F L LE NY US	falls of the Essequibo R.
29		LAU	Nectandra lambigua Meisn. — Nectandra amazonum Nees	tc: U US	Essequibo R. and Rupununi R.
30		RUB	Cephaëlis tomentosa Willd. — Psychotria poeppigiana Muell. Arg. var. poeppigiana		woods of the Essequibo R.
31	1836	STR	Melochia arenosa Benth. — Melochia arenosa Benth.	st: BM	Essequibo R.
32	1836	RUB	Chomelia angustifolia Benth. — Chomelia angustifolia Benth.	tc: BM NY US	falls of the Essequibo R.
33		EUP	Croton essequiboensis Klotzsch — Croton essequiboensis Klotzsch	tc: G U	Essequibo R.
34		FAB	Amphymenium rohrii Kunth — Pterocarpus santalinoides L'Hér, ex DC.		falls of the Essequibo R.
35		EUP	Discocarpus essequiboensis Klotzsch — Discocarpus essequiboensis Klotzsch	lt: BM; ilt: G K OXF P U W	Upper Essequibo R.
36	1836	EUP	Amanoa guianensis Aubl. — Amanoa guianensis Aubl.	st: BM F G K U	Essequibo R.
36*		CEL	Maytenus guianensis Klotzsch [n.n.] — Maytenus guyanensis Klotzsch ex Reissek	tc: U	
37	1836	APO	Thyrsanthus schomburgkii Benth. — Forsteronia acouci (Aubl.) A. DC.	BM U	
38	1836	HPC	Hippocratea aubletiana Miers — Peritassa compta Miers		
39		APO	Tabernaemontana gracilis Benth. — Malouetia gracilis (Benth.) A. DC	ht: K; it: BM F G K NY U	Upper Essequibo R.
39*		APO	Forsteronia	tc: L P US P	Upper Essequibo R.
39*		APO	Tabernaemontana undulata Vahl — Tabernaemontana undulata Vahl		Upper Essequibo R.
40	1836	EUP	Mabea schomburgkii Benth. — Mabea taquari Aubl.	lt: K; ilt: AAU BM BR E G K L P U W	Upper Essequibo R.
41	1836	BRS	Icica heptaphylla Aubl. — Protium heptaphyllum (Aubl.) Marchand subsp. heptaphyllum	BM	
41*		APO	Tabernaemontana longifolia Benth. — Tabernaemontana siphilitica (L.f.) Leeuwenb.		
42	1836	APO	Bonafousia undulata (Vahl.) A. DC. var. ovalifolia Miers — Bonafousia undulata (Vahl) A. DC.	ht: BM	Essequibo R.
43		CSL	Tachigali pubiflora Benth. ('Tachigalia') — Tachigali pubiflora Benth. ('Tachigalia')	tc: NY US	Essequibo R.
44		EUP	Croton nervosus Klotzsch var. villosus Klotzsch — Croton nervosus Klotzsch	st: G	Waraputa F.
45		CEL	Maytenus guianensis Klotzsch [n.n.] — Maytenus guyanensis Klotzsch ex Reissek	K U	Horotoko (Orotoko) F.
46		AMA	Serturnera guianensis Klotzsch [n.n.] — Pfaffia glomerata (Spreng.) Pedersen var. glomerata	NY U US	near Horotoko (Orotoko) F.
47					
48		ADI	Adiantum triangulatum Kaulf. — Adiantum latifolium Lam.	B	woods of Essequibo R. and Rupununi R.

No.		Family	Name	Accepted name	Collection	Locality
49		CSL	Martia excelsa Benth.	Martiodendron excelsum (Benth.) Gleason	st: K US	Essequibo R. and Rupununi R.
50		CHB	Licania coriacea Benth.	Licania coriacea Benth.	ht: K; it: BM BR CGE G L OXF P US	Essequibo R.
51		RUB	Psychotria fimbriata Benth.	Rudgea cornifolia (Humb. & Bonpl. ex Roem. & Schult.) Standley	ht: K; it: U US	Essequibo R.
52		FAB	Lonchocarpus densiflorus Benth.	Lonchocarpus densiflorus Benth.	tc: U	banks Upper Essequibo R. and Rupununi R.
53		PIP	Artanthe apiculata Klotzsch	Piper anonifolium (Kunth) C. DC.	ht: B; it: BM US	Kurupukari (Ouropocari) I.; Essequibo R.
53*		PIP	Artanthe berbicencis Miq.	Piper hostmannianum (Miq.) C. DC. var. berbicense (Miq.) C. DC.		Essequibo R.
54		TRG	Trigonia macrocarpa Benth.	Trigonia villosa Aubl. var. macrocarpa (Benth.) Lleras	ht: K; it: BM C CGE F G K NY US W	Essequibo R.
54*		TRG	Trigonia villosa Aubl.	Trigonia villosa Aubl. var. villosa	BM BR CGE G K L OXF TCD W	Essequibo R.
55		CMB	Combretum obtusifolium Rich.	Combretum pyramidatum Desv.		Essequibo R.
56	1836	TRG	Trigonia subcymosa Benth.	Trigonia subcymosa Benth.	lt: K; ilt: BM CGE G K NY U W	Essequibo R.
56*		TRG	Trigonia hypoleuca Griseb.	Trigonia hypoleuca Griseb.		Essequibo R.
57		MLP	Byrsonima crassifolia (L.) Kunth	Byrsonima crassifolia (L.) Kunth	U	savanna of the Parima Mts. and Kanuku Mts.
58		FAB	Neurocarpum cajanifolium C. Presl ('cajanaefolium')	Clitoria laurifolia Poir.		savanna at Annai
58*		FAB	Neurocarpum longifolium Mart. ex Benth.	Clitoria guianensis (Aubl.) Benth.	MEL US	savanna at Annai
59		CSL	Cassia flexuosa L.	Chamaecrista flexuosa (L.) Greene var. flexuosa	st: BM G K U	arid savanna
60		MLP	Byrsonima schomburgkiana Benth.	Byrsonima schomburgkiana Benth.	ht: K	savanna
60*		MLP	Byrsonima sessilifolia Benth.	Byrsonima coccolobifolia Kunth	B BM	savanna
61	1836	ONA	Jussiaea nervosa Poir. ('Jussieua')	Ludwigia nervosa (Poir.) Hara	BM	
61*	1836	ONA	Jussiaea nervosa Poir. ('Jussieua')	Ludwigia rigida (Miq.) Sandw.		
62		FAB	Crotalaria stipularia Desv.	Crotalaria stipularia Desv.		savanna
63		TRG	Trigonia villosa Aubl.	Trigonia villosa Aubl. var. villosa	BR CGE G OXF U W	Essequibo R.
63*		TRG	Trigonia subcymosa Benth.	Trigonia subcymosa Benth.		
64		CSL	Cassia lotoides Kunth	Chamaecrista hispidula (Vahl) H.S. Irwin & Barneby		savanna
65		CHB	Parinari coriaceum Benth. ('Parinarium')	Exellodendron coriaceum (Benth.) Prance	ht: K; it: BM BR CGE GH L NY OXF U US	brook Annai
66		FAB	Tephrosia brevipes Benth.	Tephrosia sessiliflora (Poir.) Hassl.	tc: U	savanna at Annai
67	1836	STR	Waltheria americana L.	Waltheria indica L.	BM	moist savanna
67*		STR	Waltheria paniculata Benth.	Waltheria paniculata Benth.	tc: BM G	moist savanna

Index 1. Robert Schomburgk's first collection series (1835–1839) – continued

No.	Year	Fam.	Original name	Current name	Collections	Locality
68	1836	MLV	Pavonia bracteosa Benth.	Peltaea trinervis (C. Presl) Krapov. & Cristóbal	st: BM U	moist savanna
69	1836	PRT	Andripetalum rubescens Pohl	Panopsis rubescens (Pohl) Pittier		brook Annai
70	1836	BOR	Tournefortia schomburgkii DC.	Tournefortia bicolor Sw. var. calycosa J.D. Sm.		Berbice
71		CSL	Bauhinia macrostachya Benth.	Bauhinia macrostachya Benth.	ht: G; it: BM K L P	woods skirting savanna
72		AST	Eupatorium conyzoides Mill.	Chromolaena odorata (L.) R.M. King & H. Rob.	tc: GH MEL US	woods at Parima Mts.
73		EUP	Euphorbia hypericifolia L. var. falciformis Klotzsch	Euphorbia hyssopifolia L. subsp. hyssopifolia	st: BM US P	
74		LYG	Lygodium venustum Sw.	Lygodium venustum Sw.		
75		VRB	Lippia microphylla Cham. & Schltdl.	Lippia origanoides Kunth	G P	woods skirting savanna
76	1836	AST	Eupatorium subvelutinum DC.	Chromolaena odorata (L.) R.M. King & H. Rob.	BM GH	on stony places in savanna
77		LYT	Cuphea antisyphilitica Kunth	Cuphea antisyphilitica Kunth var. acutifolia Benth.	K P	Rupununi savanna
77*	1836	MLV	Sida rhombifolia L.	Sida rhombifolia L.	P	moist savanna
78		FAB	Machaerium affine Benth.	Machaerium affine Benth.	ht: K; it: BM F P W	moist savanna
79		AST	Eupatorium ixodes Benth.	Ayapana amygdalina (Lam.) R.M. King & H. Rob.	tc: GH NY US	woods near Parima Mts. and Kanuku Mts.
80		CHB	Hirtella hexandra Willd.	Hirtella racemosa Lam. var. hexandra (Willd. ex Roem. & Schult.) Prance	BM BR F Fl K P US	Rupununi savanna
81	1836	PGL	Polygala appressa Benth.	Polygala appressa Benth.	tc: BM G GH K NY US W	moist savanna
82		MLS	Rhynchanthera acuminata Benth.	Rhynchanthera grandiflora (Aubl.) DC.		savanna at Annai
83		FAB	Dioclea guianensis Benth. var. guianensis	Dioclea guianensis Benth.	ht: K; it: F G U US W	shady woods at Parima Mts.
84	1836	CYP	Xyris eriophora Klotzsch	Bulbostylis lanata (Kunth) C.B. Clarke	BM	moist savanna
85		LOG	Antonia pilosa Hook.	Antonia ovata Pohl		Essequibo R.
85*		CYP	Isolepis lanata Kunth	Bulbostylis lanata (Kunth) C.B. Clarke	U	
86	1836	CSL	Cassia undulata Benth.	Senna undulata (Benth.) H.S. Irwin & Barneby	st: B † IPA K NY US	woods of the Essequibo R.
87		CMB	Combretum aurantiacum Benth.	Combretum fruticosum (Loefl.) Stuntz	tc: BM G K OXF P TCD W	Essequibo R.
87*		CMB	Combretum elegans Kunth	Combretum rotundifolium Rich.		
88		CSL	Peltogyne pubescens Benth.	Peltogyne paniculata Benth. subsp. pubescens (Benth.) M.F. Silva	st: NY US	Essequibo R.
89		CNV	Convolvulus agrestis Mart. ex Choisy	Jacquemontia agrestis (Choisy) Meisn.		
90		ADI	Adiantum triangulatum Kaulf.	Adiantum pulverulentum L.	BM	woods skirting savanna
91		MLP	Byrsonima verbascifolia (L.) DC.	Byrsonima verbascifolia (L.) DC.	B P	in woods among rocks
92		DLL	Curatella americana L.	Curatella americana L.	tc: BM K	Pirara
93		AST	Mikania convolvulacea DC.	Mikania cordifolia (L.f.) Willd.	BM	savanna
94		CSL	Cassia bacillaris L.f.	Senna bacillaris (L.f.) H.S. Irwin & Barneby var. bacillaris		woods near Parima Mts.
95	1836	RUB	Sipanea dichotoma Kunth	Sipanea pratensis Aubl. var. dichotoma (Kunth) Steyerm.	BM	moist savanna

96		FAB	Indigofera pascuorum Benth.	Indigofera lespedezioides Kunth	ht: G; it: U US	dry savanna
97		AST	Vernonia odoratissima Kunth	Vernonanthura brasiliana (L.) H. Rob.	U	rocky ground in Rupununi savanna
98	1836	XYR	Xyris	Xyris paraensis Poepp. ex Kunth var. paraensis	BM L U	moist savanna
98*		XYR	Xyris savanensis Miq.	Xyris savanensis Miq.		moist savanna
99		SCR	Buchnera rosea Kunth	Buchnera rosea Kunth		dry savanna
100	1836	RUB	Perama hirsuta Aubl.	Perama hirsuta Aubl. var. hirsuta	BM	moist savanna
100*	1836	RUB	Perama stricta Benth.	Perama hirsuta Aubl. var. stricta (Benth.) Bremek.	tc: BM NY US	moist savanna
101		EUP	Brachystachys hirta Klotzsch	Croton hirtus L'Hér.	tc: ? U	
102		DRS	Drosera dentata Benth.	Drosera sessilifolia J. St.-Hil.	ht: K; it: U US	moist savanna
103	1836	FLC	Casearia carpinifolia Benth.	Casearia sylvestris Sw. var. lingua (Cambess.) Eichl.	lt: K; ilt: BM F G K P U US	skirts of woods savanna
104		HYD	Hydrolea spinosa L.	Hydrolea spinosa L.		moist savanna
105	1836	TRN	Turnera guianensis Aubl.	Turnera guianensis Aubl.	BM G I K US	dry savanna
106		MLS	Microlicia recurva (Rich.) DC.	Acisanthera uniflora (Vahl) Gleason		moist savanna
106*		MLS	Comolia microphylla Benth.	Comolia microphylla Benth.	ht: K	moist savanna
107		ERO	Eriocaulon brevifolium Klotzsch ex Körn.	Eriocaulon tenuifolium Klotzsch ex Körn.	tc: GH K NY U	savanna
108		CYP	Scleria bracteata Cav.	Scleria bracteata Cav.	tc: K MANCH US	
109		BOR	Cordia sericicalyx A. DC.	Cordia sericicalyx A. DC.	ht: G; it: B P US	
109*		CYP	Kyllinga decora Steud.	Ascolepis brasiliensis (Kunth) Benth. ex C.B. Clarke	ht: W; it: BM K U W	
110	1836	MRT	Psidium parviflorum Benth.	Psidium striatulum DC.	tc: BM G US	
111		CHB	Licania leptostachya Benth.	Licania leptostachya Benth.	ht: K; it: CGE F G L LE NY U US W	Essequibo R. and Rupununi R.
112		CHB	Moquilea multiflora (Benth.) Walp.	Couepia multiflora Benth.	ht: K; it: BM BR CGE F GH L LE NY OXF P US	Upper Rupununi R.
113	1836	CHB	Hirtella rubra Benth.	Hirtella ciliata Mart. & Zucc.	ht: K; it: BM CGE GH LE NY OXF P U	savanna near Pirara
114		EUP	Peridium bicolor Klotzsch	Pera bicolor (Klotzsch) Muell. Arg.	ht: B †; lt: G; ilt: BM G K P U W	
115		CSL	Schnella rubiginosa (Bong.) Benth.	Bauhinia rubiginosa Bong.	MEL	Rupununi R.
116		PHT	Ancistrocarpus maypurensis Kunth	Microtea maypurensis (Kunth) G. Don	K NY U	Essequibo R.
117		AMA	Mogiphanes			
118		SLG	Selaginella pedata Klotzsch	Selaginella parkeri (Hook. & Grev.) Spring	ht: B; it: BM P K	
119	1835–36	VIO	Alsodeia pubiflora Benth.	Rinorea pubiflora (Benth.) Sprague & Sandw. var. pubiflora	BM G K OXF P US W	Essequibo R. and Aripai R.
119*	1836	VIO	Alsodeia flavescens Spreng.	Rinorea flavescens (Aubl.) Kunze	US	

Index 1. Robert Schomburgk's first collection series (1835–1839) – continued

No.	Year	Code	Name	Current name	Herbarium	Locality
120	1836	RUB	Coussarea schomburgkiana (Benth.) Benth. & J.D. Hook	Coussarea violacea Aubl.	tc: BM W	
120*		RUB	Guettarda xylosteoides Kunth	Guettarda divaricata (Roem. & Schult.) Standley	BM	
121	1836	VIO	Ionidium oppositifolium (L.) Roem. & Schult.	Hybanthus oppositifolius (L.) Taub.	BM	
122	1836	MLV	Sida glomerata Cav.	Sida glomerata Cav.	BM	Rupununi R.
123	1836	TRN	Turnera ulmifolia L.	Turnera scabra Millsp.	BM G K P US	
124	1836	MNS	Cissampelos crenata DC.	Cissampelos ovalifolia DC.	BM	
125	1836	VIO	Alsodeia laxiflora Benth.	Rinorea brevipes (Benth.) S.F. Blake	ht: K; it: BM CGE G L K OXF P US W	savanna near Pirara Rupununi R.
126	1836	CNN	Rourea revoluta Planch.	Rourea revoluta Planch. var. revoluta	ht: G; it: BM F U	Essequibo R.
127	1836	TRN	Piriqueta lanceolata Benth.	Piriqueta cistoides (L.) Griseb. subsp. cistoides	ht: K; it: BM Fl G K P U US W	Rupununi R.
127*		FAB	Desmodium benthamianum Klotzsch [n.n.]	Desmodium sclerophyllum Benth.	tc: BM G P U	
128	1836	AST	Wedelia scaberrima Benth.	Wedelia scaberrima Benth.	BM	
129	1836	AST	Baccharis leptocephala DC.	Baccharis varians Gardner		
130	1836	MRT	Eugenia nitida Benth.	Myrcia inaequiloba (DC.) D. Legrand	st: BM US	Essequibo R. and Rupununi R.
131	1836	MLV	Sida linifolia Cav.	Sida linifolia Cav.	BM	savanna at Annai
132		EUP	Caperonia angustissima Klotzsch [n.n.]	Caperonia stenophylla Muell. Arg.		
133		MLS	Chaetogastra hypericoides DC.	Desmoscelis villosa (Aubl.) Naudin		
133*	1836	STR	Melochia vestita Benth.	Melochia spicata (L.) Fryxell	st: BM	savanna
134		EUP	Traganthus sidoides Klotzsch	Bernardia sidoides (Klotzsch) Muell. Arg.	ht: ?; it: G K	Annai
135		DRY	Aspidium inerme Fée	Cyclodium inerme (Fée) A.R. Sm.	B BM K	
135*		THL	Thelypteris schomburgkii A.R. Sm.	Thelypteris schomburgkii A.R. Sm.	ht: K; it: G K	Essequibo R.
136		CHB	Licania pubiflora Benth.	Licania apetala (E. Mey.) Fritsch var. aperta (Benth.) Prance	ht: K; it: BM CGE NY OXF P	Upper Essequibo R.
137	1836	MLV	Pavonia typhalaea Cav.	Pavonia fruticosa (Mill.) Fawc. & Rendle	BM GH K MO OXF US	inlet Primos on the Upper Essequibo R.
138		PLG	Symmeria paniculata Benth.	Symmeria paniculata Benth.	st: BM G NY	Upper Essequibo R.
139	1836	RUB	Psychotria polycephala Benth.	Psychotria polycephala Benth.	lt: K; ilt: BM F G K P US W	
140	1836	FLC	Casearia laurifolia Benth.	Casearia commersoniana Cambess.	lt: L U US W	Upper Essequibo R.
141		SOL	Schwenckia grandiflora Benth. ('Schwenkia')	Schwenckia grandiflora Benth.	tc: NY U	
142	1836	THY	Goodallia guianensis Benth.	Goodallia guianensis Benth.	lt: K; ilt BM BR F Fl G K L P U US W	Curassawaka, near Rupununi R.
143	1836	FLC	Casearia densiflora Benth.	Casearia commersoniana Cambess.	BM K L US	near brook Curassawaka
144	1836	MLP	Hiraea chrysophylla A. Juss.	Hiraea faginea (Sw.) Nied.		

No.	Date	Fam	Name (Schomburgk)	Determination	Herbaria	Locality
145	1836	TIL	Vasivaea alchorneoides Baill.	Vasivaea alchorneoides Baill.	st: BM U US	
146		POR	Talinum patens (L.) Willd.	Talinum patens (L.) Willd.	BM	Upper Essequibo R.
147		SAP	Paullinia barbadensis Jacq.	Paullinia dasygonia Radlk.		
148		CSL	Mora guianensis Benth.	Mora excelsa Benth.		
149	1837	AST	Vernonia tricholepis DC. var. microphylla Benth.	Lepidaploa remotiflora (Rich.) H. Rob.	BM	Upper Essequibo R.
150		SOL	Solanum callicarpaefolium Kunth & Bouché			brook Curassawaka
151	1837	POA	Oplismenus crus-galli (L.) Kunth	Echinochloa crus-pavonis (Kunth) Schult.	BM E F L Z	Rupununi R.
152	1836	GEN	Coutoubea ramosa Aubl.	Coutoubea ramosa Aubl. var. racemosa (G. Mey.) Benth.	BM	brook Curassawaka
153	1836	AST	Trichospira menthoides Kunth	Trichospira verticillata (L.) S.F. Blake	BM US	brook Curassawaka
154		AST	Sparganophorus vaillantii Gaertn.	Struchium sparganophorum (L.) Kuntze	st: US	Kanuku Mts.
155		LOG	Strychnos toxifera R.H. Schomb. ex Benth.	Strychnos toxifera R.H. Schomb. ex Benth.		
156		LOG	Strychnos cogens Benth.	Strychnos cogens Benth.	BM G K U	Berbice R.
157		LAU	Persea	Aniba canelilla (Kunth) Mez		along brook Curassawaka
158		VRB	Petrea macrostachya Benth.	Petrea macrostachya Benth.	ht: K; it: BM BR F G GH L NY	Mt. Annai
159		BML	Tillandsia usneoides (L.) L.	Tillandsia usneoides (L.) L.	BM U	Essequibo R. and Rupununi R.
160		ARE	Astrocaryum jauari Mart.	Astrocaryum jauari Mart.	BM	Upper Rupununi R.
161	1836	RUB	Diodia barbata DC.	Diodia apiculata (Willd. ex Roem. & Schult.) K. Schum.	BM	arid savanna of Annai
162	1836	OCH	Sauvagesia sprengelii A. St.-Hil.	Sauvagesia sprengelii A. St.-Hil.	US	moist savanna
163	1836	OCH	Sauvagesia surinamensis Miq.	Sauvagesia rubiginosa A. St.-Hil.	tc: BM	arid savanna
164	Mar 1836	GEN	Oclinium clavatum Benth.	Irlbachia caerulescens (Aubl.) Griseb.	BM F G K L P W	Essequibo R. and Rupununi R.; arid savannas
165	1836	AST	Lisianthus caerulescens Aubl.	Chromolaena ivaefolia (L.) R.M. King & H. Rob.		Essequibo R. and Rupununi R.; arid savannas
166	Mar 1836	PGL	Polygala longicaulis Kunth	Polygala longicaulis Kunth		Essequibo R. and Rupununi R.; arid savannas
166*	Mar 1836	PGL	Polygala adenophora DC.	Polygala adenophora DC.	G	Essequibo R. and Rupununi R.; arid savannas
166*	Mar 1836	PGL	Polygala trichosperma L.	Polygala trichosperma L.	P	Essequibo R. and Rupununi R.; arid savannas
167	Mar 1836	PGL	Polygala longicaulis Kunth	Polygala longicaulis Kunth	W	arid savanna of the Rupununi
167*		GEN	Schuebleria coarctata Benth.	Curtia tenuifolia (Aubl.) Knobl.		
168		APO	Tabernaemontana sp.	Tabernaemontana sp.		
169		FAB	Etaballia guianensis Benth.	Etaballia dubia (Kunth) Rudd		Essequibo R.
170	1837	BIX	Bixa orellana L.	Bixa orellana L.	BM	
171	1837	CNV	Ipomoea tamnifolia L.	Jacquemontia tamnifolia (L.) Griseb.	BM	savanna
172	1837	EUP	Euphorbia dioeca Kunth	Euphorbia dioeca Kunth	BM K L P U	
173		FAB	Tephrosia toxicaria Pers.	Tephrosia sinapou (Bucholz) A. Chevalier		dry savanna of the Rupununi

Index 1. Robert Schomburgk's first collection series (1835–1839) – continued

No.	Year	Fam.	Schomburgk name	Current determination	Herbaria	Habitat
174		ANA	Tapirira guianensis Aubl.	Tapirira guianensis Aubl.		
175		CYP	Isolepis junciformis Kunth	Bulbostylis junciformis (Kunth) C.B. Clarke		abandoned fields
175*		CYP	Cyperus aurantiacus Kunth	Cyperus amabilis Vahl		savanna
176		CSL	Cassia disadena Steud.	Chamaecrista nictitans (L.) Moench var. disadena (Steud.) H.S. Irwin & Barneby	ht: K; it: BM NY US	savanna
177		SOL	Solanum juripeba Rich.	Solanum subinerme Jacq.		
178		FAB	Stylosanthes viscosa Sw.	Stylosanthes viscosa Sw.		
179	1837	RUB	Commianthus schomburgkii Benth.	Retiniphyllum schomburgkii (Benth.) Muell. Arg.		
180		ACA	Aphelandra pectinata Willd. ex Nees	Aphelandra scabra (Vahl) Sm.		
181		FAB	Aeschynomene paniculata Willd. ex Vogel	Aeschynomene paniculata Willd. ex Vogel	E GH US	savanna
182	1837	CLU	Vismia guianensis (Aubl.) Choisy	Vismia guianensis (Aubl.) Choisy	BM GL US	dry savanna
183		ERO	Eriocaulon	Paepalanthus bifidus (Schrad.) Kunth	BM	dry savanna
184	1837	AST	Pectis elongata Kunth	Pectis elongata Kunth var. elongata	BM US	
185	1837	AST	Wulffia platyglossa (Cass.) DC.	Tilesia baccata (L.) Pruski	BM	dry savanna
186		CSL	Cassia viscosa Kunth	Chamaecrista viscosa (Kunth) H.S. Irwin & Barneby var. major (Benth.) H.S. Irwin & Barneby	US	savanna
187		FAB	Aeschynomene conferta Benth.	Aeschynomene histrix Poir. var. histrix	tc: F GH US	
188		LAM	Hyptis lantanifolia Poit. ('lantanaefolia')	Hyptis lantanifolia Poit. ('lantanaefolia')		
189	1837	TRN	Piriqueta villosa Aubl.	Piriqueta cistoides (L.) Griseb. subsp. cistoides	BM BR F G K P	savanna
190	1837	CSL	Cassia ramosa Vogel	Chamaecrista ramosa (Vogel) H.S. Irwin & Barneby var. ramosa	BM F K MICH US	sands of the Essequibo R. Mahaica-Berbice Region
191	1837	MRT	Psidium aquaticum Benth.	Psidium striatulum DC.	ht: K; it: BM CGE P G US	
192		ACA	Dicliptera ciliaris Juss.	Dicliptera ciliaris Juss.	G	
193		ORC	Schomburgkia crispa Lindl.	Schomburgkia marginata Lindl.		
194	1837	RUB	Geophila reniformis Cham. & Schltdl.	Psychotria herbacea Jacq.	BM	Courantyne R.
195		ORC	Epidendrum graniticum Lindl.	Encyclia granitica (Lindl.) Schlecht.	tc: BM US	Courantyne R.
195*		ORC	Epidendrum pictum Lindl.	Encyclia picta (Lindl.) Hoehne	P	Courantyne R.
196		VRB	Lantana tilliaefolia var. scabra Schauer	Lantana camara L.	tc: L	Courantyne R.
197	1837	MLP	Tetrapterys discolor (G. Mey.) DC.	Tetrapterys discolor (G. Mey.) DC.	BM K L US	
198	1837	STR	Helicteres guazumaefolia Kunth	Helicteres guazumaefolia Kunth	BM	
199	1837	RUB	Coffea crassiloba Benth.	Rudgea crassiloba (Benth.) B.L. Rob.	tc: BM US	Essequibo R.
200		MLS	Spennera indecora (Bonpl.) DC.	Aciotis laxa (DC.) Cogn.		
201		MLS	Mouriria guianensis Aubl.	Mouriria guianensis Aubl.		abandoned fields
202	1837	RUB	Siderodendron laxiflorum Benth. ('Siderodendron')	Ixora graciliflora Benth.	tc: BM	Berbice R.

No.	Year	Code	Name	Accepted name	Herbaria	Locality
203	1837	STR	Melochia ulmifolia Benth.	Melochia ulmifolia Benth.	it: BM	savanna
204	1837	CPP	Physostemon intermedium Moric.	Cleome guianensis Aubl.	BM	Courantyne R.
205	1837	STR	Byttneria divaricata Benth. ('Büttneria')	Byttneria divaricata Benth.	tc: BM P U US	
206	1837	AST	Sparganophorus vaillantii Gaertn.	Struchium sparganophorum (L.) Kuntze	BM US	Courantyne R.
207		VRB	Lantana annua L.	Lippia alba (Mill.) N.E. Br.	BM	
208	1837	RUB	Genipa americana L.	Genipa americana L.	U	
209		MRS	Conomorpha magnoliifolia Mez	Cybianthus fulvopulverulentus (Mez) G. Agostini subsp. magnoliifolius (Mez) Pipoly		
210		BMN	Burmannia bicolor Mart.	Burmannia bicolor Mart.	BM K P	
211	1837	BOM	Quararibea guianensis Aubl.	Quararibea guianensis Aubl.	BM	
212	1837	MLV	Paritium tiliaceum (L.) A. Juss.	Hibiscus tiliaceus L.	BM	
213	1837	TRN	Turnera ulmifolia L.	Turnera scabra Millsp.	lt: ?; ilt: BM G K	Courantyne R.; post Orealla
214		VIO	Calyptrion aubletii DC.	Corynostylis arborea (L.) S.F. Blake	tc: BM US	
214*		VIO	Calyptrion nitidum Benth.	Corynostylis arborea (L.) S.F. Blake		
215		LAM	Marsypianthes hyptoides Mart. ex Benth.	Marsypianthes chamaedrys (Vahl) Kuntze		
216		ERO	Paepalanthus humboldtii Kunth	Syngonanthus humboldtii (Kunth) Ruhland	U	moist savanna
217		FAB	Desmodium incanum (Sw.) DC.	Desmodium incanum (Sw.) DC.		
217*		FAB	Desmodium rubiginosum Benth.	Desmodium distortum (Aubl.) J.F. Macbr.		
218		FAB	Stenolobium coeruleum Benth.	Calopogonium coeruleum (Benth.) Sauv.		
219		CYR	Cyrilla antillana Michx.	Cyrilla racemiflora L.	G	
220		CHB	Chrysobalanus pellocarpus G. Mey.	Chrysobalanus icaco L.	BM K OXF P US	sand bank of the Essequibo R. savanna
221		BIG	Amphilophium			
222	1837	MLP	Heteropterys macrostachya A. Juss.	Heteropterys macrostachya A. Juss.	st: BM K L US	Essequibo R.
223		PLG	Triplaris surinamensis Cham.	Triplaris weigeltiana (Rchb.) Kuntze		Long John Creek on Upper Essequibo R.
224	1837	DSC	Dioscorea truncata Miq.	Dioscorea megacarpa Gleason	tc: BM	
225	1837	FLC	Homalium racoubea Sw.	Homalium guianense (Aubl.) Oken	BM BR FFI G K NY US W	
226		MIM	Inga corymbifera Benth.	Inga nobilis Willd.	ht: K; it: NY P US	
227		CYP	Becquerelia merkeliana Nees	Becquerelia cymosa Brongn. subsp. merkeliana (Nees) T. Koyama		moist savanna
227*		CYP	Scleria tuberculata Boeck.	Becqueleria tuberculata (Boeck.) H. Pfeiff.	st: US	moist savanna
228		VRB	Amasonia erecta L.f.	Amasonia campestris (Aubl.) Moldenke		Rupununi savanna
229		PHT	Microtea debilis Sw.	Microtea debilis Sw.		
230		PLP	Taenitis desvauxii Klotzsch	Dicranoglossum desvauxii (Klotzsch) Proctor	B P	
231	1837	CSL	Cynometra bauhiniifolia Benth. ('bauhiniaefolia')	Cynometra bauhiniifolia Benth. var. bauhiniifolia	BM F G K NY US W	Berbice R.; Essequibo R.
232		ASC	Tassadia propinqua Decne.	Tassadia propinqua Decne.	lt: P	
233		CYP	Rhynchospora longibracteata Boeck.	Rhynchospora longibracteata Boeck.		wet savanna

Index 1. Robert Schomburgk's first collection series (1835–1839) – continued

No.	Year	Fam.	Determination	Accepted name	Herbaria	Locality
234		CYP	Kyllinga ('Kyllingia')	Kyllinga brevifolia Rottb. ('Kyllingia')	tc: BM W	
235	1837	OCH	Sauvagesia elata Benth.	Sauvagesia elata Benth.	tc: U US	
236		ERO	Paepalanthus subulatus Klotzsch [n.n.]	Paepalanthus subtilis Miq.		savanna at Pirara
237		CYP	Rhynchospora polycephala Wydler	Rhynchospora holoschoenoides (Rich.) Herter		
238		FAB	Lonchocarpus floribundus Benth.	Lonchocarpus floribundus Benth.		
239		FAB	Centrosema brasilianum (L.) Benth.	Centrosema brasilianum (L.) Benth.		
239*		FAB	Neurocarpum ellipticum Desv.	Clitoria falcata Lam.		
240	1837	FAB	Stylosanthes gracilis Kunth	Stylosanthes guianensis (Aubl.) Sw. var. guianensis	NY	dry savanna
241		EUP	Geiseleria chamaedrifolia Klotzsch	Croton trinitatis Millsp.		
242		CNV	Batatas cissoides Choisy	Merremia cissoides (Lam.) Hallier		
243		MLS	Leiostegia vernicosa Benth.	Comolia vernicosa (Benth.) Triana	ht: K	dry savanna
244		CYP	Fuirena			moist savanna savanna
245		FAB	Eriosema crinitum (Kunth) G. Don	Eriosema crinitum (Kunth) G. Don		savanna
245*		FAB	Eriosema pulchellum (Kunth) G. Don			
246	1837	MEL	Guarea aubletti A. Juss.	Guarea guidonia (L.) Sleumer	BM	
247	1837	AST	Latreillea glabrata Benth.	Ichthyothere terminalis (Spreng.) S.F. Blake	st: NY US	dry savanna
248		MLS	Chaetogastra glomerata (Rottb.) Benth.	Pterolepis glomerata (Rottb.) Miq.		
249		CRY	Polycarpaea brasiliensis Cambess.	Polycarpaea corymbosa (L.) Lam.		dry savanna
250	1837	RUB	Borreria suaveolens G. Mey.	Borreria capitata (Ruiz & Pav.) DC. var. suaveolens (G. Mey.) Steyerm.	BM	dry savanna
251	1837	RUB	Psychotria cornigera Benth.	Psychotria bahiensis DC. var. cornigera (Benth.) Steyerm.	tc: BM G US	
252		MLS	Tibouchina aspera Aubl.	Tibouchina aspera Aubl. var. aspera		dry savanna
252*		MLS	Tibouchina aspera Aubl.	Tibouchina aspera Aubl. var. asperrima Cogn.		dry savanna
253	1837	MLV	Pavonia speciosa Kunth	Peltaea speciosa (Kunth) Standley	tc: BM GH	Rupununi savanna
254	1837	AST	Centratherum muticum Less.	Centratherum punctatum Cass.	BM	sandy savanna
255	1837	AIZ	Mollugo verticillata L.	Mollugo verticillata L.	BM	savanna
256		SOL	Solanum foetidum Ruiz & Pav.			edges of savanna
257		FAB	Zornia latifolia DC.	Zornia pardina Mohlenbr. var. vichadana Killip ex Mohlenbr.		savanna
258	1837	AST	Vernonia scorpioides (Lam.) Pers.	Cyrtocymura scorpioides (Lam.) H. Rob.	BM	
259		MLS	Miconia holosericea (L.) DC. var. obtusifolia Benth.	Miconia albicans (Sw.) Triana		skirts of savanna
260	1837	PGL	Polygala mollis Kunth	Polygala hebeclada DC.	BM F G K L P U	
261		FAB	Collaea rosea Benth.	Galactia jussiaeana Kunth var. jussiaeana	tc: U US	arid savanna of Annai savanna

262		VRB	Stachytarpheta cayennensis (Rich.) Vahl ('Stachytarpha')	Stachytarpheta cayennensis (Rich.) Vahl ('Stachytarpha')	st: BM Fl G GH K L P U US W	savanna
263	1837	FLC	Casearia carpinifolia Benth.	Casearia sylvestris Sw. var. lingua (Cambess.) Eichl.	BM	
264	1837	RUB	Palicourea rigida Kunth	Palicourea rigida Kunth	BM	savanna
265	Mar 1837	GEN	Lisianthus uliginosus Griseb. var. uliginosus	Irlbachia purpurascens (Aubl.) Maas	US	moist savanna
266		AQF	Ilex celastroides Klotzsch [n.n.]	Ilex vismiifolia Reissek	BM G K NY P	Rupununi
267		FAB	Drepanocarpus ferox Mart. ex Benth.	Machaerium ferox (Mart. ex Benth.) Ducke	BM F Fl K L P	
268	1837	RUB	Coccocypselum tontanea Kunth	Coccocypselum guianense (Aubl.) K. Schum.	BM	savanna
269		ORC	Galeandra juncea Lindl.	Galeandra stylomisantha (Vell.) Hoehne	tc ?: BM US	savanna near Berbice R.
269*		CSL	Cassia hispida Collad.	Chamaecrista hispidula (Vahl) H.S. Irwin & Barneby		
270	1837	HUM	Humiria guianensis Benth. ('Humirium')	Humiria balsamifera (Aubl.) J. St.-Hil. var. guianensis (Benth.) Cuatrec.	ht: P; it: K L MO NY P U US	savanna
271		TIL	Triumfetta eriocarpa A. St.-Hil.	Triumfetta rhomboidea Jacq.		
272		CMB	Cacoucia coccinea Aubl.	Combretum cacoucia (Baill.) Exell ex Sandw.	BM CGE K OXF P TCD W	Essequibo R.
273	1837	MLV	Pavonia cancellata (L.) Cav.	Pavonia cancellata (L.) Cav.	BM	
274		SCZ	Schizaea incurvata Schkuhr	Actinostachys pennula (Sw.) Hook.	B	dry sandy savanna
275		CMM	Commelina platyphylla Klotzsch ex Seub. ('Commelyna')	Commelina platyphylla Klotzsch ex C.B. Clarke ('Commelyna')		
276	1837	SYM	Symplocos ciponima L'Hér.	Symplocos guianensis (Aubl.) Guerke	BM	
277	1837	APO	Thenardia corymbosa Benth.	Forsteronia schomburgkii DC.	tc: BM	
278		FAB	Stylosanthes viscosa Sw.	Stylosanthes viscosa Sw.		savanna
279	1837	MLP	Heteropterys cristata Benth.	Heteropterys cristata Benth.	ht: K; it: BM CGE G K L US	Essequibo R.
280	1837	BOM	Catostemma fragrans Benth.	Catostemma fragrans Benth.	ht: K; it: US	Berbice R.
281	1837	RUB	Isertia hypoleuca Benth.	Isertia hypoleuca Benth.	ht: K; it: BM L NY US	R. Casiquiare [BRA]
282	1837	AST	Vernonia tricholepis DC. var. tricholepis	Lepidaploa remotiflora (Rich.) H. Rob.	BM	
283	1837	EUP	Stillingia prunifolia Baill. ex Muell. Arg.	Sapium glandulosum (L.) Morong	ht: K; it: BM G L P US	savanna
284		MIM	Pithecellobium trapezifolium (Vahl) Benth. ('Pithecolobium')	Abarema jupunba (Willd.) Britton & Killip var. trapezifolia (Vahl) Barneby & J.W. Grimes		
285	1837	DSC	Dioscorea piperifolia Humb. & Bonpl. ex Willd.	Dioscorea piperifolia Humb. & Bonpl. ex Willd.	BM	
286	1837	LCY	Lecythis grandiflora Aubl.	Lecythis corrugata Poit. subsp. corrugata	ht: U; it: BM G GH K L P OXF U	Berbice R.
287	1837	CEC	Coussapoa cuneata Miq.	Coussapoa microcephala Trécul	ht: K; it: BM NY US	Essequibo R.
288	1837	MLS	Tococa subnuda Benth.	Tococa subciliata (DC.) Triana	BM	
289	1837	PAS	Cieca appendiculata Kunth	Passiflora auriculata Kunth	US	Essequibo R.
290		RUB	Cephaëlis rosea Benth.	Psychotria rosea (Benth.) Muell. Arg.	US	Essequibo R.
291	1837	TRN	Turnera aurantiaca Benth.	Turnera aurantiaca Benth.	ht: BM; it: G US	Essequibo R. and Rupununi R.

Index 1. Robert Schomburgk's first collection series (1835–1839) – continued

292	1837	APO	Tabernaemontana longifolia Benth.	Tabernaemontana siphilitica (L.f.) Leeuwenb.	lt: K; ilt: BM CGE E F G L OXF P SING TCD UPS US W	
293	1837	API	Eryngium foetidum L.	Eryngium foetidum L.		
294		AST	Clibadium erosum DC.	Clibadium sylvestre (Aubl.) Baill.	BM	
295		PDS	Lacis fluviatilis J.G. Gmel.	Mourera fluviatilis Aubl.		
296	1837	CSL	Campsiandra comosa Benth.	Campsiandra comosa Benth. var. comosa	lt: K; ilt: BM GH NY TCD US W	Essequibo R.
297	1837	STR	Sterculia ivira Sw.	Sterculia pruriens (Aubl.) K. Schum. var. pruriens	BM	
298		GEN	Lisianthus schomburgkii Griseb.	Irlbachia alata (Aubl.) Maas subsp. alata	tc ?: L US	
299		HMP	Trichomanes schomburgkianum J.W. Sturm	Trichomanes pinnatum Hedw.	st: B BR K P US	Berbice R.
300	1837	ADI	Adiantum phyllitidis J. Sm.	Adiantum phyllitidis J. Sm.	ht: B; it: B BM K L US	Berbice R.
301	1837	PLP	Polypodium persicariifolium Schrad. ('persicariaefolium')	Microgramma persicariifolia (Schrad.) C. Presl	B	
302		PLP	Polypodium lycopodioides L.	Microgramma lycopodioides (L.) Copeland	B	
303		OLN	Polypodium rivulare Vahl	Nephrolepis rivularis (Vahl) Mett. ex Krug	BM G K	
304		CTH	Alsophila subaculeata Splitg.	Cyathea surinamensis (Miq.) Domin	G TCD	Bartica
304*		CTH	Hemitelia multiflora (Sm.) Spreng. var. hostmannii Baker	Cyathea cyatheoides (Desv.) K.U. Kramer		
305	1837	ACA	Leptostachya martiana (L.) Nees	Justicia comata (L.) Lam.	tc: G U US	Berbice R.
306		MRS	Badula schomburgkiana A. DC.	Stylogyne schomburgkiana (DC.) Mez	tc: U US	
307		PLP	Polypodium ciliatum Willd.	Microgramma reptans (Cav.) A.R. Sm.	B	
308	1837	ONA	Jussiaea affinis DC. ('Jussieua')	Ludwigia affinis (DC.) Hara	B BM	Essequibo R. and Rupununi R.
309		APO	Odontadenia speciosa Benth.	Odontadenia macrantha (Roem. & Schult.) Markgr.		Berbice R.
310	1837	AST	Mikania parkeriana DC.	Mikania congesta DC.	BM US	
311	1837	APO	Echites tubulosa Benth.	Mesechites trifida (Jacq.) Muell. Arg.	ht: G; it: NY US	
312		PLP	Polypodium percussum Cav.	Pleopeltis percussa (Cav.) Hook. & Grev.	B	
312*		RHZ	Cassipourea serrata Benth.	Cassipourea serrata Benth.		
313	1837	MTX	Amphidesmium blechnoides (Sw.) Klotzsch	Metaxya rostrata (Kunth) C. Presl	B BM G K L	Berbice R.
314	1837	RUB	Chomelia tenuiflora Benth.	Chomelia tenuiflora Benth.	ht: K; it: BM U US	
315		ARA	Pistia stratiotes L.	Pistia stratiotes L.		
316	1837	DRY	Aspidium confertum Kaulf.	Cyclodium meniscioides (Willd.) C. Presl var. meniscioides	BM G K P	Berbice R.
317	1837	CLU	Havetia flavida Benth.	Havetiopsis flavida (Benth.) Planch. & Triana	tc: B BM U US	
318		MLS	Salpinga parviflora DC.	Macrocentrum cristatum (DC.) Triana var. parviflorum (DC.) Cogn.		

No.	Year	Family	Name	Current name	Specimens	Locality
319		MRT	Eugenia nitida Benth.	Myrcia inaequiloba (DC.) D. Legrand	st: BM	
320	1837	MIM	Inga pubiramea Steud.	Macrosamanea pubiramea (Steud.) Barneby & J.W. Grimes var. pubiramea		
321	1837	AST	Mikania denticulata DC.	Mikania microptera DC.	BM US	
322		VTT	Taenitis angustifolia Spreng.	Vittaria costata Kunze	B G OXF P TCD	
323	1837	ASL	Asplenium schomburgkianum Klotzsch	Asplenium serratum L.	st: B BM BR G K OXF P TCD US	Berbice R.
324	1837	PLP	Polypodium costatum Kunze	Polypodium nitidum Kaulf.	st: BM NY US	Berbice R.
325		RUB	Euosmia corymbosa Benth. ('Evosmia')	Bothriospora corymbosa Hook.f.	ht: K; it: NY	
326		MLS	Comolia veronicaefolia Benth.	Comolia villosa (Aubl.) Triana	ht: K; it: BM BR F MICH NY W	
327		MIM	Machaerium schomburgkii Benth.	Mimosa annularis Benth. var. odora Barneby	G K W	
327*		FAB	Drepanocarpus inundatus Mart. ex Benth.	Machaerium inundatum (Mart. ex Benth.) Ducke	B	
328		GMM	Polypodium taxifolium L.	Terpsichore taxifolia (L.) A.R. Sm.	ht: P; it: US	Berbice R.
329		APO	Echites macrostoma Benth.	Rhabdadenia macrostoma (Benth.) Muell. Arg.	BM	
330	1837	RUB	Posoqueria longiflora Aubl.	Posoqueria longiflora Aubl.	BM US	
331	1837	AST	Eclipta erecta L.	Eclipta prostrata (L.) L.		
332	1837	SAP	Cupania vouarana Cambess.	Matayba camtoneura Radlk.	ht: K; it: BM G K P U	Berbice R.
333	1837	CNN	Connarus schomburgkii Planch.	Connarus lambertii (DC.) Sagot	B BM	Berbice R.
334	1837	BEG	Begonia guyanensis A. DC.	Begonia semiovata Liebm.	BM	
335	1837	SCR	Torenia parviflora Ham.	Torenia thouarsii (Cham. & Schltdl.) Kuntze		
336	1837	VIO	Alsodeia pubiflora Benth.	Rinorea pubiflora (Benth.) Sprague & Sandw. var. pubiflora	BM CGE F G GH K L NY OXF P W	Berbice R.
336*		VIO	Alsodeia flavescens Spreng.	Rinorea flavescens (Aubl.) Kunze	BR CGE F G GH K US W	
337		RUB	Palicourea riparia Benth.	Palicourea croceoides Desv.	tc ?: NY US	
338		CYP	Nemochloa	Pleurostachys		
339		CYP	Cyperus elegans L.	Cyperus miliifolius Poepp. & Kunth		
340		ASL	Asplenium cuneatum Lam.	Asplenium cuneatum Lam.	B BM G K NY OXF P TCD	Berbice R.
341		CPP	Crateva acuminata DC. ('Crataeva')	Crateva tapia L.		Berbice R.
342	1837	FLC	Casearia laurifolia Benth.	Casearia commersoniana Cambess.	st: B † BM F Fl G GH K L P U US	
343		PLP	Polypodium crassifolium L.	Niphidium crassifolium (L.) Lellinger	B	
344		LOR	Loranthus		BM	
345	1837	SMR	Simaba guianensis Aubl.	Simaba multiflora A. Juss.	ht: B	
346		DST	Lindsaea rufescens Kunze	Lindsaea portoricensis Desv.	BM F K OXF P TCD U W	Berbice R.
346*		DST	Lindsaea guianensis (Aubl.) Dryand.	Lindsaea guianensis (Aubl.) Dryand. subsp. guianensis	B BM G K L OXF P TCD US	Berbice R.
347	1837	DST	Lindsaea trapeziformis Dryand.	Lindsaea lancea (L.) Bedd. var. lancea	US	
347*	1837	DST	Lindsaea divaricata Klotzsch	Lindsaea divaricata Klotzsch		

Index 1. Robert Schomburgk's first collection series (1835–1839) – continued

No.	Year	Code	Original name	Current name	Herbarium	Locality
348		PHT	Seguieria macrophylla Benth.	Seguieria macrophylla Benth.	st: K	
349	1837	ADI	Adiantum tomentosum Klotzsch	Adiantum tomentosum Klotzsch	st: B K	Berbice R.
350		APO	Echites brachystachya Benth.	Mandevilla scabra (Hoffmanns. ex Roem. & Schult.) K. Schum.	tc: US	
350*		APO	Echites rugosa Benth.	Mandevilla scabra (Hoffmanns. ex Roem. & Schult.) K. Schum.	tc: BM L US	
351	1837	BOR	Heliotropium helophilum Mart.	Heliotropium filiforme Lehmann	K P	Berbice R.; sandy savanna
352	1837	HPC	Pristimera apiculata Miers	Cuervea kappleriana (Miq.) A.C. Sm.	tc: BM G GH K L W	
353	1837	SML	Smilax globifera G. Mey.	Smilax cumanensis Willd.	BM P	
354		VTT	Taeniopis lineata J. Sm.	Vittaria lineata (L.) Sm.	B BM G K OXF P TCD	Berbice R.
355	1837	ASC	Roulinia guianensis Decne.	Cynanchum blandum (Decne.) Sundell	ht: ?: it: L NY	
356		PLP	Polypodium aureum L.	Phlebodium aureum (L.) J. Sm.	B	
357		URT	Pilea guyanensis Wedd.	Pilea pubescens Liebm.		
358		POA	Paspalum repens Bergius	Paspalum repens Bergius		
359	1837	MEL	Trichilia brachystachya Klotzsch ex C. DC.	Trichilia pallida Sw.	lt: BM; ilt: BR G GH K NY P U	
360		OCH	Ochna superba Kunze	Ouratea superba Engl.	tc: BM	
361	1837	OCH	Gomphia rupununiensis Klotzsch [n.n.]	Ouratea rupununiensis Engl.	st: BM	
362	1837	STR	Melochia lanceolata Benth.	Melochia lanceolata Benth.	tc: BM	
363	1837	RUB	Coffea crassiloba Benth.	Rudgea crassiloba (Benth.) B.L. Rob.	st: BM	Berbice R.
364		MIM	Inga floribunda Benth.	Inga splendens Willd.	ht: K; it: NY P US	
365	1837	BIG	Spathodea schomburgkii DC.	Memora schomburgkii (DC.) Miers	tc: BM K L	
366	1837	STR	Melochia melissifolia Benth.	Melochia melissifolia Benth.	st: BM	
367		POA	Panicum pallens Sw.	Ichnanthus pallens (Sw.) Munro ex Benth.		
368		CMM	Commelina guianensis Klotzsch [n.n.] ('Commelyna')	Commelina rufipes Seub. var. glabrata (D.R. Hunt) Faden & D.R. Hunt		wet savanna
369	1837	FLC	Casearia spinosa (L.) Willd.	Casearia aculata Jacq.	BM F Fl G GH K L P S U US W	
370		PLG	Polygonum acuminatum Kunth	Polygonum acuminatum Kunth	GH	
371		CYP	Oxycaryum schomburgkianum Nees	Oxycaryum cubense (Poepp. & Kunth) Palla	P U	moist savanna
372		EUP	Phyllanthus congesta Benth. ex Muell. Arg.	Jablonskia congesta (Benth. ex Muell. Arg.) G.L. Webster		
373		FAB	Centrosema vexillatum Benth.	Centrosema vexillatum Benth.		
374		ASC	Tassadia guianensis Decne.	Tassadia trailiana (Benth.) Fontella		
375		CSL	Vouapa staminea (G. Mey.) DC.	Macrolobium bifolium (Aubl.) Pers.	st: UPS	Berbice R.
376		OCH	Sauvagesia erecta L.	Sauvagesia erecta L.		Essequibo R.

No.	Year	Family	Name	Accepted name	Herbarium codes	Locality
377	1837	ACA	Dipteracanthus canescens Nees	Ruellia geminiflora Kunth var. angustifolia (Nees) Griseb.	st: G K P U US	
378	1837	GLC	Mertensia pectinata Willd.	Dicranopteris pectinata (Willd.) Underw.	B BM G US	
379		ADI	Adiantum villosum Willd.	Adiantum villosum Willd.	P	
380	1837	AST	Unxia camphorata L.f.	Unxia camphorata L.f.	BM US	Rupununi savanna
381		CHB	Licania crassifolia Benth.	Licania incana Aubl.		Rupununi savanna
382	1837	BOR	Cordia ulmifolia Benth.	Cordia polycephala (Lam.) I.M. Johnst.	B BM G K L P	savanna
383	1837	SYM	Symplocos ciponima L'Hér.	Symplocos guianensis (Aubl.) Guerke	BM	
384	1837	BOR	Varronia polycephala Lam.	Cordia polycephala (Lam.) I.M. Johnst.	BM	
384*		RUB	Malanea sarmentosa Aubl.	Malanea sarmentosa Aubl.		
385		AQF	Ilex martiniana D. Don	Ilex martiniana D. Don	tc: BM G P U W	
386	1837	FLC	Casearia densiflora Benth.	Casearia commersoniana Cambess.	st: BM Fl G GH K P U US W	
387		CMM	Commelina			
388		CHB	Licania crassifolia Benth.	Licania incana Aubl.	ht: K; it: CGE L NY OXF P US W	Rupununi savanna
389		PON	Heteranthera	Eichhornia heterosperma Alexander		Berbice R.
390	1837	ADI	Gymnogramma calomelanos (L.) Kaulf.	Pityrogramma calomelanos (L.) Link	ht: K; it: B G P TCD	Berbice R.
391		PON	Heteranthera grandiflora Klotzsch [n.n.]	Eichhornia diversifolia (Vahl) Urb.	ht: B; it: BM G K	Berbice R.
392		MLS	Chaenopleura hypoleuca Benth.	Miconia hypoleuca (Benth.) Triana	ht: K	
393		MLS	Miconia rufescens (Aubl.) DC.	Miconia rufescens (Aubl.) DC.	U	savanna
394	1837	RUB	Mitracarpus puberulum Benth.	Borreria ocymoides (Burm.f.) DC.	tc: BM US	
395		LYC	Lycopodium cernuum L.	Lycopodiella cernua (L.) Pic.-Ser.	B	
396		CYP	Isolepis junciformis Kunth	Bulbostylis junciformis (Kunth) C.B. Clarke	U	
397		CYP	Isolepis conifera Kunth	Bulbostylis conifera (Kunth) C.B. Clarke	K U	
398	1837	MLS	Miconia macrothyrsa Benth.	Miconia macrothyrsa Benth.	ht: K; it: NY	Berbice R.: savanna
399		LYG	Lygodium volubile Sw.	Lygodium volubile Sw.	ht: B; it: BM G P	
400		PLG	Coccoloba excelsa Benth.	Coccoloba excelsa Benth.	st: GH NY	
401	1837	CSL	Cassia cultrifolia Kunth	Chamaecrista diphylla (L.) Greene	BM F K NY	Berbice R.
402		MLS	Clidemia rariflora Benth.	Clidemia bullosa DC.	ht: K	
403		MLS	Henriettea succosa (Aubl.) DC.	Clidemia bullosa DC.		
404		VRB	Aegiphila arborescens Vahl	Aegiphila integrifolia (Jacq.) B.D. Jacks.	ht: K; it: P US	savanna
405	1837	CLU	Vismia macrophylla Kunth	Vismia macrophylla Kunth	BM CGE F Fl OXF W	Berbice R.
406	1837	BOR	Cordia schomburgkii A. DC.	Cordia schomburgkii A. DC.	ht: G; it: BM L U US W	
407		POA	Panicum parvifolium Lam.	Panicum parvifolium Lam.		
408	1837	BOR	Cordia umbraculifera DC.	Cordia tetrandra Aubl.	ht: G; it: B BM G K P	
409	1837	RUB	Mitracarpus rudis Benth.	Mitracarpus hirtus (L.) DC.	tc: BM G US	
410	1837	HPC	Hippocratea aubletiana Miers	Peritassa compta Miers	ht: K; it: BM BR G U	savanna
411	1837	LAU	Oreodaphne	Ocotea puberula (Rich.) Nees	US	Essequibo R. and Rupununi R.

Index 1. Robert Schomburgk's first collection series (1835–1839) – continued

No.	Family	Year	Name	Current name	Herbarium	Locality
412	SOL		Markea coccinea Rich.	Markea coccinea Rich.		
413	SMR		Quassia amara L.	Quassia amara L.		
414	POA		Setaria macrostachya Kunth	Setaria tenax (Rich.) Desv.		
415	RUB		Psychotria inundata Benth.	Psychotria capitata Ruiz & Pav. subsp. inundata (Benth.) Steyerm.	BM	Berbice R.
415*	RUB		Psychotria arcuata Benth.	Psychotria capitata Ruiz & Pav. subsp. inundata (Benth.) Steyerm.	tc: US	Berbice R.
416	OLN	1837	Aspidium pendulum Splitg.	Oleandra pilosa Hook.	tc: B E G K OXF P	Berbice R.
417	AQF		Ilex macoucoua Pers.	Ilex guianensis (Aubl.) Kuntze	BM P U	
418	MLS		Miconia ciliata (Rich.) DC.	Miconia ciliata (Rich.) DC.		savanna
419	SCR		Buchnera palustris (Aubl.) Spreng.	Buchnera palustris (Aubl.) Spreng.		moist savanna
420	EUP		Phyllanthus microphyllus Kunth	Phyllanthus stipulatus (Raf.) G.L. Webster		savanna
421	LNT		Polypompholyx schomburgkii Klotzsch [n.n.]	Utricularia longeciliata A. DC.	ht: G; it: BM CGE G K L P	moist savanna
422	LNT		Polypompholyx schomburgkii Klotzsch [n.n.]	Utricularia longeciliata A. DC.		moist savanna
422*	LNT	1837	Utricularia angulosa Poir.	Utricularia juncea Vahl		moist savanna
423	ORC		Epidendrum fragrans Sw.	Encyclia aemula (Lindl.) Carnevali & Ramirez	BM P	
424	ORC	1837	Epidendrum raniferum Lindl.	Epidendrum cristatum Ruiz & Pav.	BM P	
425	ORC		Aspasia variegata Lindl.	Aspasia variegata Lindl.	BM	
426	ORC	1837	Epidendrum imatophyllum Lindl.	Epidendrum flexuosum G. Mey.	BM	
427	ORC	1837	Stelis argentata Lindl.	Stelis argentata Lindl.	tc: BM US	
428	ORC		Brassavola angustata Lindl.	Brassavola martiana Lindl.	BM P	
429	ORC	1837	Epidendrum bicornutum Hook.	Caularthron bicornutum (Hook.) Raf.	BM P	
430	ORC	1837	Rodriguezia secunda Kunth	Rodriguezia lanceolata Ruiz & Pav.	BM	
431	PDS			Tristicha trifaria (Bory ex Willd.) Spreng.		
431*	PDS			Oserya sphaerocarpa Tul. & Wedd.		
432	PDS			Tristicha trifaria (Bory ex Willd.) Spreng.		
433	PDS		Mourera	Weddellina squamulosa Tul.	tc: ?: US	
434	PDS			Apinagia richardiana (Tul.) van Royen	tc: US	
435	PDS			Rhyncholacis hydrocichorium Tul.		
436	PDS		Mourera	Apinagia corymbosa (Tul.) Engl. var. corymbosa	tc: U	
436*	PDS			Apinagia richardiana (Tul.) van Royen	B BM CGE US	
437	PDS			Apinagia longifolia (Tul.) van Royen	tc: U	
438	ONA	1837	Jussiaea nervosa Poir. ('Jussieua')	Ludwigia nervosa (Poir.) Hara	BM	Berbice R.
439	HMP		Trichomanes coriaceum Kuntze	Trichomanes arbuscula Desv.	B BM K	swampy savanna
439*	HMP		Neurophyllum pinnatum C. Presl	Trichomanes pinnatum Hedw.	P	Berbice R.

No.	Year	Fam.	Name	Name	Specimens	Locality
439**		HMP	Neurophyllum hostmannianum Klotzsch	Trichomanes hostmannianum (Klotzsch) Kunze	K P US	
440		HMP	Trichomanes kraussii Hook. & Grev.	Trichomanes kraussii Hook. & Grev.	BR G L P PRC U US W	
440*		HMP	Hymenophyllum polyanthos (Sw.) Sw.	Hymenophyllum polyanthos (Sw.) Sw.	B K	
441		GMM	Grammitis serrulata Sw.	Cochlidium serrulatum (Sw.) L.E. Bishop	B K	Berbice R.
442		HMP	Trichomanes crispum L.	Trichomanes martiusii C. Presl	ht: L; it: K P US	Berbice R.
442*		HMP	Trichomanes schomburgkii Bosch	Trichomanes crispum L.	B	
442**		HMP	Trichomanes laxum Klotzsch	Trichomanes cristatum Kaulf.	B	sandy hills
443		SCZ	Schizaea flabellum Mart.	Schizaea elegans (Vahl) Sw.	B BM E G K OXF P	Berbice R.; sandy hills
444	1837	OLN	Nephrolepis ensifolia (Schkuhr) C. Presl	Nephrolepis biserrata (Sw.) Schott	B BM K	Berbice coast; swampy savanna
445	1837	BLE	Blechnum serrulatum Rich.	Blechnum serrulatum Rich.	tc: BM E K MPU OXF W ht: BM; it: B BM E K OXF P US	Berbice R.
446	1837	LOM	Acrostichum squamosum (Sw.) J. Sm.	Elaphoglossum plumosum (Fée) T. Moore	st: B BM BR E G K OXF P US	Berbice R.
447	1837	LOM	Acrostichum glabellum J. Sm.	Elaphoglossum glabellum J. Sm.	st: B BM E G K L OXF P US	Berbice R.
448	1837	LOM	Acrostichum flaccidum Fée	Elaphoglossum flaccidum (Fée) T. Moore	st: B BM E G K OXF P US W	Berbice R.
449	1837	LOM	Acrostichum alatum Fée	Elaphoglossum pteropus C. Chr.	st: B BM E G K OXF P US W	
450		LOM	Elaphoglossum schomburgkii (Fée) T. Moore	Elaphoglossum luridum (Fée) Christ	st: BM E G K P	
450*	1837	LOM	Acrostichum brevipes Kunze ex Fée	Elaphoglossum luridum (Fée) Christ	L	
450**		LOM	Elaphoglossum pteropus C. Chr.	Elaphoglossum pteropus C. Chr.	BM K OXF TCD	
451		ASL	Asplenium salicifolium L.	Asplenium salicifolium L.	B BM K OXF TCD	
451*		ASL	Asplenium integerrimum Spreng.	Asplenium juglandifolium Lam.	ht: K; it: B + BM BR frag F G NY OXF P U W	
452		MLS	Macairea pachyphylla Benth.	Macairea pachyphylla Benth.	BM	swampy savanna
453		CYP	Dichromena radicans Schltdl. & Cham.	Rhynchospora radicans (Schltdl. & Cham.) H. Pfeiff.	BM US	
454		MRT	Myrcia splendens (Sw.) DC.	Myrcia splendens (Sw.) DC.		
455	1837	AST	Bidens bipinnata L.	Bidens cynapiifolia Kunth		
456		MLS	Spennera aquatica Mart.	Nepsera aquatica (Aubl.) Naudin		
457		MLS	Clidemia pustulata DC.	Clidemia capitellata (Bonpl.) D. Don var. dependens (D. Don) J.F. Macbr.		
458	1837	MLS	Tococa aristata Benth.	Tococa aristata Benth.	ht: K; it: NY	sides of creeks
459	1837	THL	Meniscium serratum Cav.	Thelypteris serrata (Cav.) Alston	BM G P	Berbice R.
459*		DRY	Cyclodium confertum C. Presl	Cyclodium meniscioides (Willd.) C. Presl var. meniscioides		
460		PLP	Goniophlebium neriifolium J. Sm.	Polypodium neriifolium Schkuhr		
461		CYP	Cyperus elegans L.	Cyperus laxus Lam.		
462		CYP	Mariscus elatus Kunth	Cyperus aggregatus (Willd.) Endl.	st: US	

Index 1. Robert Schomburgk's first collection series (1835–1839) – continued

No.	Year	Fam.	Schomburgk name	Current name	Herbarium	Habitat
463		CHB	Licania divaricata Benth.	Licania divaricata Benth.	ht: K; it: BM CGE G L NY OXF P	
464		LAU	Oreodaphne schomburgkiana Nees	Ocotea schomburgkiana (Nees) Mez	st: B U	
465		CYP	Cyperus ligularis L.	Cyperus ligularis L.		
466		ANN	Guatteria schomburgkiana Mart.	Guatteria schomburgkiana Mart.	BM K P	
467	1837	RUB	Pagamea guianensis Aubl.	Pagamea guianensis Aubl.	BM	Warrewarrema
467*	1837	RUB	Siderodendrum macrophyllum Benth. ('Siderodendron')	Ixora schomburgkiana Benth.	ht: K; it: BM NY	Warrewarrema
468		APO	Aspidosperma excelsum Benth.	Aspidosperma excelsum Benth.		
469		MLP	Byrsonima spicata (Cav.) DC.	Byrsonima spicata (Cav.) DC.		
470		PHT	Phytolacca icosandra L.	Phytolacca icosandra L.		
471		MIM	Pithecellobium glomeratum Benth. ('Pithecolobium')	Zygia cataractae (Kunth) L. Rico		
471*		MIM	Inga bourgoni (Aubl.) DC.	Inga bourgoni (Aubl.) DC.	G K US	
472	1837	VTT	Antrophyum guayanense Hieron.	Antrophyum guayanense Hieron.	B BM E K OXF P TCD	
472*		VTT	Antrophyum cajenense (Desv.) Spreng. ('Antrophium')	Antrophyum cajenense (Desv.) Spreng.		
473	1837	AST	Elephantopus mollis Kunth	Elephantopus mollis Kunth	BM	
474	1837	AST	Trinchinettia caleoides Endl. ex Walp.	Calea caleoides (DC.) H. Rob.	ht: US; BM GH NY US	
475		VRB	Clerodendron fragrans (Vent.) Willd. var. pleniflorum Schauer	Clerodendron fragrans (Vent.) Willd. var. pleniflora	K	
476		MLV	Sida rhombifolia L.	Sida rhombifolia L.	BM	
477		DCH	Chailletia pedunculata DC.	Dichapetalum pedunculatum (DC.) Baill.	BM CGE F G GH K L P US W	
478		MLS	Clidemia campestris Benth.	Miconia campestris (Benth.) Triana	ht: K	moist savanna
479	1837	AST	Mikania hookeriana DC.	Mikania hookeriana DC.	BM US	
480	1837	AST	Mikania racemulosa Benth.	Mikania psilostachya DC.	tc: BM GH US	
481		POA	Panicum pilosum Sw.	Panicum pilosum Sw.	US	
482		FAB	Machaerium leiophyllum (DC.) Benth.	Machaerium leiophyllum (DC.) Benth.	F FI K US	
483		MLS	Diplochita parviflora Benth.	Miconia pubipetala Miq.	ht: K; it: US	
484		MRS	Weigeltia guianensis Klotzsch [n.n.]	Cybianthus surinamensis (Spreng.f.) G. Agostini	U	
485		CHB	Moquilea bracteosa Walp.	Couepia bracteosa Benth.	ht: K; it: BM CGE GH L OXF US	sandy savanna
486	1837	MRT	Calyptranthes obtusa O. Berg	Marlierea montana (Aubl.) Amshoff	tc: BM P US	

No.	Year	Fam.	Name	Accepted name	Collections	Locality
487		MIM	Pithecellobium lasiopus Benth. ('Pithecolobium')	Zygia latifolia (L.) Fawc. & Rendle var. lasiopus (Benth.) Barneby & J.W. Grimes	ht: K; it: US	
488	1837	RUB	Psychotria chlorantha Benth.	Psychotria anceps Kunth var. anceps	tc: BM G US	sandy hills
489		MLS	Diplochita fothergilla DC.	Miconia mirabilis (Aubl.) L.O. Williams	tc: BM F G K L MO U US W	Berbice coast
490	1837	EUP	Amanoa guianensis Aubl.	Amanoa guianensis Aubl.	tc: BM U	
491	1837	CNV	Lysiostyles scandens Benth.	Lysiostyles scandens Benth.		
492		FAB	Ecastophyllum monetaria (L.f.) DC.	Dalbergia monetaria L.f.	ht: K; it: NY	
493		MLS	Miconia eriopoda Benth.	Miconia acinodendron (L.) Sweet		
494		PIP	Artanthe olfersiana Klotzsch	Piper hostmannianum (Miq.) C. DC.	K	
495	1837	MLV	Sida althaeifolia Sw. var. aristosa DC.	Sida cordifolia L.	BM	
496	1837	CSL	Mora guianensis Benth.	Mora excelsa Benth.	tc: NY US	Berbice R.
497	1837	RUB	Palicourea guianensis Aubl.	Palicourea guianensis Aubl.	BM	
498		MIM	Pentaclethra filamentosa Benth.	Pentaclethra macroloba (Willd.) Kuntze	tc: NY US	
499	1837	MLV	Sida glomerata Cav.	Sida glomerata Cav.	BM	
500		CUC	Anguria multiflora Miq.	Gurania spinulosa (Poepp. & Endl.) Cogn.	lt: BM L	swampy savanna
501		CYP	Hypolytrum macrophyllum Boeck.	Mapania macrophylla (Boeck.) H. Pfeiff.	st: BM	
502		ORC	Zygopetalum rostratum Hook.	Zygosepalum labiosum (Rich.) C. Schweinf.	BM US	
503		ORC	Promenaea graminea Lindl.	Koellensteinia graminea (Lindl.) Rchb.f.	BM	
504		PLP	Goniophlebium distans J. Sm.			
505	1837	SPT	Chrysophyllum schomburgkianum A. DC.	Pradosia schomburgkiana (A. DC.) Cronq. subsp. schomburgkiana	ht: G; it: BM BR F K NY OXF P U US W	
506	1837	ORC	Epidendrum orchidiflorum Salzm. ex Lindl.	Epidendrum orchidiflorum Salzm. ex Lindl.	BM	sandy savanna
507	1837	MLS	Miconia myriantha Benth.	Miconia myriantha Benth.	ht: K; it: NY US	Berbice coast
508		ORC	Dichaea graminoides Lindl.	Dichaea rendlei Gleason		
509		HMP	Hymenophyllum schomburgkii C. Presl ex Sturm	Hymenophyllum decurrens (Jacq.) Sw.	st: B L US	
510		RUB	Pagamea guianensis Aubl.	Strychnos guianensis (Aubl.) Mart.		Essequibo R. and Rupununi R.
511		CSL	Vouapa staminea (G. Mey.) DC.	Macrolobium vuapa J.F. Gmel.	BM K	Essequibo R. and Rupununi R.
512		CCR	Anthodiscus trifoliatus G. Mey.	Anthodiscus trifoliatus G. Mey.		
513		MLS	Spennera dichotoma Benth.	Aciotis dichotoma (Benth.) Cogn.	ht: K; it: U	Upper Essequibo R.
514	1838	CLU	Calophyllum lucidum Benth.	Calophyllum lucidum Benth.	tc: BM NY US	Essequibo R. and Rupununi R.
515	1838	CSL	Eperua falcata Aubl.	Eperua rubiginosa Miq. var. rubiginosa	BM U	Essequibo R. and Rupununi R.
516		SCR	Vandellia diffusa L.	Lindernia diffusa (L.) Wettst.	lt: K	Essequibo R. and Rupununi R.
517	1837	CSL	Parivoa grandiflora Aubl.	Eperua schomburgkiana Benth.		Kumut (Cunuti) Mts.
518		SPT	Sideroxylon cuspidatum A. DC.	Pouteria cuspidata (A. DC.) Baehni subsp. cuspidata	ht: G; it: BM BR F G K NY U W	
519	1838	STR	Melochia arenosa Benth.	Melochia arenosa Benth.	st: BM	Rupununi R.
520		FAB	Drepanocarpus inundatus Mart. ex Benth.	Machaerium inundatum (Mart. ex Benth.) Ducke	BM GH P W	

Index 1. Robert Schomburgk's first collection series (1835–1839) – continued

No.	Year	Fam.	Name as collected	Determination	Herbaria	Locality
521	1838	CSL	Outea acaciifolia Benth. ('acaciaefolia')	Macrolobium acaciifolium (Benth.) Benth.	ht: K; it: US	Essequibo R. and Rupununi R.
522		CSL	Cassia multijuga Rich.	Senna multijuga (Rich.) H.S. Irwin & Barneby var. multijuga	US	Essequibo R. and Rupununi R.
523		CLU	Garcinia macrophylla Mart.	Rheedia benthamiana Planch. & Triana	tc: US	Waraputa F.
524		CSL	Aldina insignis (Benth.) Endl.	Aldina insignis (Benth.) Endl. var. insignis	ht: K; it: NY US	Upper Essequibo R. and Rupununi R.
525	1838	MLP	Byrsonima ceranthera Benth.	Byrsonima gymnocalycina A. Juss.	ht: K; it: BM CGE FG K L MO P	Essequibo R. and Rupununi R.
526		FAB	Leptolobium nitens Vogel	Acosmium nitens (Vogel) Yakovlev		falls of the Essequibo R. and Rupununi R.
527		RHZ	Cassipourea serrata Benth.	Cassipourea guianensis Aubl.		Essequibo R. and Rupununi R.
528		SCR	Beyrichia ocimoides Cham. & Schltdl.	Achetaria ocimoides (Cham. & Schltdl.) Wettst.		sands of the Essequibo R. and Rupununi R.
529		EUP	Phyllanthus guianensis Klotzsch	Phyllanthus carolinensis Walter subsp. guianensis (Klotzsch) G.L. Webster	st: G K	
530		MIM	Pithecellobium multiflorum (Kunth) Benth. ('Pithecolobium')	Albizia subdimidiata (Splitg.) Barneby & J.W. Grimes var. subdimidiata		
531		PLG	Coccoloba ovata Benth.	Coccoloba ovata Benth.		
532	1838	SCR	Bacopa aquatica Aubl.	Bacopa aquatica Benth.	BM	Essequibo R. and Rupununi R.
533	1838	DST	Lindsaea reniformis Dryand.	Lindsaea reniformis Dryand.	B BM BR K L NY P TCD US W	Essequibo R. and Rupununi R.
534	1838	MIM	Inga platycarpa Benth.	Inga pilosula (Rich.) J.F. Macbr.	ht: K; it: BM	Essequibo R. and Rupununi R.
535		CHB	Parinari campestris Aubl. ('Parinarium campestre')	Parinari campestris Aubl. ('Parinarium campestre')	BM GH K NY	Essequibo R. and Rupununi R.
536		LOG	Spigelia humilis Benth.	Spigelia humilis Benth.	st: L	Kwitaro R.
537	1838	MEL	Trichilia guianensis Klotzsch ex C. DC. var. parvifolia C. DC.	Trichilia rubra C. DC.	ht: K; it: BM	Kwitaro R.
538	1838	RUB	Sabicea glabrescens Benth.	Sabicea glabrescens Benth.	tc: BM G NY US	abandoned Indian settlement on Kwitaro R.
539		POA	Panicum zizanioides Kunth	Acroceras zizanioides (Kunth) Dandy		
540		CYP	Scirpus longifolius Rich.	Hypolytrum longifolium (Rich.) Nees		
541		CYP	Diplasia karataefolia Rich.	Diplasia karataefolia Rich.	K	
542		CYP	Rhynchospora cephalotes (L.) Vahl	Rhynchospora cephalotes (L.) Vahl		
543	1838	HUM	Humiria densiflorum Benth. ('Humirium')	Schistostemon densiflorum (Benth.) Cuatrec.	ht: K; it: BM L MO P U US	Kwitaro R.
544						

No.	Year	Family	Name	Revised name	Herbaria	Locality
545	1838	ERX	Erythroxylum roraimae Klotzsch ('Erythroxylon')	Erythroxylum roraimae Klotzsch ex O.E. Schulz	lt: K; ilt: BM	Kwitaro R.
546	1838	MRT	Eugenia xylopifolia DC.	Eugenia biflora (L.) DC.	BM	Kwitaro R.
547	1838	MRT	Eugenia quitarensis Benth.	Myrcia quitarensis (Benth.) Sagot	ht: K; it: BM MO US	Kwitaro R.
548	1838	MRT	Myrcia hebepetala (Mart.) DC.	Myrcia calycampa Amshoff	st: BM US	Kwitaro R.
549	1838	MRT	Eugenia vismeaefolia Benth.	Myrciaria vismiifolia (Benth.) O. Berg	tc: BM US	
550		APO	Echites rugosa Benth.	Mandevilla rugosa (Benth.) Woodson	BM Fl G GH K L U US	
551		LCS	Lacistema myricoides Sw.	Lacistema aggregatum (Bergius) Rusby		
552	1838	POA	Cenchrus parviflorus Poir.	Setaria parviflora (Poir.) Kerguélen	ht: K; it: BM G GF L NY P US	
553		RUB	Buena triflora Benth.	Cosmibuena grandiflora (Ruiz & Pav.) Rusby		falls of the Kwitaro R.
554		LOR	Viscum pennivenium DC.	Phoradendron pennivenium (DC.) Eichl.	BM	Kwitaro R.
555	1838	ACA	Teliostachya alopecuroides (Vahl) Nees	Lepidagathis alopecuroidea (Vahl) R. Br. ex Griseb.		Kwitaro R.
556		APO	Thyrsanthus schomburgkii Benth.	Forsteronia acouci (Aubl.) A. DC.	tc: BM NY US	Kwitaro R.
557	1838	APO	Thyrsanthus schomburgkii Benth.	Forsteronia acouci (Aubl.) A. DC.		Kwitaro R.
558		RUB	Euosmia corymbosa Benth. ('Evosmia')	Bothriospora corymbosa Hook.f.		Kwitaro R.
559		FAB	Dipteryx oppositifolia (Aubl.) Willd.	Taralea oppositifolia Aubl.	BM K P	Kwitaro R.
560	1838	ANN	Xylopia salicifolia Kunth	Xylopia discreta (L.f.) Sprague & Hutch.	ht: K; it: BM E F G L OXF P S TCD US	Kwitaro R.
561	1838	ANN	Duguetia quitarensis Benth.	Duguetia quitarensis Benth.	BM US	
562		LMC	Hydrocleis commersonii Rich.	Hydrocleys nymphoides (Humb. & Bonpl. ex Willd.) Buchenau		
563		ALI	Alisma subalatum Mart.	Echinodorus subalatus (Mart.) Griseb. subsp. subalatus	BM G K	Kwitaro R.
564		FAB	Deguelia scandens Aubl.	Derris pterocarpa (DC.) Killip	tc: US	Kwitaro R.
565		CSL	Schnella brachystachya Benth.	Bauhinia glabra Jacq.	tc: BM K P US W	Kwitaro R.
566		TIL	Mollia glabrescens Benth.	Mollia glabrescens Benth.		
566*		TIL	Mollia speciosa Mart. & Zucc.	Mollia speciosa Mart. & Zucc.	tc: U US	Kwitaro R.
567		LAU	Acrodiclidium guianense Nees	Licaria polyphylla (Nees) Kosterm.	tc: B E F GH	
568		LAU	Nectandra leucantha Nees	Nectandra paucinervia Coe-Teixeira	ht: G; it: K P	
568*		LAU	Nectandra schomburgkii Meisn.	Nectandra hihua (Ruiz & Pav.) Rohwer		
569		EUP	Sagotia racemosa Baill. var. brachysepala Muell. Arg.	Sagotia brachysepala (Muell. Arg.) R. Secco		
570	1838	ONA	Jussiaea accuminata Sw. ('Jussieua')	Ludwigia hyssopifolia (G. Don) Exell	B BM	Kwitaro R.
571		BOR	Tournefortia obscura A. DC.	Tournefortia cuspidata Kunth	ht: G; it: B BM G K L	Kwitaro R.
572	1838	POR	Portulaca pilosa L.	Portulaca pilosa L.	BM	
573	1838	AST	Gnaphalium schomburgkii Sch. Bip. [n.n.]	Gnaphalium polycaulon Pers.	BM US	
573*	1838	VIO	Alsodeia pubiflora Benth.	Rinorea pubiflora (Benth.) Sprague & Sandw. var. pubiflora	ht: K; it: BM G HH NY OXF P US	

Index 1. Robert Schomburgk's first collection series (1835–1839) – continued

No.	Year	Fam.	Name as published	Current name	Herbaria	Locality
574	1838	VIO	Alsodeia brevipes Benth.	Rinorea brevipes (Benth.) S.F. Blake	ht: K; it: BM CGE F HH K OXF P US W	Kwitaro R.
575	1838	BIG	Lundia schomburgkii Klotzsch [n.n.]	Callichlamys latifolia (Rich.) K. Schum.	BM F G L UPS	
576	1838	BIG	Bignonia aequinoctialis L.	Cydista aequinoctialis (L.) Miers	BM L	
577		SAP	Serjania paucidentata DC.	Serjania paucidentata DC.		
578		FAB	Swartzia microstylis Benth.	Swartzia dipetala Willd. ex Vogel	ht: K; it: L U US	Kwitaro R.
579		FAB	Machaerium nervosum Vogel	Machaerium quinata (Aubl.) Sandw.	BM F Fl G K L P W	
580		FAB	Ormosia coccinea (Aubl.) Jackson	Ormosia smithii Rudd	ht: K; it: BM	Kwitaro R.
581	1838	ORC	Epidendrum ibaguense Kunth var. confluans (Lindl.) C. Schweinf.	Epidendrum calanthum Rchb.f. & Warsz.		Mt. Ataraipu
582		MIM	Calliandra stipulacea Benth.	Calliandra stipulacea Benth.	ht: K; it: GH NY US	Kwitaro R.
583	1838	CLU	Caraipa laxiflora Benth.	Caraipa densifolia Mart.	tc: BM L NY P U	Kwitaro R.
584		RHM	Gouania virgata Reissek	Gouania virgata Reissek	st: P	
585	1838	BML	Pitcairnia kegeliana Schltdl.	Pitcairnia caricifolia Mart. ex Schult.f. var. carcifolia	K	
586		AMA	Chamissoa macrocarpa Kunth	Chamissoa altissima (Jacq.) Kunth var. altissima	U	
587		FAB	Andira laurifolia Benth.	Andira surinamensis (Bondt) Splitg. ex Amshoff		
588		POA	Echinolaena polystachya Kunth	Pseudechinolaena polystacha (Kunth) Stapf	BM K P US	Kwitaro R.
589		CSL	Martia excelsa Benth.	Martiodendron excelsum (Benth.) Gleason	lt: ?; ilt: US	Kwitaro R.
590		ERX	Erythroxylum citrifolium A. St.-Hil. ('Erythroxylon')	Erythroxylum citrifolium A. St.-Hil.	st: GH US	Kwitaro R.
591		EUP	Alchornea schomburgkii Klotzsch	Alchornea schomburgkii Klotzsch	tc: BM G K US	
592	1838	FLC	Casearia javitensis Kunth	Casearia javitensis Kunth	BM BR F Fl G GH K L NY P US W	Kwitaro R.
593		CHB	Licania aperta Benth.	Licania apetala (E. Mey.) Fritsch var. aperta (Benth.) Prance	ht: K; it: BM BR CGE GH L NY OXF P US	brook Curassawaka
594		SOL	Solanum pensile Sendtn.	Solanum pensile Sendtn.	tc: B † G	
595		MIM	Inga sapida Kunth sensu Benth.	Inga	K	
596		STY	Styrax guianensis A. DC.	Styrax guianensis A. DC.	tc: B MO US	
597		MRT	Eugenia subobliqua Benth.	Myrcia subobliqua (Benth.) Nied.	ht: G; it: BM MO US	Kwitaro R.
598		LAM	Hyptis parkeri Benth.	Hyptis parkeri Benth.		
599		APO	Secondatia densiflora A. DC.	Secondatia densiflora A. DC.		sands of the Essequibo R.
600		BOR	Heliophytum indicum DC.	Heliotropium indicum L.	K	Curassawaka
600*		SOL	Physalis pubescens L.	Physalis pubescens L.		
601	1838	BOR	Cordia bicolor A. DC.	Cordia bicolor A. DC.	BM	
602		ULM	Sponia mollis Decne.	Trema micrantha (L.) Blume	B	

No.	Fam.	Year	Name (as published)	Current name	Collections	Locality
603	FAB		Aeschynomene sensitiva Sw.	Aeschynomene sensitiva Sw.	F US	
604	MIM		Entada polyphylla Benth.	Entada polyphylla Benth.	st: NY US	
605	LAM	1838	Hyptis recurvata Poit.	Hyptis recurvata Poit.		Kwitaro R. sand banks
606	PAS	1838	Passiflora glandulosa Cav.	Passiflora glandulosa Cav.	BM P	
607	CLU	1838	Vismia cayennensis (Jacq.) Pers.	Vismia cayennensis (Jacq.) Pers.	BM CGE G GL K S US	
608	APO	1838	Thyrsanthus gracilis Benth.	Forsteronia gracilis (Benth.) Muell. Arg.	tc: BM NY US	Curassawaka
609	ANN	1838	Xylopia grandiflora A. St.-Hil.	Xylopia aromatica (Lam.) Mart.	BM K P	Annai
610	EUP	1838	Dalechampia scandens L.	Dalechampia scandens L.	BM	
611	ASL		Asplenium schomburgkianum Klotzsch	Asplenium serratum L.	G	
611*	ASL		Asplenium angustum Sw.	Asplenium angustum Sw.	BM K	
612	AST		Elephantosis angustifolia DC.	Orthopappus angustifolius (Sw.) Gleason		
613	MLS		Miconia fallax DC.	Miconia fallax DC.		
614	CSL		Cassia bacillaris L.f.	Senna bacillaris (L.f.) H.S. Irwin & Barneby var. bacillaris		
615	CYP		Rhynchospora barbata (Vahl) Kunth	Rhynchospora barbata (Vahl) Kunth		dry savanna
616	SAP		Cupania	Cupania rubiginosa (Poir.) Radlk.		
617	ORC		Cyrtopodium parviflorum Lindl.	Cyrtopodium parviflorum Lindl.	tc: BM G	
617*	LYT		Cuphea antisyphylitica Kunth	Cuphea antisyphilitica Kunth var. acutifolia Benth.	P	
618	RUB	1838	Borreria verticillata (L.) G. Mey.	Borreria verticillata (L.) G. Mey.	BM	
619	EUP	1839	Euphorbia pilulifera L.	Euphorbia hirta L.	BM K U	
620	CYP		Rhynchospora cephalotes (L.) Vahl	Rhynchospora cephalotes (L.) Vahl	tc: NY	savanna
621	CSL		Cassia polystachya Benth.	Chamaecrista polystachya (Benth.) H.S. Irwin & Barneby	BM	dry savanna
622	SCR	1838	Scoparia dulcis L.	Scoparia dulcis L.	BM	
623	CNV	1839	Evolvulus sericeus Sw.	Evolvulus sericeus Sw.	BM	dry savanna near Pirara
624	ASC		Oxypetalum capitatum Mart. & Zucc.	Oxypetalum capitatum Mart. & Zucc.		
625	CNV	1839	Ipomoea juncea Choisy	Merremia aturensis (Kunth) Hallier	BM	Rupununi savanna
626	TRN	1838	Turnera benthamiana M.R. Schomb.	Turnera velutina Benth.	tc: BM G K US	savanna at Annai
627	ERX	1838	Erythroxylum campestre A. St.-Hil. ('Erythroxylon')	Erythroxylum suberosum A. St.-Hil.	BM U	
628	ORC	1839	Cyrtopodium cristatum Lindl.	Cyrtopodium cristatum Lindl.	ht: BM; it: G	Mt. Annai
629	FAB	1838	Dioclea guianensis Benth. var. villosior Benth.	Dioclea guianensis Benth.	ht: K; it: NY US	
630	RUB	1838	Richardsonia divergens (Pohl) DC.	Richardia sp.	BM	near brook Curassawaka
631	CYP		Scleria cyperina Willd. ex Kunth	Scleria cyperina Willd. ex Kunth	BR F K	dry savanna
632	PAS	1838	Dysosmia foetida (L.) Roem.	Passiflora foetida L. var. foetida	BM	
633	PAS	1838	Passiflora pedata L.	Passiflora pedata L.	BM P	
634	MRT	1838	Eugenia subalterna Benth.	Eugenia punicifolia (Kunth) DC.	tc: BM F G US	Rupununi savanna
635	MLS		Miconia alata (Aubl.) DC.	Miconia alata (Aubl.) DC.		

Index 1. Robert Schomburgk's first collection series (1835–1839) – continued

No.	Date	Fam.	Name	Determination	Herbaria	Habitat
636		MRT	Psidium polycarpon Lamb.	Psidium guineense Sw.		arid savanna near Pirara and Rupununi R.
637	1838	SAP	Serjania bignonioides Klotzsch [n.n.]	Serjania caracasana (Jacq.) Willd.		Mt. Annai
638	1838	FAB	Phaseolus gracilis Poepp. ex Benth.	Phaseolus gracilis Poepp. ex Benth.		savanna at Annai
639		SAP	Urvillea berteriana (DC.) Radlk.	Urvillea ulmacea Kunth		
640		FAB	Bowdichia major Mart.	Bowdichia virgilioides Kunth		
641		ZIN	Renealmia aromatica (Aubl.) Griseb.	Renealmia aromatica (Aubl.) Griseb.	tc ?: BM CGE K L	
642		FAB	Rhynchosia violacea DC.	Eriosema violaceum (Aubl.) G. Don		
643		MLS	Clidemia sericea D. Don	Clidemia sericea D. Don		
644	1838	STR	Byttneria scabra L. ('Büttneria')	Byttneria scabra L.	K US	Curassawaka
645	1838	STR	Byttneria ramosissima Pohl ('Büttneria')	Byttneria genistella Triana & Planch.	BM	Curassawaka
646		POA	Echinolaena scabra Kunth	Echinolaena inflexa (Poir.) Chase	BM	
647		CYP	Scleria distans Poir.	Scleria distans Poir.		
648		FAB	Desmodium cajanifolium (Kunth) DC. ('cajanaefolium')	Desmodium cajanifolium (Kunth) DC. ('cajanaefolium')	F K US	
649		FAB	Galactia velutina Benth.	Galactia latisiliqua Desv.	tc: NY US	dry savanna
650	Jun 1838	AST	Wedelia discoidea Less.	Eleutheranthera ruderalis (Sw.) Sch. Bip.	US	
651	1838	FAB	Eriosema lanceolatum Benth.	Eriosema simplicifolium (Kunth) G. Don var. simplicifolium	ht: K; it: BM F G GH LE NY US	dry savanna
652						
653		FAB	Rhynchosia schomburgkii Benth.	Rhynchosia schomburgkii Benth.	U	dry savanna
654		POA	Saccharum cayennense (P. Beauv.) Benth.	Eriochrysis cayennensis P. Beauv.		moist savanna
655		CYP	Rhynchospora globosa (Kunth) Roem. & Schult.	Rhynchospora globosa (Kunth) Roem. & Schult.		moist savanna
656		POA	Panicum pilcomayense Hack.	Panicum bergii Arechav.	BM K	dry savanna
657		FAB	Desmodium benthamianum Klotzsch [n.n.]	Desmodium sclerophyllum Benth.		dry savanna
657*	1839	CYP	Hemicarpha schomburgkii Friedland	Lipocarpha schomburgkii (Friedland) G. Tucker	ht: NY; it: G L P	dry savanna
658		AST	Clibadium asperum DC.	Clibadium surinamense L.		
659		EUP	Discocarpus essequiboensis Klotzsch	Discocarpus essequiboensis Klotzsch	st: BM E F G K L MANCH OXF P U W	
660		CYP	Scleria microcarpa Nees	Scleria microcarpa Nees	tc: BR F K U US	
661		PHT	Seguieria foliosa Benth.	Seguieria americana L.	tc: B BM F G K	
662	1838	BIG	Memora nobilis Miers	Arrabidaea bilabiata (Sprague) Sandw.	BM K	
663	1838	AST	Acanthospermum xanthioides DC.	Acanthospermum australe (Loefl.) Kuntze	BM US	
664	1839	PAS	Decaloba hemicycla Roem.	Passiflora vespertilio L.	BM	

No.	Year	Fam.	Name (recorded)	Name (accepted)	Codes	Habitat
665	1839	POA	Imperata arundinacea Cirillo var. americana Anderss.	Imperata brasiliensis Trin.	st: BM US	moist savanna
666		POA	Andropogon	Andropogon virgatus Desv.	st: US	moist savanna
667		CYP	Psilocarya rufa Nees	Rhynchospora rufa (Nees) Boeck.	BM F G GH P	moist savanna
668	1838	MEL	Moschoxylum cipo A. Juss.	Trichilia cipo (A. Juss.) C. DC.	G K L P US	
669		MNM	Siparuna guianensis Aubl.	Siparuna guianensis Aubl.	US	sandy savanna
670	1838	PGL	Polygala paludosa A. St.-Hil.	Polygala leptocaulis Torrey & Gray		
671		LOG	Spigelia anthelmia L.	Spigelia anthelmia L.		
672		POA	Olyra latifolia L.	Olyra latifolia L.		
673		POA	Trachypogon plumosus (Willd.) Nees	Trachypogon spicatus (L.f.) Kuntze	BM	sandy swamps
674	1838	SCR	Gerardia hispidula Mart.	Agalinis hispidula (Mart.) D'Arcy	tc: BM K P W	moist savanna
675		TIL	Corchorus argutus Kunth	Corchorus orinocensis Kunth	st: BM	
676		CYP	Rhynchospora sylvatica Nees	Rhynchospora comata (Link) Roem. & Schult.		
677		MNS	Cissampelos fasciculata Benth.	Cissampelos fasciculata Benth.		
677*	1838	CYP	Fimbristylis limosa Poepp. & Kunth	Fimbristylis limosa Poepp. & Kunth		near brook Akalauri, on the Upper Rupununi R.
678		FAB	Tephrosia penicillata Benth.	Tephrosia adunca Benth		
679		ARS	Aristolochia surinamensis Willd.	Aristolochia surinamensis Wild.	ht: G; it: BM F K L US	Curassawaka
680	1838	RUB	Psychotria capitellata DC.	Psychotria officinalis (Aubl.) Sandw.	BM	sandy swamps
681	1839	PGL	Polygala hygrophila Kunth	Polygala hygrophila Kunth	tc: BM F G K P US W	
682	1838	MLV	Gaya subtriloba Kunth	Herissantia crispa (L.) Brizicky	BM	
683	1838	CSL	Cassia aeschynomene DC.	Chamaecrista nictitans (L.) Moench subsp. patellaria (Collad.) H.S. Irwin & Barneby		
684		POA	Panicum molle Sw.	Urochloa mollis (Sw.) Morrone & Zuloaga	P U	dry savanna
685		CYP	Psilocarya candida Nees	Rhynchospora candida (Nees) Boeck.		dry savanna
686		LAM	Hyptis paludosa A. St.-Hil. ex Benth.	Hyptis recurvata Poit.		moist savanna
687		VIO	Ionidium itoubou Kunth	Hybanthus calceolaria (L.) G.K. Schulze	BM	dry savanna
688	1839	ORC	Epidendrum rigidum Jacq.	Epidendrum rigidum Jacq.	BM L US	
689		AST	Leria nutans DC.	Chaptalia nutans (L.) Pol.		sand banks of the in Rupununi R.
690		MLS	Mouriria brevipes Hook.	Mouriri brevipes Hook.	ht: K; it: NY	
691	1838	MRT	Eugenia polystachya Rich.	Eugenia polystachyoides Amshoff	tc: BM	
692	1839	CNV	Ipomoea schomburgkii Choisy	Ipomoea schomburgkii Choisy	tc: BM	
693	1838	RUB	Faramea longifolia Benth.	Faramea sessilifolia (Kunth) A. DC.	tc: BM BR NY US	Curassawaka
694	1838	VRB	Cryptocalyx nepetifolia Benth.	Lippia betulifolia Kunth	st: BM K US	Curassawaka
695	1838	AST	Trichospira menthoides Kunth	Trichospira verticillata (L.) S.F. Blake	BM US	
695*		MRS	Rapanea guianensis Aubl.	Myrsine guianensis (Aubl.) Kuntze	U	Pirara
696	1839	PIP	Artanthe peduncularis Miq.	Piper pedunculare (Miq.) C. DC.	ht: G; it: B BM K NY P	Curassawaka

Index 1. Robert Schomburgk's first collection series (1835–1839) – continued

No.	Date	Family	Name	Accepted name	Specimens	Locality
697		AMA	Pupalia densiflora Mart.	Cyathula achyranthoides (Kunth) Moq.	ht: BM	
698		ORC	Goodyera guyanensis Lindl.	Brachystele guyanensis (Lindl.) Schltr.	BM P	
699		ORC	Cattleya superba R.H. Schomb. ex Lindl.	Cattleya violaceae (Kunth) Rolfe	BM	
700	1839	AML	Hippeastrum solandriflorum Herb. ('solandraeflorum')	Hippeastrum elegans (Spreng.) H.E. Moore	BM	north of Rupununi R. sandy savanna
701	1839	CNV	Batatas edulis Choisy	Ipomoea batatas (L.) Poir.	BM	
701*	1839	CNV	Batatas paniculata (L.) Choisy	Ipomoea mauritiana Jacq.	BM	Pirara
702		AMA	Amaranthus bahiensis Mart.	Amaranthus viridis L.	U	
703	May 1838	MRT	Eugenia schomburgkii Benth.	Eugenia lambertiana DC. var. lambertiana	tc: BM K L US	Curassawaka; near Rupununi R.
704	1838	FLC	Casearia stipularis Vent.	Casearia arborea (Rich.) Urb.	BM U	Pirara
705	1838	AST	Wulffia platyglossa (Cass.) DC.	Tilesia baccata (L.) Pruski	BM US	
706		FAB	Etaballia guianensis Benth.	Etaballia dubia (Kunth) Rudd	st: NY US	Rupununi
706*	1838	EUP	Discocarpus essequiboensis Klotzsch	Discocarpus essequiboensis Klotzsch	st: BM G K L P U W	
707	1838	RUB	Diodia rigida Cham. & Schltdl.	Borreria ocymifolia (Roem. & Schult.) Bacigalupo & E.L. Cabral	st: BM US	arid savanna near Pirara and Rupununi R.
708	Apr 1838	CNV	Ipomoea guyanensis Choisy	Jacquemontia guyanensis (Aubl.) Meisn.	BM	Annai and Curassawaka
709	1838	AST	Baccharis erioptera Benth.	Pterocaulon alopecuroides (Lam.) DC.	tc: BM NY US W	dry savannas near Upper Rupununi
710		HPC	Hippocratea schomburgkii Klotzsch [n.n.]	Prionostemma aspera (Lam.) Miers		
711	1839	MYS	Myristica sebifera Sw.	Virola sebifera Aubl.	BM W	dry savanna
712		MLP	Byrsonima crassifolia (L.) Kunth	Byrsonima crassifolia (L.) Kunth	BM F G	Pirara; savanna
713		APO	Haemadictyon marginatum Benth.	Prestonia marginata (Benth.) Woodson		dry savanna
714		PGL	Securidaca latifolia Benth.	Securidaca pubiflora Benth.	tc: GH US	Pirara
714*	1839	PGL	Securidaca longifolia Poepp. & Endl.	Securidaca longifolia Poepp. & Endl.	BM G K P W	Pirara
714**	1839	PGL	Catocoma lucida Benth.	Bredemeyera lucida (Benth.) Klotzsch ex Hassk.	BM	Pirara
715		MIM	Mimosa schomburgkii Benth.	Mimosa schomburgkii Benth.	ht: K: it: BM E F G K N Y US W	Pirara
716	1838	EUP	Dactylostemon schomburgkii Klotzsch [n.n.]	Actinostemon schomburgkii (Klotzsch) Hochr.	ht: K: it: BM G MO	
717	May 1838	PGL	Catocoma lucida Benth.	Bredemeyera lucida (Benth.) Klotzsch ex Hassk.	tc: BM FFl G K L P US W	
718	1838	FLC	Casearia petraea Benth.	Casearia ulmifolia Vahl ex Vent.	ht: K: it: BM F G K L P US W	Pirara
719		MLS	Chaetogastra hypericoides DC.	Desmoscelis villosa (Aubl.) Naudin	NY	Pirara; stony savanna
720		CSL	Cassia parkeriana DC.	Chamaecrista glandulosa (L.) Greene var. swartzii H.S. Irwin & Barneby		Lake Amuku

No.	Date	Fam.	Name	Accepted name	Herbaria	Locality
720*		CSL	Cassia disadena Steud.	Chamaecrista nictitans (L.) Moench var. disadena (Steud.) H.S. Irwin & Barneby	BM K US	
721	1838	MLS	Chaetogastra divaricata (Bonpl.) DC.	Perogastra divaricata (Bonpl.) Naudin	tc: BM GH US	high banks of Rupununi R.
722	1838	STR	Waltheria involucrata Benth.	Waltheria involucrata Benth.	tc ?: BM NY US	Pirara
723		RUB	Declieuxia chiococcoides Kunth	Declieuxia fruticosa (Willd. ex Roem. & Schult.) Kuntze	ht: K; it: BR CGE Fl G GH P US W	Pirara; dry savanna
724		FAB	Swartzia latifolia Benth.	Swartzia latifolia Benth. var. latifolia	ht: K; it: F M NY US W	savanna
725		MIM	Mimosa camporum Benth.	Mimosa camporum Benth.		savanna
725*		MIM	Mimosa casta L.	Mimosa casta L.	ht: G; it: BM US	Rupununi R.
726	1838	MRT	Eugenia incanescens Benth.	Eugenia incanescens Benth.	BM	
727	1838	STR	Helicteres althaeifolia Lam.	Helicteres baruensis Jacq.	lt: K; ilt: BM BR CGE GH L NY OXF P US	Pirara
728	May 1838	CHB	Licania incana Benth.	Licania kunthiana Hook.f	BM	
728*	May 1838	MLP	Heteropterys lessertiana A. Juss.	Heteropterys macradena (DC.) W.R. Anderson	BM K US	Pirara
729	1838	MLP	Heteropterys lessertiana A. Juss.	Heteropterys macradena (DC.) W.R. Anderson		Pirara
730	May 1838	VRB	Lantana canescens Kunth	Lantana canescens Kunth	ht: K; it: BM G GH K	Pirara
731	1838	MLV	Pavonia angustifolia Benth.	Pavonia angustifolia Benth.	st: US	Pirara
732		MRT	Myrcia prunifolia DC.	Myrcia tomentosa (Aubl.) DC.	tc: BM US	Pirara
733	1838	MRT	Eugenia dipoda DC. var. brachypoda DC.	Eugenia punicifolia (Kunth) DC.	tc: BM GH K P U US W	Pirara
734		TIL	Apeiba tibourbou Aubl.	Apeiba schomburgkii Szyszyl.	tc: G K NY	Pirara
735	1839	RUB	Coffea tenuiflora Benth.	Morinda tenuiflora (Benth.) Steyerm.	BM	
736	1839	CNV	Batatas cissoides Choisy	Merremia cissoides (Lam.) Hallier	ht: K; it: BM CGE G GH K L NY P	Pirara
737	Apr 1838	MLP	Stigmaphyllon purpureum Benth.	Stigmaphyllon sinuatum (DC.) A. Juss.		dry savanna near Pirara
738		APO	Echites coriacea Benth.	Odontadenia geminata (Roem. & Schult.) Muell. Arg.	st: BM US	Pirara
738*	1838	STR	Melochia fasciculata Benth.	Melochia parvifolia Kunth var. fasciculata (Benth.) Hassl.	ht: K	Pirara
739	May 1838	MLS	Clidemia miconioides Benth.	Miconia ibaguensis (Bonpl.) Triana		
740		MIM	Pithecellobium pubescens Benth. ('Pithecolobium')	Pithecellobium roseum (Vahl) Barneby & J.W. Grimes		
740*	Jun 1838	MIM	Inga brevipes Benth.	Inga brevipes Benth.	ht: K; it: G US	Pirara
741		SAP	Schmidelia velutina Turcz.	Allophylus racemosus Sw.	tc: K L U	savanna
742	May 1838	MLP	Bunchosia mollis Benth.	Bunchosia mollis Benth.	ht: K; it: BM CGE G GH K L NY P	Pirara
743		SAP	Cardiospermum halicacabum L.	Cardiospermum halicacabum L.		savanna
744	1839	STR	Ayenia tomentosa L.	Ayenia tomentosa L.	BM	Pirara
745		FAB	Lonchocarpus rufescens Benth.	Lonchocarpus rufescens Benth.		
746		LOR	Psittacanthus guianensis Klotzsch [n.n.]	Psittacanthus cordatus (Hoffmanns.) Blume		savanna
747	1839	RHM	Gouania velutina Reissek	Gouania velutina Reissek	tc: P US W	Pirara

Index 1. Robert Schomburgk's first collection series (1835–1839) – continued

No.	Date	Fam.	Name (as collected)	Current name	Herbaria	Locality
748		MRN	Calathea villosa Lindl.	Calathea villosa Lindl.	U US	skirts of woods on savanna
749	1838	BOR	Tournefortia spigeliiflora A. DC. ('spigeliaeflora')	Tournefortia paniculata Cham. var. spigeliiflora (A. DC.) I.M. Johnst.	ht: G; it: BM K L	near Pirara
750	1838	TEO	Clavija ornata D. Don	Clavija macrophylla (Link ex Roem. & Schult.) Radlk.	BM G K L OXF P W	
750*		MIM	Pithecellobium pubescens Benth. ('Pithecolobium')	Pithecellobium roseum (Vahl) Barneby & J.W. Grimes		
751		MIM	Neptunia plena Benth.	Neptunia plena (L.) Benth.	st: NY	savanna
752	1838	DSC	Dioscorea microura Knuth	Dioscorea microura Knuth	tc: B BM US	
753		SOL	Solanum radula Vahl	Solanum asperum Rich.		
754	1838	TRN	Turnera opifera Mart.	Turnera caerulea Moç. & Sessé ex DC. var. surinamensis (Urb.) Arbo	BM G K	Pirara
755		APO	Prestonia latifolia Benth.	Prestonia tomentosa R. Br.	tc: US	Pirara; savanna
756		ALI	Sagittaria rhombifolia Cham.	Sagittaria rhombifolia Cham.	BM G	
757		ORC	Bonatea pauciflora Lindl.	Habenaria trifida Kunth	BM US	dry savanna
758		SAP	Paullinia bipinnata Poir.	Paullinia leiocarpa Griseb.		savanna
759		POA	Paspalum densum Poir.	Paspalum densum Poir.		moist savanna
760		CYP	Holoschoenus elatior Nees	Rhynchospora stricta Boeck. var. demerarensis (C.B. Clarke) Kük.		moist savanna
761		POA	Andropogon bicornis L.	Andropogon bicornis L.		moist savanna
762		POA	Andropogon contortus L.	Heteropogon contortus (L.) P. Beauv. ex Roem. & Schult.	U US	dry savanna
763	1838	RUB	Borreria alata (Aubl.) DC.	Borreria alata (Aubl.) DC.	BM	Pirara
764	May 1838	ERX	Erythroxylum testaceum Peyr. ('Erythroxylon')	Erythroxylum suberosum A. St.-Hil.	lt: ?; ilt: BM	Pirara; stony savanna
765		CYP	Abildgaardia monostachya Vahl	Abildgaardia ovata (Burm.f.) Kral		dry savanna
766		ERX	Erythroxylum mucronatum Benth. ('Erythroxylon')	Erythroxylum mucronatum Benth.		Pirara
767		APO	Tabernaemontana grandiflora Jacq.	Stemmadenia grandiflora (Jacq.) Miers		savanna at Pirara
768	1839	POA	Panicum zizanioides Kunth	Acroceras zizanioides (Kunth) Dandy		dry savanna
769		POA	Ischaemum guianense Hack. var. schomburgkii Hack.	Ischaemum guianense Kunth ex Hack.	ht: K; it: US	dry savanna
770		ASC	Ditassa pauciflora Decne.	Ditassa pauciflora Decne.		
771	1838	RUT	Monnieria trifolia L.	Ertela trifolia (L.) Kuntze	BM	dry savanna
772	Apr 1838	VRB	Aegiphila laxiflora Benth.	Aegiphila laxiflora Benth.	tc: BM BR G K NY US	
773	May 1838	FLC	Casearia brevipes Benth.	Casearia spinescens (Sw.) Griseb.	ht: K; it: BM K	dry savanna near Pirara
774		POA	Streptostachys asperifolia Desv.	Streptostachys asperifolia Desv.		dry savanna

No.	Date	Fam.	Name	Current name	Herbarium	Locality
775		RUB	Randia hebecarpa Benth.	Randia hebecarpa Benth.	tc: US st: BM L US	moist savanna
776	1838	CNV	Aniseia ensifolia Choisy var. minor Choisy	Aniseia cernua Moric.		
777						
778	1838	RUB	Guettarda macrantha Benth.	Guettarda macrantha Benth.	BM	dry savanna
779	1838	APO	Malouetia	Malouetia pubescens Markgr.	BM	Kanuku Mts.
780	Aug 1838	MRT	Eugenia salzmanni Benth.	Myrciaria floribunda (West ex Willd.) O. Berg	tc: BM US	R. Branco
781	Aug 1838	SAP	Koernickea guianensis Klotzsch [n.n.]	Paullinia anisoptera Turcz.		
782		APO	Thyrsanthus schomburgkii Benth.	Forsteronia acouci (Aubl.) A. DC.	BM	sandy savanna
783	July 1838	SAP	Cupania affinis Klotzsch [n.n.]	Matayba macrostylis Radlk.	sc: G GH U	
784	Aug 1838	LAU	Goeppertia polyantha Meisn.	Endlicheria anomala (Nees) Mez	st: K MO P US	falls of the R. Branco
785		CHB	Parinari brachystachya Benth. ('Parinarium brachystachyum')	Parinari excelsa Sabine	ht: K; it: BM BR CGE GH L NY OXF P US	
786	Aug 1838	MLP	Byrsonima schomburgkiana Benth.	Byrsonima schomburgkiana Benth.	lt: K; ilt: BM CGE G GH K L P US	Fort São Joaquim
787	Aug 1838	CSL	Cassia filipes Benth.	Chamaecrista rotundifolia (Pers.) Greene var. grandiflora (Benth.) H.S. Irwin & Barneby	ht: K; it: GH NY US	savanna near Fort São Joaquim
788	Aug 1838	FAB	Crotalaria leptophylla Benth.	Crotalaria maypurensis Kunth	ht: K; it: US	Rupununi savanna
789	Aug 1838	GEN	Schultesia benthamiana Klotzsch [n.n.]	Schultesia benthamiana Klotzsch ex Griseb.	st: US	R. Surumu
790	Aug 1838	HAE	Xiphidium fockeanum Miq.	Schiekia orinocensis (Kunth) Meisn. subsp. orinocensis		Fort São Joaquim
791	Aug 1838	CSL	Peltogyne pubescens Benth.	Peltogyne paniculata Benth. subsp. pubescens (Benth.) M.F. Silva	st: NY US	R. Branco
792	Aug 1838	LCY	Lecythis schomburgkii O. Berg	Lecythis schomburgkii O. Berg	lt: BM; ilt: CGE E G INPA K L NY OXF P U	skirts of savanna
793	July 1838	ASC	Metastelma stenolobium Decne.	Blepharodon nitidus (Vell.) J.F. Macbr.	st: NY US	R. Branco
794		RUB	Euosmia corymbosa Benth. ('Evosmia')	Bothriospora corymbosa Hook.f.	ht: K; it: BM K NY P US	R. Branco
795	Aug 1838	MIM	Inga stenoptera Benth.	Inga stenoptera Benth.	BM	R. Branco
796		RUB	Genipa caruto Kunth	Genipa americana L.		R. Branco
797	Aug 1838	CSL	Outea multijuga DC.	Macrolobium multijugum (DC.) Benth. var. multijugum		R. Branco
798	1838	AST	Ooclinium villosum DC.	Praxelis pauciflora (Kunth) R.M. King & H. Rob.	BM US	
799		POA	Aristida capillacea Lam.	Aristida capillacea Lam.		sandy savanna
800		TIL	Carpodiptera schomburgkii Baill.	Christiana africana DC.	tc: BM K P U	
801	Jun 1838	FAB	Triptolemea ovata Mart.	Dalbergia variabilis Vogel		Pirara
802		EUP	Croton nervosus Klotzsch var. pubescens Klotzsch	Croton nervosus Klotzsch	tc: F GH P US	Takutu R.
803		FAB	Aeschynomene interrupta Benth.	Aeschynomene interrupta Benth.	ht: K; it: F GH US	R. Branco
804		CYP	Isolepis junciformis Kunth	Bulbostylis junciformis (Kunth) C.B. Clarke		sandy savanna
805	Aug 1838	STR	Melochia graminifolia A. St.-Hil.	Melochia graminifolia A. St.-Hil.	BM	R. Branco
806		CYP	Cyperus surinamensis Rottb.	Cyperus surinamensis Rottb.		sandy savanna

Index 1. Robert Schomburgk's first collection series (1835–1839) – continued

No.	Family	Date	Name as labelled	Current name	Herbaria	Locality
807	CYP		Kyllinga ('Kyllingia')			sandy savanna
808	LYT		Cuphea micrantha Kunth	Cuphea micrantha Kunth		savanna at Pirara
809	CYP		Cyperus simplex Kunth	Cyperus simplex Kunth	K	
810	CYP		Cyperus schomburgkianus Nees	Cyperus schomburgkianus Nees	ht: K; it: US	dry savanna
811	RUB	1839	Faramea crassifolia Benth.	Faramea crassifolia Benth.	ht: K; it: BM NY US	Pirara and Rupununi
812	AST	1839	Wedelia hispida Kunth	Sphagneticola brachycarpa (Baker) Pruski	BM US	Annai
813	SOL	July 1838	Solanum schomburgkii Sendtn.	Solanum schomburgkii Sendtn.	tc: L U W	R. Branco
814	ORC		Habenaria schomburgkii Lindl. ex Benth.	Habenaria schomburgkii Lindl. ex Benth.	tc: BM US	R. Branco
815	LYT		Cuphea melvilla Lindl.	Cuphea melvilla Lindl.		
816	PGL		Polygala camporum Benth.	Polygala violacea Aubl. emend. Marques	tc: US	R. Branco
816*	PGL	1839	Polygala monticola Kunth	Polygala monticola Kunth	BM K P	R. Branco; dry savanna
817	BOR	Aug 1838	Cordia	Cordia grandifolia (Desv.) Kunth	BM G K	R. Branco
818	PAS	July 1838	Cieca discolor Roem.	Passiflora misera Kunth	BM U	R. Branco
819	ASC	July 1838	Sarcostemma cumanense Kunth	Philibertia cumanensis Hemsl.		R. Branco
820	LNT		Utricularia myriocysta A. St.-Hil. & Girard	Utricularia myriocysta A. St.-Hil. & Girard		
821	FAB	Aug 1838	Centrosema pascuorum Mart. ex Benth.	Centrosema pascuorum Mart. ex Benth.	tc: G	dry savanna
822	FAB		Aeschynomene mucronulata Benth.	Aeschynomene histrix Poir. var. histrix		dry savanna
823	PGL	July 1838	Polygala galioides Kunth	Polygala galioides Poir.	K L P G	R. Branco
824	CYP		Cyperus cuspidatus Kunth	Cyperus cuspidatus Kunth		
825	PLG		Coccoloba grandis Benth.	Coccoloba latifolia Lam.		R. Branco
825*	MIM		Acacia westiniana DC.	Acacia riparia Kunth		R. Branco
826	MNY		Limnanthemum humboldtianum Griseb.	Nymphoides indica (L.) Kuntze		
827	CNN	1839	Connarus incomptus Planch.	Connarus incomptus Planch.	ht: K; it: BM G GH K P U W	
828	FAB	1839	Eriosema rufum (Kunth) G. Don	Eriosema rufum (Kunth) G. Don	BM F G GH NY US	
829	BIG	1839	Tanaecium albiflorum DC.	Tanaecium jaroba Sw.	ht: K; it: BM	
830	MIM		Pithecellobium multiflorum (Kunth) Benth. ('Pithecolobium')	Albizia subdimidiata (Splitg.) Barneby & J.W. Grimes var. subdimidiata		
831	VRB	Sep 1838	Stachytarpheta mutabilis (Jacq.) Vahl ('Stachytarpha')	Stachytarpheta sprucei Moldenke		Serra Grande or Caruma Mts.
832	AMA		Iresine polymorpha Mart.	Iresine diffusa Humb. & Bonpl. ex Willd.	U	
833	MNS	1839	Anomospermum schomburgkii Miers	Orthomene schomburgkii (Miers) Barneby & Krukoff	ht: BM; it: B BM US	Pirara
834	ORC		Epidendrum schomburgkii Lindl.	Epidendrum macrocarpum (Rich.) Garay	BM	Kanuku Mts.
835	PHT		Ancistrocarpus maypurensis Kunth	Microtea maypurensis (Kunth) G. Don		
836	MRT	1839	Psidium polycarpon Lamb.	Psidium guineense Sw.	BM	
837	HYD		Hydrolea multiflora Mart. ex Choisy	Hydrolea elatior Schott		

No.	Date	Family	Name	Accepted name	Codes	Location
838	Sep 1838	MLV	Fugosia campestris Benth.	Cienfuegosia phlomidifolia (A. St.-Hil.) Garcke	tc: GH NY US	R. Branco; dry savanna
839		CSL	Copaifera pubiflora Benth.	Copaifera pubiflora Benth.	tc: NY US	
840		CSL	Cassia prostrata L.	Chamaecrista serpens (L.) Greene var. serpens	US	dry savanna
841		CYP	Cyperus luzulae Rottb. var. microphyllus Nees	Cyperus luzulae Rottb. ex Retz.		dry savanna
842		CMM	Aneilema schomburgkianum Kunth	Murdannia schomburgkiana (Kunth) Brückner		
843		CSL	Cassia obtusifolia (L.) var. uniglandulosa Vogel	Senna obtusifolia (L.) H.S. Irwin & Barneby	US	
844		MLP	Banisteria schomburgkiana Benth.	Banisteriopsis muricata (Cav.) Cuatrec.	ht: K; it: BM F G K L NY P US W	R. Branco
845		MLV	Abutilon spicatum Kunth	Briquetia spicata (Kunth) Fryxell		R. Branco
846		FAB	Aeschynomene densiflora Benth.	Aeschynomene histrix Poir. var. densiflora (Benth.) Rudd	tc: US W	dry savanna
847		ASC	Metastelma campanulatum Decne.	Metastelma parviflorum (Sw.) R. Br. ex Schult.	ht: K; it: BM NY US	R. Branco
848		MIM	Mimosa floribunda Willd.	Mimosa sensitiva L.		R. Branco
849		MLV	Abutilon lucianum Sweet	Wissadula contracta (Link) R.E. Fr.	BM	R. Branco
850		FAB	Cymbosema roseum Benth.	Cymbosema roseum Benth.	ht: K; it: US	R. Branco
851		CYP	Cyperus infucatus Kunth	Cyperus filifolius Willd. ex Kunth		R. Branco
852		MIM	Piptadenia peregrina (L.) Benth.	Adenanthera peregrina (L.) Speg.		R. Branco
853		LAM	Marsypianthes hyptoides Mart. ex Benth.	Marsypianthes chamaedrys (Vahl) Kuntze		R. Branco
854		CMB	Bucida buceras L.	Buchenavia suaveolens Eichl.		R. Branco
855		CYP	Fimbristylis brizoides Nees var. microstachya Nees	Fimbristylis sp.		dry savanna
856	1839	RUB	Mitracarpus scabrellus Benth.	Mitracarpus microspermus K. Schum.	tc: BM US	R. Branco
857		XYR	Abolboda pulchella Humb. & Bonpl.	Abolboda pulchella Humb. & Bonpl.	BM	dry savanna
858		CYP	Kyllinga cruciformis Schrad. ('Kyllingia')	Kyllinga brevifolia Rottb. ('Kyllingia')		R. Negro
859		EUP	Pogonophora schomburgkiana Miers ex Benth.	Pogonophora schomburgkiana Miers ex Benth.	tc: G	R. Negro
860		PHT	Petiveria hexaglochin Fisch. & C.A. Mey.	Petiveria hexaglochin Fisch. & C.A. Mey.	tc: BM US	R. Negro
861		RUB	Psychotria bracteata DC.	Psychotria pseudinundata Wernham	BM	Pedrero on the R. Negro
862		CPP	Cleome aculeata L.	Cleome aculeata L.		Pedrero on the R. Negro
863	1839	MLS	Spennera circaeifolia DC.	Aciotis circaeifolia (Bonpl.) Triana	BM	
864		RUB	Borreria alata (Aubl.) DC.	Borreria alata (Aubl.) DC.		
865		LCY	Lecythis micrantha O. Berg	Eschweilera micrantha (O. Berg) Miers	lt: K; ilt: BM CGE F GH L NY U US	
866		MIM	Piptadenia peregrina (L.) Benth.	Adenanthera peregrina (L.) Speg.	st: NY US	
867		AST	Latreillea glabrata Benth.	Ichthyothere terminalis (Spreng.) S.F. Blake		R. Negro
868		VOC	Erisma floribundum Rudge	Erisma floribundum Rudge var. floribundum		R. Negro
869		CSL	Heterostemon mimosoides Desf.	Heterostemon mimosoides Desf. var. mimosoides	US	R. Negro
870	1839	STR	Theobroma bicolor Humb. & Bonpl.	Theobroma bicolor Humb. & Bonpl.	BM	R. Negro

Index 1. Robert Schomburgk's first collection series (1835–1839) – continued

No.	Date	Fam.	Name as collected	Current name	Herbarium/type	Locality
871		MLS	Melastoma aplostachyum Humb. & Bonpl.	Miconia aplostachya (Bonpl.) DC.		
871*		CHB	Licania floribunda Benth.	Licania apetala (E. Mey.) Fritsch var. apetala		
872	1839	MRT	Eugenia fallax Rich.	Myrcia fallax (Rich.) DC.	BM	Pedrero on the R. Negro
872*		MRT	Eugenia multiflora Lam.	Myrcia multiflora (Lam.) DC.	BM	Pedrero on the R. Negro
873		CHB	Licania heteromorpha Benth.	Licania heteromorpha Benth. var. heteromorpha	ht: K; it: BM F GH L NY OXF P US	Pedrero on the R. Negro
874		MIM	Pithecellobium schomburgkii Benth. ('Pithecolobium')	Enterolobium schomburgkii (Benth.) Benth.	tc: US	Pedrero on the R. Negro
875		FAB	Aeschynomene ciliata Vogel	Aeschynomene hispida Willd.	B † F K US	Pedrero on the R. Negro
876		CYP	Scleria melaleuca Rchb. ex Cham. & Schltdl.	Scleria melaleuca Rchb. ex Cham. & Schltdl.	K L P UPS W	R. Negro
877		BIG	Cuspidaria	Anemopaegma jucundum Bureau & K. Schum.		R. Branco
878		CYP	Cyperus sphacelatus Rottb.	Cyperus sphacelatus Rottb.		Pedrero on the R. Negro
879	1839	RUB	Psychotria amplectens Benth.	Psychotria amplectens Benth.	BM	R. Branco
880		DSC	Dioscorea consanguinea Kunth	Dioscorea amazonum Mart. var. consanguinea (Kunth) Uline	lt: BM; ilt P US	Pedrero on the R. Negro
881		CNV	Ipomoea umbellata (L.) G. Mey.	Merremia umbellata (L.) Hallier	BM	Pedrero on the R. Negro
882		CYP	Cyperus compressus L.	Cyperus compressus L.	K	Pedrero and at falls of R. Branco
882*		CYP	Cyperus sphacelatus Rottb.	Cyperus sphacelatus Rottb.		Pedrero and at falls of R. Branco
883		EUP	Alchornea latifolia Sw.	Aparisthmium cordatum (A. Juss.) Baill.		R. Branco
884	Oct 1839	MLS	Clidemia desmantha Benth.	Miconia rhytidophylla Naudin	ht: K; it: NY US	Pedrero on the R. Negro
885	Oct 1839	MRS	Conomorpha laxiflora A. DC.	Cybianthus spicatus (Kunth) G. Agostini	st: GH US	R. Negro
886	Oct 1839	CHB	Hirtella eriandra Benth.	Hirtella eriandra Benth.	ht: K; it: BM CGE GH NY P	Pedrero on the R. Negro
887		MLS	Tococa barbata Benth.	Tococa guianensis Aubl.	ht: K; it: BM NY US	Pedrero on the R. Negro
888		VRB	Stachytarpheta cayennensis (Rich.) Vahl ('Stachytarpha')	Stachytarpheta cayennensis (Rich.) Vahl ('Stachytarpha')		Pedrero on the R. Negro
889	Nov 1839	MLV	Malachra capitata L.	Malachra alceifolia Jacq. var. alceifolia	BM	Pedrero on the R. Negro
890		AST	Spilanthes poeppigii DC.	Acmella uliginosa (Sw.) Cass.	US	
891		APO	Rauvolfia polyphylla Benth. ('Rauwolfia')	Rauvolfia polyphylla Benth. ('Rauwolfia')		R. Negro
892	1839	ANA	Thyrsodium schomburgkianum Benth.	Tapirira obtusa (Benth.) J.D. Mitchell	ht: K; it: BM K NY	R. Negro
893		PLG	Coccoloba ovata Benth.	Coccoloba ovata Benth.	st: NY U	R. Negro
894		CSL	Cassia moschata Kunth	Cassia leiandra Benth.		R. Negro
895	Nov 1839	CSL	Cassia trinitatis Rchb. ex DC.	Senna spectabilis (DC.) H.S. Irwin & Barneby var. spectabilis		Pedrero on the R. Negro

No.	Date	Fam.	Name 1	Name 2	Type info	Locality
896		FAB	Diplotropis nitida Benth.	Diplotropis nitida Benth.	tc: US	R. Negro
897		CHB	Licania floribunda Benth.	Licania apetala (E. Mey.) Fritsch var. apetala	ht: K; it: BM CGE F GH L OXF P U US	
898		APO	Tabernaemontana rupicola Benth.	Tabernaemontana rupicola Benth.	ht: K; it: BM CGE E F FI G L NY P SING TCD US W	Pedrero on the R. Negro
899		THY	Lasiadenia rupestris Benth.	Lasiadenia rupestris Benth.	tc: NY	Pedrero on the R. Negro
900		RUB	Sphinctanthus rupestris Benth.	Sphinctanthus striiflorus (DC.) Hook.f.	tc: BM NY US	rocks on R. Negro
901		PIP	Artanthe guianensis Klotzsch	Piper guianense (Klotzsch) C. DC.	tc: B K NY P W	R. Branco
902	Oct 1839	BIG	Bignonia tubulosa Klotzsch [n.n.]	Pyrostegia dichotoma Miers ex K. Schum.	K	R. Branco
903		MIM	Schrankia brachycarpa Benth.	Mimosa diplotricha C. Wright ex Sauv.	ht: K; it: US	R. Negro
904		PLP	Polypodium lepidopteris Sodiro	Polypodium lepidopteris Sodiro	B	Pedrero on the R. Negro
905		MLP	Lophanthera kunthiana A. Juss.	Lophanthera longifolia (Kunth) Griseb.	BM	Barcelos on R. Negro
906	Nov 1839	CHB	Licania pendula Benth.	Licania apetala (E. Mey.) Fritsch var. apetala	ht: K; it: BM CGE F GH LE NY OXF P US W	lagoons in R. Negro
907		FAB	Triptolemea riparia Mart. ex Benth.	Dalbergia riparia Benth.	tc: NY US	R. Negro
908		CSL	Peltogyne paniculata Benth.	Peltogyne paniculata Benth. subsp. paniculata	ht: K; it: BM CGE G GH K MO P	lagoons in R. Negro
909	Nov 1839	MLP	Byrsonima staminea Griseb.	Byrsonima japurensis A. Juss.		R. Negro
910		CHB	Licania mollis Benth.	Licania mollis Benth.	ht: K; it: BM CGE F G GH L NY OXF US W	R. Negro
911	1839	BOR	Cordia scabrifolia A. DC.	Cordia scabrifolia A. DC.	tc: B BM G K L P US	Barcelos on R. Negro
912	Nov 1839	EUP	Microstachys guianensis Klotzsch	Sebastiana corniculata (Vahl) Muell. Arg.	ht: K; it: BM G U US	Barcelos on R. Negro
913	Nov 1839	CYP	Scleria stipularis Nees	Scleria stipularis Nees	tc: K US	Barcelos on R. Negro
914	Nov 1839	FAB	Tounatea grandifolia (Bong. ex Benth.) Taub.	Swartzia grandifolia Bong. ex Benth.	lt: K; it F FI G GH P US W	
915		CYP	Calyptrostylis longirostris Nees	Rhynchospora amazonica Poepp. & Kunth subsp. guianensis (Kük.) T. Koyama	tc: NY US	
916		CPP	Cleome pungens Willd.	Cleome spinosa Jacq.	ht: K	R. Negro
917	Nov 1839	MIM	Entada myriadena Benth.	Mimosa myriadenia (Benth.) Benth.	BM	R. Negro
918		EUP	Schismatopera distichophylla Klotzsch	Pera distichophylla (Mart.) Baill.	tc: L US	R. Negro
919		APO	Tabernaemontana laxa Benth.	Molongum laxum (Benth.) Pichon	ht: K; it: F FI G GH K L P US W	Pedrero on the R. Negro
920		FLC	Carpotroche paludosa Benth.	Lindackeria paludosa (Benth.) Gilg		Pedrero on the R. Negro
921	Nov 1839	MRT	Eugenia leptantha Benth.	Eugenia inundata DC.	tc: BM	Pedrero on the R. Negro
922	Nov 1839	ANN	Guatteria inundata Mart.	Guatteria obovata R.E. Fr.	tc: B BM F K L P U US	Pedrero on the R. Negro
923	Nov 1839	RHM	Gouania	Ruprechtia tenuifolia Benth.	st: GH NY U US	Pedrero on the R. Negro
924	Nov 1839	PLG	Ruprechtia tenuifolia Benth.	Miconia argyrophylla DC.		Pedrero on the R. Negro
925	Nov 1839	MLS	Miconia argyrophylla DC.	Spigelia humilis Benth.	st: BM L	Pedrero on the R. Negro
926	Nov 1839	LOG	Spigelia humilis Benth.			R. Negro

Index 1. Robert Schomburgk's first collection series (1835–1839) – continued

No.	Date	Fam.	Name in first collection	Current name	Herbaria	Locality
927	Nov 1839	EUP	Phyllanthus piscatorum Kunth	Phyllanthus piscatorum Kunth		Barcelos on R. Negro
928	Nov 1839	SPT	Myrsine schomburgkiana Miq.	Elaeoluma schomburgkiana (Miq.) Baill.		Barcelos on R. Negro
929		MLS	Jucunda tomentosa Benth.	Miconia tomentosa (Rich.) D. Don ex DC.	BM	Barcelos on R. Negro
930		CRY	Drymaria cordata (L.) Willd. ex Roem. & Schult.			Barcelos on R. Negro
931		MIM	Schrankia leptocarpa DC.	Mimosa quadrivalvis L. var. leptocarpa (DC.) Barneby		R. Negro
932	Nov 1839	MLS	Spenmera dysophylla Benth.	Aciotis annua (DC.) Triana	ht: K; it: GH NY	Barcelos on R. Negro
933	Nov 1839	CUC	Melothria pendula L.	Melothria pendula L.	st: BM	
934	Nov 1839	AMA	Gomphrena sessilis L.	Alternanthera sessilis (L.) R. Br. ex DC.		Barcelos on R. Negro
935	1839	CLU	Caraipa leiantha Benth.	Haploclathra leiantha (Benth.) Benth.	tc: BM NY P US	R. Negro
936	Dec 1839	ELC	Dasynema laurifolium Benth.	Sloanea laurifolia (Benth.) Benth.	tc: BM GH US	R. Negro
937		LAU	Oreodaphne guianensis Aubl.	Ocotea guianensis Aubl.	st: US	
938	Dec 1839	PAS	Passiflora coccinea Aubl.	Passiflora coccinea Aubl.	BM P U	Barcelos
939		PIP	Piper pellucidum L.	Peperomia pellucida (L.) Kunth	BM P	R. Negro
940		EUP	Amanoa oblongifolia Muell. Arg.	Amanoa oblongifolia Muell. Arg.	BM K L P U W	Barcelos on R. Negro
941	1839	OCH	Elvasia calophyllea DC.	Elvasia calophyllea DC.	BM	
942	1839	RUB	Psychotria polycephala Benth.	Psychotria polycephala Benth.	st: BM	R. Negro
943	1839	MRT	Eugenia egensis DC.	Eugenia egensis DC.	BM	R. Negro
944	Dec 1839	EUP	Croton suavis Kunth	Croton suavis Kunth	P	Barcelos on R. Negro
945	Dec 1839	RUB	Psychotria longistipula Benth.	Psychotria	BM	Barcelos on R. Negro
946	Dec 1839	MLV	Sida urens L.	Sida urens L.	BM	Barcelos on R. Negro
947		VIO	Alsodeia racemosa Mart. & Zucc.	Rinorea sprucei (Eichl.) Kuntze	CGE F G GH K L M P US W	Barcelos on R. Negro
948		LOG	Strychnos rouhamon Benth.	Strychnos guianensis (Aubl.) Mart.	BM U	
949		ORC	Oncidium iridifolium Kunth	Psychomorchis pusilla (L.) Dodson & Dressler	tc: B BM	R. Negro
950		ANN	Guatteria heteropetala Benth.	Heteropetalum brasiliense Benth.	tc: L	Barcelos on R. Negro
951	Dec 1839	APO	Tabernaemontana odorata Vahl	Malouetia tamaquarina (Benth.) A. DC.	tc: BM BR NY US	Barcelos on R. Negro
952	Dec 1839	RUB	Remijia tenuiflora Benth.	Morinda tenuiflora (Benth.) Steyerm.	tc: BM NY US	Barcelos on R. Negro
953	1839	APO	Thenardia laurifolia Benth.	Forsteronia laurifolia (Benth.) A. DC.		R. Negro
954		PIP	Heckeria peltata Kunth	Piper peltatum L.	B	
955		SMR	Simaba obovata Klotzsch [n.n.]	Simaba cedron Planch.	BM	
956	1839	PAS	Passiflora nitida Kunth	Passiflora nitida Kunth	st: GH L U	
957		PLG	Ruprechtia tenuifolia Benth.	Ruprechtia tenuifolia Benth.	tc: BM US	Pedrero on the R. Negro
958		MRT	Eugenia divaricata Benth.	Myrciaria dubia (Kunth) McVaugh	st: B BM F G K L P	R. Negro
959		HPC	Tontelea longifolia Miers	Tontelea attenuata Miers		

No.	Year	Family	Name	Accepted name	Specimens	Locality
960		MLS	Phyllopus martii Mart.	Henriettea martii Naudin	ht: BM: it: B G K L P U U S	sand banks of the R. Negro
961	1839	MLS	Clidemia campestris Benth. var. pauciflora Benth.	Miconia biglomerata (Bonpl.) DC.	ht: K: it: F NY US W	R. Negro
962		ANN	Guatteria elongata Benth.	Guatteria elongata Benth.	US	R. Negro
963		RUB	Psychotria remota Benth.	Psychotria remota Benth.		R. Negro
964		MRT	Eugenia pyrifolia Hamilton	Myrcia pyrifolia (Hamilton) Nied.		R. Negro
965		APO	Echites elegans Benth.	Odontadenia geminata (Roem. & Schult.) Muell. Arg.		R. Negro
966	1839	MEL	Guarea affinis A. Juss.	Guarea pubescens (Rich.) A. Juss. subsp. pubiflora (A. Juss.) T.D. Penn.	BM	
967		MEL	Guarea	Elaeoluma schomburgkiana (Miq.) Baill.	ht: U: it: BM US	R. Negro
968		SPT	Myrsine schomburgkiana Miq.	Ferdinandusa rudgeoides (Benth.) Wedd.	tc: BM G US	islands in the R. Negro
969	1839	RUB	Aspidanthera rudgeoides Benth.	Emmotum acuminatum (Benth.) Miers	ht: K: it: P US	R. Negro
970		ICC	Pogopetalum acuminatum Benth.	Rhynchospora schomburgkiana (Boeck.) T. Koyama	tc: US	islands in the R. Negro
971		CYP	Leptoschoenus prolifer Nees	Psychotria subundulata Benth.	tc: NY US	R. Negro
972		RUB	Psychotria subundulata Benth.	Parkia discolor Spruce ex Benth.		R. Negro
973		MIM	Parkia	Chlorocardium rodiei (Rob. Schomb.) Rohwer, Richter, & van der Werff	tc: NY US	
974		LAU	Nectandra rodiei R.H. Schomb.	Ocotea aciphylla (Nees) Mez	ht: B: it: GH U	
974*		LAU	Nectandra rhynchophylla Meisn.	Cybianthus guianensis (A. DC.) Miq. subsp. guianensis	tc: US	
975		MRS	Conomorpha guyanensis A. DC.		tc: B K	
976		LAU	Oreodaphne gracilis Meisn.	Ocotea gracilis (Meisn.) Mez	ht: K: it: BM CGE G L OXF P	
977		CHB	Licania parviflora Benth.	Licania parviflora Benth.	tc: K U	R. Negro
978		EUP	Podocalyx loranthoides Klotzsch	Podocalyx loranthoides Klotzsch	tc: B LG US	R. Negro; dry savanna
979		SLG	Lycopodium sp. nov.	Selaginella puberula Spring	BM K P U US	R. Padauiri
979*		SLG	Selaginella coarctata Spring	Selaginella coarctata Spring	tc: BM	R. Padauiri
980	1839	MLS	Tococa truncata Benth.	Tococa truncata Benth.	tc: G NY	R. Negro
980*	1839	MLS	Tococa coronata Benth.	Tococa coronata Benth.	B	R. Negro
981		VRB	Aegiphila mollis Kunth		BM K P	
982		SLG	Lycopodium concinnum Sm.	Selaginella breynii Spring	ht: ?; it: BM NY	R. Padauiri
982*		SLG	Selaginella guianensis Spring	Selaginella falcata (P. Beauv.) Spring	B BM K P	R. Padauiri
983		ARE	Lepidocaryum tenue Mart.	Lepidocaryum tenue Mart. var. gracile (Mart.) Henderson	tc: BM US	
984		BOR	Cordia nodosa Lam.	Cordia nodosa Lam.	US	
985	1839	RUB	Pagamea guianensis Aubl.	Pagamea guianensis Aubl.		
985*		RUB	Ferdinandusa goudotiana K. Schum.	Ferdinandusa goudotiana K. Schum.		
986		SAP	Toulicia guianensis Aubl.	Porocystis toulicioides Radlk.		
987		LAU	Oreodaphne costulata Nees	Nectandra cuspidata Nees	st: B BM G K P U	R. Padauiri
988		HUG	Roucheria calophylla Planch.	Roucheria calophylla Planch.		R. Padauiri

Index 1. Robert Schomburgk's first collection series (1835–1839) – continued

No.	Date	GEN	Collector's name	Determined name	Herbaria	Locality
989			Coutoubea ramosa Aubl.	Coutoubea ramosa Aubl. var. ramosa	US	sands R. Negro near Manaus (Barra)
990		OCH	Godoya gemmiflorus Mart.	Blastemanthus gemmiflorus Planch.		
991	1839	CLU	Tovomita umbellata Benth.	Tovomita umbellata Benth.		
992	1839	PLP	Phymatodes schomburgkiana J. Sm.	Micrgramma megalophylla (Desv.) de la Sota	ht: K; it: BM G K U US	R. Orinoco
993	1839	FLC	Tetracocyne angustifolia Turcz.	Ryania angustifolia (Turcz.) Monach.	ht: B	R. Negro; near São Gabriel
994	1839	RUB	Coffea subsessilis Benth.	Psychotria erecta (Aubl.) Standley & Steyerm.	ht: CW; it: BM Fl G K L NY P U US W	R. Negro
995	1839	ANN	Guatteria foliosa Benth.	Guatteria foliosa Benth.	BM US	R. Negro
996	1839	FLC	Casearia javitensis Kunth	Casearia javitensis Kunth	ht: B; it: BM K L P NY U US	R. Negro
997		EUP	Phyllanthus guyanensis Muell. Arg.	Phyllanthus attenuatus Miq.	BM	expedition to R. Orinoco
998		MLS	Clidemia radulaefolia Benth.	Miconia radulaefolia (Benth.) Naudin	st: K P	R. Negro
999		MLS	Spennera viscida Benth.	Aciotis indecora (DC.) Triana	ht: K	R. Negro
1000		CMM	Dithyrocarpus schomburgkianus Kunth	Aneilema umbrosum (Vahl) Kunth subsp. ovato-oblongum (P. Beauv.) J.K. Morton	tc: NY US	R. Negro
1001		VRB	Stachytarpheta elatior Schrad. ex Schult & Roem. ('Stachytarpha')	Stachytarpheta angustifolia (Mill.) Vahl	K P	Upper Rupununi R.
1002	Mar 1839	MRS	Conomorpha laxiflora A. DC.	Cybianthus spicatus (Kunth) G. Agostini	tc: US	near São Isabel
1003	1839	AST	Pectis elongata Kunth	Pectis elongata Kunth var. elongata	BM US	Fort São Gabriel
1004		POA	Eragrostis maypurensis (Kunth) Steud.	Eragrostis maypurensis (Kunth) Steud.		São Gabriel on the R. Negro
1005	1839	AST	Clibadium sylvestre (Aubl.) Baill.	Clibadium sylvestre (Aubl.) Baill.	BM	
1006		LAM	Hyptis spicata Poit.	Hyptis mutabilis (Rich.) Briq.		abandoned village of São José
1007	1839	MLV	Urena americana L.	Urena lobata L.	BM	R. Padauiri
1008		EUP	Palanostigma crotonoides Mart. ex Klotzsch ('Palamostigma')	Croton palanostigma Klotzsch	st: G P US	near São Gabriel; R. Padauiri
1009	Mar 1839	VRB	Vitex	Vitex calothyrsa Sandw.		
1010		CNV	Dicranostyles scandens Benth.	Dicranostyles scandens Benth.	U	R. Padauiri; R. Negro
1011		RUB	Calycophyllum coccineum DC.	Warszewiczia coccinea (Vahl) Klotzsch	tc: BM U	Camanow
1012	1839	CNV	Evolvulus alsinoides (L.) L.	Evolvulus alsinoides (L.) L.	BM U	São Gabriel on the R. Negro
1013		OXL	Oxalis casta Zucc.	Oxalis casta Zucc.		São Isabel on the R. Negro
1014	1839	AST	Eupatorium schomburgkii Benth.	Baccharis brachylaenoides DC. var. brachylaenoides	BM	
1015		MLS	Macairea rigida Benth.	Macairea rigida Benth.		near Padamo R.; Mt. Maravaca
1016	Feb 1839	AST	Vernonia opaca Benth.	Piptocarpha opaca (Benth.) Baker subsp. opaca		Mt. Maravaca
1017		LYC	Lycopodium linifolium L.	Huperzia linifolia (L.) Trevis.		Serra Mey
1018	1839	ERI	Cavendishia duidae A.C. Sm.	Cavendishia callista J.D. Sm.	BM K	R. Padauiri

No.	Date	Code	Name	Current name	Herbaria	Locality
1019	Mar 1839	CAM	Centropogon surinamensis (L.) C. Presl	Centropogon cornutus (L.) Druce	BM	R. Padauiri
1020		MLS	Miconia schomburgkii Benth.	Miconia lepidota DC.	ht: K	R. Padauiri
1021	1839	ELC	Sloanea massoni Sw.	Sloanea brevipes Benth.	tc: BM GH L MO U US	R. Negro
1021*		ELC	Sloanea sinemariensis Aubl.	Sloanea sinemariensis Aubl.		São Isabel
1022		URT	Urtica caracasana Jacq.	Urera caracasana (Jacq.) Griseb.		R. Padauiri
1023	Mar 1839	CYP	Trichelostylis stricta Nees	Bulbostylis schomburgkiana (Steud.) M.T. Strong	ht: CGE; it: BM K US	
1024	1839	BOR	Heliotropium inundatum Sw.	Heliotropium procumbens Mill.	BM K G	Roraima
1025		ORC	Stelis ophioglossoides Sw.	Stelis schomburgkii Fawc. & Rendle	ht: BM	Roraima
1026		ORC	Masdevallia guianensis Lindl.	Masdevallia guianensis Lindl.	tc: BM	Barcelos on R. Negro
1027		ERX	Erythroxylum amplum Benth. ('Erythroxylon')	Erythroxylum macrophyllum Cav.		R. Negro
1028		BIG	Tabebuia fluviatilis (Aubl.) DC.	Tabebuia fluviatilis (Aubl.) DC.		R. Negro; near Toupeaging Mts.
1029		LAM	Hyptis dilatata Benth.	Hyptis dilatata Benth.		Serra Mey
1030	Feb 1839	HMP	Hymenostachys diversifrons Bory	Trichomanes trollii Bergdolt		
1031		GMM	Polypodium discolor Hook.	Grammitis discolor (Hook.) C.V. Morton		
1032		SOL	Schwenckia hirta Klotzsch ex Schltdl. ('Schwenkia')	Schwenckia americana L. var. hirta (Klotzsch) Carvalho	tc: K	dry savanna near Roraima
1033	Oct 1838	SCR	Herpestis gratioloides Benth.	Bacopa gratioloides (Benth.) Chodat	BM	Pakaraima Mts.
1034		LAM	Hyptis membranacea Benth.	Hyptidendron arboreum (Benth.) Harley		Roraima
1035	Nov 1838	AST	Vernonia ehretifolia Benth.	Lepidaploa ehretifolia (Benth.) H. Rob.	tc: BM	Roraima
1036	Oct 1838	CNV	Evolvulus strictus Benth.	Evolvulus glomeratus Nees & Mart.	tc: BM	dry savanna near Roraima
1037		IXO	Ochthocosmus roraimae Benth.	Ochthocosmus roraimae Benth.	ht: K	Roraima
1038	1839	ERI	Notopora schomburgkii Hook.f.	Notopora schomburgkii Hook.f.	BM	Roraima
1039		GLC	Mertensia pedalis Kaulf.	Dicranopteris pedalis (Kaulf.) Looser		open savanna
1040		MLS	Rhexia taxifolia A. St.-Hil.	Marcetia taxifolia (A. St.-Hil.) DC.		near Roraima
1041	1839	CLU	Mahurea exstipulata Benth.	Mahurea exstipulata Benth.	tc: BM	Pakaraima Mts.
1042	1839	AST	Achyrocline flaccida DC.	Achyrocline flaccida DC.	BM	Roraima
1043	Nov 1838	ASC	Ditassa taxifolia Decne.	Ditassa taxifolia Decne.	tc: BM	Roraima
1044	1839	ELC	Sloanea usurpatrix Sprague & Riley	Sloanea tuerckheimii J.D. Sm.	BM MO	
1045		CSL	Dimorphandra macrostachya Benth.	Dimorphandra macrostachya Benth.	ht: K; it: BM G GH NY US	Roraima
1046	Oct 1838	BNN	Bonnetia sessilis Benth.	Bonnetia sessilis Benth.	lt ?: BM	Roraima
1047	1839	VOC	Qualea schomburgkiana Warm.	Qualea schomburgkiana Warm.	ht: K	Roraima
1048		AQF	Ilex retusa Klotzsch [n.n.]	Ilex retusa Klotzsch ex Reissek	L	
1049	Oct 1838	BIG	Digomphia laurifolia Benth.	Digomphia laurifolia Benth.		Roraima
1050		SAR	Heliamphora nutans Benth.	Heliamphora nutans Benth.	ht: K; it: BM CGE F NY	Roraima
1051		CHB	Hirtella scabra Benth.	Hirtella scabra Benth.	K P	near Mt. Roraima
1052		CYP	Hypolytrum pungens (Vahl) Kunth	Hypolytrum pulchrum (Rudge) H. Pfeiff.		near Mt. Roraima

Index 1. Robert Schomburgk's first collection series (1835–1839) – continued

No.	Date	Code	Name	Accepted name	Type	Locality
1053	Oct 1838	APO	Echites angustifolia Benth.	Mandevilla benthamii (A. DC.) K. Schum.	ht: G; it: L	Mt. Roraima
1054	Oct 1838	XYR	Xyris involucrata Nees	Xyris involucrata Nees	lt: K; ilt: BM K L	Roraima
1055	1839	RPT	Amphiphyllum schomburgkii Maguire	Marahuacaea schomburgkii (Maguire) Maguire	ht: NY; it: BM K NY	near Mt. Roraima
1055*		RUB	Chiococca nitida Benth.	Chiococca nitida Benth.	tc: NY	near Mt. Roraima
1056		VRB	Vitex umbrosa Sw.	Vitex compressa Turcz.		
1057						
1058		ZIN	Xyris guianensis Steud.	Xyris guianensis Steud.	lt: K; ilt: BM K	
1059	1839	ORC	Sobralia liliastrum Lindl.	Sobralia liliastrum Lindl.	BM P	Roraima
1060		ERO	Paepalanthus schomburgkii Klotzsch [n.n.]	Syngonanthus longipes Gleason		
1061		MLS	Miconia revoluta Benth.	Miconia prasina (Sw.) DC.		
1062		MLS	Meisneria cordifolia Benth.	Siphanthera cordifolia (Benth.) Gleason	ht: K	Roraima
1063		MLS	Miconia holosericea (L.) DC. var. obtusifolia Benth.	Miconia albicans (Sw.) Triana		
1063*		MLS	Miconia fallax DC.	Miconia fallax DC.		
1064		LNT	Utricularia humboldtii R.H. Schomb.	Utricularia humboldtii R.H. Schomb.	st: BM K	Mt. Roraima
1065		ERI	Befaria sprucei Meisn.	Bejaria sprucei Meisn.		

INDEX 2. ROBERT SCHOMBURGK'S ADDITIONAL COLLECTION SERIES (1835–1839).

Ro add. ser.	Year	Fam.	Original name	Current name	Herbarium	Location
1.S						
2.S						
3.S		MLV	Pavonia speciosa Kunth	Peltaea speciosa (Kunth) Standley		arid savanna
4.S						south chain of the Kanuku Mts.; arid savanna
5.S		ACA	Dipteracanthus canescens Nees	Dipteracanthus humilis Nees var. diffusus Nees		
6.S		FAB	Phaseolus linearis Kunth	Vigna linearis (Kunth) Maréchal, Mascherpa & Stainier		Upper Rupununi R.; moist savanna
7.S		LAM	Hyptis brevipes Poit.	Hyptis lanceolata Poir.		
8.S		LNT	Utricularia angustifolia Benj.	Utricularia hispida Lam.		Kanuku Mts.; arid savanna
9.S		FAB	Zornia reticulata Sm.	Zornia reticulata Sm.		Rupununi savanna
10.S		CNV	Ipomoea juncea Choisy	Merremia aturensis (Kunth) Hallier		Upper Rupununi; savanna
11.S		POA	Echinolaena scabra Kunth	Echinolaena inflexa (Poir.) Chase		near Mt. Roraima; moist savanna
12.S	1836	GEN	Schultesia benthamiana Klotzsch ex Griseb.	Schultesia benthamiana Klotzsch ex Griseb.		savanna
13.S	1836	PGL	Polygala monticola Kunth	Polygala monticola Kunth	K	
14.S		TIL	Vasivaea alchorneoides Baill.	Vasivaea alchorneoides Baill.	st: K P U US W	Pirara
15.S						
16.S						
17.S		SAP	Matayba opaca Radlk.	Matayba opaca Radlk.		
18.S		GEN	Coutoubea ramosa Aubl.	Coutoubea ramosa Aubl.		
19.S		CSL	Cassia uniflora Spreng. var. ramosa (Vogel) Benth.	Chamaecrista ramosa (Vogel) H.S. Irwin & Barneby var. ramosa	K	Upper Rupununi R.; sandy savanna
20.S						
21.S		TIL	Corchorus argutus Kunth	Corchorus orinocensis Kunth		
22.S		MLV	Hibiscus furcellatus Desr.	Hibiscus furcellatus Desr.	K	
23.S	1836	PGL	Polygala variabilis Kunth	Polygala trichosperma L.		
24.S		VOC				Amai
25.S		LNT	Genlisea filiformis A. St.-Hil.	Genlisea filiformis A. St.-Hil.	W	Annai
26.S		TIL	Luehea speciosa Willd.	Luehea speciosa Willd.	lt: K	banks of brooks and rivers
27.S		ONA	Jussiaea latifolia Benth. ('Jussieua')	Ludwigia latifolia (Benth.) Hara		

Index 2. Robert Schomburgk's additional collection series (1835–1839) – continued

No.	Year	Family	Name	Name	Herb.	Locality
28.S						
29.S	1836					
30.S		ANA	Anacardium occidentale L.	Anacardium occidentale L.		banks of brooks and rivers
31.S		PGL	Securidaca marginata Benth.	Securidaca marginata Benth.	tc: K US	
32.S		EUP	Phyllanthus conami Sw.	Phyllanthus brasiliensis (Aubl.) Poir		
33.S		PDS	Lacis			
34.S		PDS	Lacis			
35.S		MRT	Campomanesia poiteaui O. Berg	Campomanesia grandiflora (Aubl.) Sagot	K	
36.S		FAB	Vigna sinensis Endl. ex Hassk.			
37.S						
38.S		MLP	Bunchosia mollis Benth.	Bunchosia mollis Benth.	K	
39.S		FAB	Vatairea guianensis Aubl.	Vatairea guianensis Aubl.	K	
40.S						
41.S		ORC	Stenorrhynchos orchioides (Sw.) Rich.	Sacoila lanceolata (Aubl.) Garay		Rupununi savanna
42.S		CNV	Ipomoea schomburgkii Choisy	Ipomoea schomburgkii Choisy		
43.S		EUP	Dalechampia scandens L.	Dalechampia scandens L.		
44.S						
45.S						
46.S		CNV	Operculina pterodes Choisy	Operculina hamiltonii (G. Don) Austin & Staples		
47.S		MLS	Clidemia sericea D. Don	Clidemia sericea D. Don		
48.S						
49.S	1837	CLU	Clusia cuneata Benth.	Clusia cuneata Benth.	tc: BM	
50.S						
51.S		MIM	Parkia nitida Miq.	Parkia nitida Miq.		
52.S	1837	RUB	Posoqueria trinitatis DC.	Posoqueria trinitatis DC.	BM	
53.S	1837	CLU	Clusia insignis Mart.	Clusia grandiflora Splitg.	BM	
54.S						
55.S						
56.S						
57.S						
58.S						
59.S	1837	SMR	Simaba trichilioides A. St.-Hil.	Simaba trichilioides A. St.-Hil.	BM	
60.S		TEA	Ternstroemia crassifolia Benth.	Ternstroemia crassifolia Benth.		
61.S	1837	VRB	Aegiphila cuspidata Mart.	Aegiphila racemosa Vell.		near Roraima
62.S						

No.	Year	Fam.	Name	Name	Code	Locality
63.S	1837	BIG	Anemopaegma	Mimosa annularis Benth. var. odora Barneby	BM	
64.S		MIM	Mimosa paniculata Benth.	Paragonia pyramidata (Rich.) Bureau	BM	
65.S	1837	BIG	Tabebuia pyramidata DC.			
66.S				Caraipa	BM	
67.S	1837	CLU	Caraipa	Arrabidaea bilabiata (Sprague) Sandw.	BM	
68.S	1837	BIG	Memora bilabiata Sprague	Merremia macrocalyx (Ruiz & Pav.) O'Donell	BM	
69.S	1837	CNV	Batatas glabra (Aubl.) Benth.			
70.S				Anthodiscus trifoliatus G. Mey.	CGE	
71.S		CCR	Anthodiscus trifoliatus G. Mey.			
72.S				Lundia erionema DC.	BM	
73.S	1837	BIG	Lundia erionema DC.	Tamonea spicata Aubl.		
74.S		VRB	Tamonea spicata Aubl.	Eriosema violaceum (Aubl.) G. Don		
75.S		FAB	Rhynchosia violacea DC.	Orthopappus angustifolius (Sw.) Gleason		
76.S		AST	Elephantosis angustifolia DC.	Maxillaria superflua Rchb.f.	BM	
77.S*	1837	ORC	Maxillaria superflua Rchb.f.	Posoqueria trinitatis DC.	BM	
78.S		RUB	Posoqueria trinitatis DC.			
79.S						
80.S				Hibiscus bifurcatus Cav.	BM	Essequibo R.
81.S	1837	MLV	Hibiscus bicornis G. Mey.	Clitoria arborescens R. Br.	K	
82.S	1837	FAB	Clitoria arborescens R. Br.			Upper Essequibo R.
83.S				Senna latifolia (G. Mey.) H.S. Irwin & Barneby		
84.S		CSL	Cassia latifolia G. Mey.			
85.S				Psygmorchis glossomystax (Rchb.f.) Dodson & Dressler	BM	
86.S	1837	ORC	Oncidium glossomystax Rchb.f.	Pleurothallis picta Lindl.	BM	
87.S		ORC	Pleurothallis picta Lindl.			
88.S						
89.S				Microstachys corniculata (Vahl) Griseb.	BM	
90.S	1838	EUP	Microstachys guianensis Klotzsch	Alibertia latifolia (Benth.) K. Schum.	tc: F K	
91.S		RUB	Cordiera latifolia Benth.	Appunia calycina (Benth.) Sandw.	type	
92.S		RUB	Coffea calycina Benth.	Exellodendron coriaceum (Benth.) Prance	BM	
93.S		CHB	Parinari coriaceum Benth. ('Parinarium')	Ornithocephalus ciliatus Lindl.	BM	
94.S	1838	ORC	Ornithocephalus ciliatus Lindl.			
95.S						
96.S				Polylychnis radicans (Nees) Wasshausen comb. nov. ined	tc: K	Hyacon cataract
97.S		ACA	Stemonacanthus humboldtianus Nees	Miconia prasina (Sw.) DC.	ht: K	
98.S		MLS	Miconia pteropoda Benth.			
99.S						

Index 2. Robert Schomburgk's additional collection series (1835–1839) – continued

No.	Date	GEN	Schomburgk name	Accepted name	Herb.	Locality
100.S						
101.S	Nov	GEN	Coutoubea reflexa Benth.	Coutoubea reflexa Benth.	type	near Mt. Roraima; moist savanna
102.S		PTR	Pteris pungens Willd.	Pteris pungens Willd.		
103.S		XYR	Xyris guianensis Steud.	Xyris guianensis Steud.	CAY	Mt. Ataraipu
104.S		DSC	Dioscorea crotalariifolia Uline	Dioscorea crotalariifolia Uline		
105.S		MRS	Conomorpha fulvopulverulenta Mez.	Cybianthus fulvopulverulenta (Mez) Agostini		falls of Essequibo R.
106.S		FAB	Clitoria	Clitoria brachycalyx Harms		
107.S		CNV	Operculina pterodes Choisy	Operculina hamiltonii (G. Don) Austin & Staples		
108.S						
109.S	May	EUP	Caperonia paludosa Klotzsch	Caperonia paludosa Klotzsch	ht: K	Annai
110.S	Oct	ACA	Beloperone schomburgkiana Nees	Justicia schomburgkiana (Nees) V.A.W. Graham	ht: K; it: BM	Pirara; moist savanna
111.S						falls of Essequibo R.
112.S		RUB	Cordiera acuminata Benth.	Alibertia acuminata (Benth.) Sandw. var. acuminata	tc: K	
113.S						
114.S		ARS	Aristolochia rumicifolia M.R. Schomb. [n.n.]	Aristolochia rugosa Lam.	BM K	
115.S						
116.S						
117.S	May	ORC	Bonatea macilenta Lindl.	Habenaria macilenta (Lindl.) Rchb.f.	type	Pirara
118.S	Jun 1838	GSN	Gesneria guianensis Benth.	Sinningia schomburgkiana (Kunth & Bouché) Chautems	ht: ?; it: BM K	
119.S	Jun 1838	HAE	Xiphidium fockeanum Miq.	Schiekia orinocensis (Kunth) Meisn. var. orinocensis		Kanuku Mts.; rocky places
120.S						
121.S						
122.S		RUB	Psychotria cordifolia Kunth	Psychotria cordifolia Kunth		
123.S		AST	Porophyllum latifolium Benth.	Porophyllum latifolium Benth.	BM	Kanuku Mts.
124.S						Upper Rupununi R.; savanna
125.S		MLS	Miconia brevipes Benth.	Miconia brevipes Benth.		
126.S					ht: K	Roraima
127.S						
128.S						
129.S	1837	VRB	Vitex umbrosa Sw.	Vitex compressa Turcz.		
130.S		ORC	Cyrtopodium andersonii (Lamb.) R. Br.	Cyrtopodium andersonii (Lamb.) R. Br.	BM	R. Miejore (Miejone)
131.S		GEN	Voyria uniflora Lam.	Voyria aphylla (Jacq.) Pers.	K	Serra Mey
132.S		CHB	Hirtella bullata Benth.	Hirtella bullata Benth.	ht: K; it: BM	near Mt. Arawogany
133.S						

No.	Year	Fam.	Name	Full name	Herb.	Location
134.S		CNN	Rourea grosourdyana Baill.	Rourea grosourdyana Baill. var. grosourdyana	K	
135.S						
136.S		ACA	Thyrsacanthus bracteolatus Nees	Odontonema bracteolatum (Jacq.) Kuntze	K	R. Merewari
137.S		MIM	Inga sertulifera DC.	Inga sertulifera DC.	U	R. Parima near Meretani Mts.
138.S		FAB	Swartzia triphylla Willd.	Swartzia arborescens (Aubl.) Pittier		
139.S						
140.S						
141.S		URT	Urtica aestuans L.	Laportea aestuans (L.) Chew		
142.S		ORC	Zygopetalum makaii Hook.	Galeottia burkei (Rchb.f.) Dressler & Christenson		R. Emekuni
143.S		ARS	Aristolochia hians Willd.	Aristolochia hians Willd.		Serra Mey
144.S		GEN	Tachia gracilis Benth.	Tachia gracilis Benth.	ht: K; it: NY	near Roraima
145.S		XYR	Abolboda rigida Gleason	Abolboda macrostachya Spruce ex Malme var. robustior Steyerm.		
146.S						
147.S		GEN	Lisianthus breviflorus Benth.	Irlbachia nemorosa (Willd. ex Roem. & Schult.) Merr.	ht: K; it: BM	Mt. Warima
148.S		LYC	Lycopodium aristatum Humb. & Bonpl. ex Willd.	Lycopodium aristatum Humb. & Bonpl. ex Willd.		Serra Mey
149.S		BMN	Dictyostega schomburgkii Miers	Dictyostega orobanchioides (Hook.) Miers subsp. orobanchioides	ht: K; it: GH	Serra Mey
150.S		FLC	Casearia spinosa (L.) Willd.	Casearia aculata Jacq.	K	R. Parima
151.S		MLS	Microlicia benthamiana Triana ex Cogn.	Microlicia benthamiana Triana ex Cogn.	ht: K	Mt. Warima
152.S		RUB	Esenbeckia pilocarpoides Kunth subsp. pilocarpoides	Esenbeckia pilocarpoides Kunth subsp. pilocarpoides	K	
153.S						
154.S		LYT	Cuphea dactylophora Koehne	Cuphea dactylophora Koehne	ht: K	R. Parima near Meretani Mts.
155.S		LYT	Cuphea rigidula Benth.	Cuphea rigidula Benth.	ht: K	
155.S*						
156.S					BM K	
157.S	1839	ERI	Vaccinium	Vaccinium		near Roraima
158.S		AST	Gongylolepis benthamiana R.H. Schomb.	Gongylolepis benthamiana R.H. Schomb.	BM	Roraima
158.S*		RUB	Retiniphyllum scabrum Benth.	Retiniphyllum scabrum Benth. var. scabrum	BM	Roraima
159.S	1839	RUB				
160.S		fern	Polypodium phlegmaria J. Sm.	Lellingeria phlegmaria (J. Sm.) A.R. Sm. & R.C. Moran	tc: BM K	Mt. Roraima
161.S		GMM				
162.S						
163.S		ALI	Alisma bolivianum Rusby	Echinodorus bolivianus (Rusby) Holm-Niels.	K	Roraima
164.S		MLS	Chaetogastra callichaeta Benth.	Pterolepis glomerata (Rottb.) Miq.	ht: K	
165.S						

Index 2. Robert Schomburgk's additional collection series (1835–1839) – continued

No.	Date	Family	Name (as collected)	Current name	Herbarium	Locality
166.S		MLS	Chaetogastra gracilis (Bonpl.) DC.	Tibouchina gracilis (Bonpl.) Cogn.		near Roraima
167.S						
168.S		MLS	Macairea parvifolia Benth.	Macairea parvifolia Benth.		Roraima
169.S	1839	AST	Calea divaricata Benth.	Calea divaricata Benth.	lt: BM US frag	near Roraima
170.S						
171.S	1839	OCH	Poecilandra retusa Tul.	Poecilandra retusa Tul.	BM	
172.S	1837	MRT	Myrcia ferruginea (Poir.) DC.	Marlierea ferruginea (Poir.) McVaugh	BM	Roraima
173.S		MLS	Clidemia capitata Benth.	Clidemia capitata Benth.	ht: K	Roraima
174.S						
175.S						
176.S		MLP	Tetrapterys glaberrima Benth.	Tetrapterys mucronata Cav.		
177.S		LAM	Hyptis salzmanni Benth.	Hyptis salzmanni Benth.		
178.S		ERI	Befaria sprucei Meisn.	Bejaria sprucei Meisn.		Roraima
179.S						
180.S		fern				
181.S						
182.S		MLS	Clidemia umbonata DC.	Clidemia strigillosa (Sw.) DC.		
183.S		APO	Echites subcarnosa Benth.	Mandevilla subcarnosa (Benth.) Woodson		Roraima
184.S		MLS	Arthrostema bonplandii DC.	Monochaetum bonplandii (Kunth) Naudin		Roraima
184.S*		TEA	Ternstroemia schomburgkiana Benth.	Ternstroemia schomburgkiana Benth.		
185.S		MLS	Miconia maculata Benth.	Miconia ibaguensis (Bonpl.) Triana		Roraima
186.S						R. Negro
187.S		AST	Eupatorium scabrum L.f.	Chromolaena pharcidodes (B.L. Rob.) R.M. King & H. Rob.	ht: K	southern slopes Roraima Mts.
188.S		MLS	Macairea multinervia Benth.	Macairea multinervia Benth.	tc: GH frag	Roraima
189.S						
190.S						
191.S						
192.S						
193.S						
194.S	1839	AST	Verbesina schomburgkii Sch. Bip. [n.n.]	Verbesina schomburgkii Sch. Bip. ex Klatt	st: BM US frag	
195.S						
196.S		MLS	Melastoma dodecandra Desv.	Miconia dodecandra (Desv.) Cogn.		
197.S						
198.S						

No.	Year	Code	Name	Accepted name	Herb.	Locality
199.S		MLS	Chaetogastra lasiophylla Benth.	Macairea lasiophylla (Benth.) Wurdack	ht: K	Roraima
200.S		KRM	Krameria spartioides Klotzsch ex O. Berg	Krameria spartioides Klotzsch ex O. Berg	K	Roraima
201.S						
202.S		RPT	Rapatea friderici-augusti R.H. Schomb.	Saxofridericia regalis R.H. Schomb.	ht: K	Roraima
203.S						
204.S						
205.S						
206.S		VRB	Vitex multiflora Schauer	Vitex orinocensis Kunth var. multiflora (Miq.) Huber		
207.S						
208.S	1839	RUB	Psychotria schomburgkii Benth.	Psychotria schomburgkii Benth.	tc: BM	
209.S						
210.S						R. Branco
211.S						R. Padauiri
212.S						
213.S		BIG	Bignonia uncata Andrews	Macfadyena uncata (Andrews) Sprague & Sandw.		São Isabel on the R. Negro
214.S		VIO	Amphirrhox longifolia (A. St.-Hil.) Spreng.	Amphirrhox longifolia (A. St.-Hil.) Spreng.	tc: B	R. Negro
215.S		PRT	Roupala obtusata Klotzsch	Roupala obtusata Klotzsch		Barcelos on R. Negro
216.S		SCR	Herpestis chamaedryoides Kunth	Bacopa chamaedryoides (Kunth) Wettst.		R. Branco
217.S		PLP	Lepicystis incana J. Sm.	Polypodium incanum Sw.		Sierra Caraman; R. Branco
218.S		PTR	Doryopteris palmata J. Sm.	Pteris palmata Willd.		Pirara
219.S		ORC	Habenaria seticauda Lindl.	Habenaria seticauda Lindl.	lt: K	Fort São Joaquim
220.S		ALI	Echinodorus paniculatus Micheli	Echinodorus paniculatus Micheli		Fort São Joaquim
221.S	1839	VRB	Amasonia erecta L.f.	Amasonia campestris (Aubl.) Moldenke		
222.S						
223.S	1839	BRS	Trattinickia burserifolia Mart.	Trattinickia burserifolia Mart.	BM	Barcelos on R. Negro
224.S		MIM	Mimosa micracantha Benth.	Mimosa rufescens Benth. var. amnis-nigri Barneby	ht: K	Pedrero on R. Negro
225.S		MLS	Tococa planifolia Benth.	Tococa subciliata (DC.) Triana	P	
226.S		FAB	Drepanocarpus lunatus (L.f.) G. Mey.	Machaerium lunatum (L.f.) Ducke		R. Branco
227.S		RUB	Richardsonia divergens (Pohl) DC.	Richardia sp.		Pedrero on R. Negro
228.S	1839	STR	Byttneria obliqua Benth. ('Büttneria')	Byttneria obliqua Benth.	tc: BM	
229.S						
230.S		HMP	Trichomanes ankersii Parker ex Hook. & Grev.	Trichomanes ankersii Parker ex Hook. & Grev.	K	R. Padauiri
231.S		ICC	Pogopetalum orbiculatum Benth.	Emmotum orbiculatum (Benth.) Miers	tc: K	R. Padauiri
232.S		CNV	Maripa densiflora Benth.	Maripa densiflora Benth.	ht: K	R. Padauiri
233.S						R. Padauiri
234.S						R. Padauiri

INDEX 3. ROBERT SCHOMBURGK'S SECOND COLLECTION SERIES (1841–1844).

Rob. ser. 2	=Rich.	Year	Fam.	Original name	Current name	Herbarium	Location
1			CYP	Diplasia karataefolia Rich.			
2			ALI	Sagittaria			
3			GSN				
4			ARS	Aristolochia trilobata L.	Aristolochia trilobata L.		
5			MRT	Campomanesia glabra Benth.	Calycolpus goetheanus (Mart. ex. DC.) O. Berg		
6			PIP	Artanthe anonaefolia Miq.	Piper anonifolium (Kunth) C. DC		
7			EUP	Amanoa latifolia Benth.	Amanoa guianensis Aubl.		
8			APO	Echites			
9			SCR	Alectra brasiliensis Benth.	Melasma melampyroides (Rich.) Pennell	G NY	
10		1841	CTH	Hemitelia parkeri Hook.	Cyathea surinamensis (Miq.) Domin	BM G P U W	Roraima
11		1841	FLC	Casearia javitensis Kunth	Casearia javitensis Kunth	BM	
12	2	1841	RUB	Coffea	Rudgea hostmanniana Benth.	BM	
13	36	1841	OLC	Jussiaea surinamensis Miq. ('Jussieua')	Ludwigia leptocarpa (Nutt.) Hara		
14			LYT	Cuphea melvilla Lindl.	Cuphea melvilla Lindl.		
15			FAB	Nicolsonia cayennensis DC.	Desmodium barbatum (L.) Benth. & Oerst.		
16			CNV	Batatas cissoides Choisy	Merremia cissoides (Lam.) Hallier f.		
17			CAM	Centropogon surinamensis (L.) C. Presl	Centropogon cornutus (L.) Druce		
18			AST	Elephantopus carolinianus Willd.	Elephantopus mollis Kunth		
19			PLP	Polypodium ciliatum Willd.	Microgramma reptans (Cav.) A.R. Sm.	BM	
20			OCH	Sauvagesia elata Benth.	Sauvagesia elata Benth.		
21	44		MLS	Spennera aquatica (Aubl.) Mart.	Nepsera aquatica (Aubl.) Naudin		
22	13		LAU	Oreodaphne costulata Nees	Ocotea aciphylla (Nees) Mez		
23			MLS	Miconia erythropila Steud.	Miconia ceramicarpa (DC.) Cogn. var. ceramicarpa		
24	4		POA	Panicum pilosum Sw.	Panicum pilosum Sw.		
25			MLS	Clidemia surinamensis Miq.	Clidemia pustulata DC.	ht: P	
26			MTX	Amphidesmium rostratum J. Sm.	Metaxya rostrata (Kunth) C. Presl	BM F G W	
27			DST	Lindsaea trapeziformis Salisb.	Lindsaea schomburgkii Klotzsch	BM G	
28	5		MRN				
29	22	1841	AST	Eupatorium macrophyllum L.	Hebeclinium macrophyllum (L.) DC.	BM	
30		1841	GEN	Coutoubea ramosa Aubl.	Coutoubea ramosa Aubl.		
31		1841	MIM	Inga corymbifera Benth.	Inga nobilis Willd.	P K	
32			POA	Panicum trichoides Sw.	Panicum trichoides Sw.		

No.	Coll.	Year	Code	Original name	Accepted name	Notes	Loc.
33			CTH	Alsophila	Cyclodium confertum C. Presl	st: BM	
34		1841	CLU	Tovomita myriandra Benth.	Clusia myriandra (Benth.) Planch. & Triana		
35			MLS	Miconia racemosa DC.	Miconia racemosa DC.		
36		1841	RUB	Oldenlandia herbacea DC.	Oldenlandia corymbosa L.	BM	
37			ANA	Spondias lutea L.	Spondias mombin L.		
38	154		PAS	Decaloba hemicycla Roem.	Passiflora vespertilio L.		
39	7		FAB	Nicolsonia cayennensis DC.	Desmodium barbatum (L.) Benth. & Oerst.		
40			HAE	Xiphidium floribundum Sw.	Xiphidium caeruleum Aubl.		
41			MLS	Clidemia heteroclita Naudin	Clidemia japurensis DC. var. japurensis		
41*	72		FAB	Machaerium leiophyllum (DC.) Benth.	Machaerium leiophyllum (DC.) Benth.		
42			HMP	Trichomanes crispum L.	Trichomanes crispum L.	P	
42*		1841	HMP	Trichomanes plumula C. Presl	Trichomanes martiusii C. Presl	P	
43		1841	MRN	Maranta divaricata Roscoe			
44			DST	Lindsaea rufescens Kunze	Lindsaea portoricensis Desv.	BM K OXF P	Roraima
44*	19		RPT	Rapatea		P	
45		1841	TEC	Aspidium macrophyllum Sw.	Tectaria incisa Cac.	P	
46	8		GEN	Lisianthus			
47			DST	Lindsaea rufescens Kunze	Lindsaea portoricensis Desv.	BM K P	
47*		1841	DST	Lindsaea dubia Spreng.	Lindsaea dubia Spreng.	B BM P US	
48	19		MLS	Miconia barbigera DC.	Miconia ciliata (Rich.) DC.		
49	156		CSL	Cassia multijuga Rich.	Senna multijuga (Rich.) H.S. Irwin & Barneby var. multijuga		
50		1841	SCR	Scoparia dulcis L.	Scoparia dulcis L.	BM ht: K; it: US frag	
51			GMM	Polypodium confusum J. Sm.	Grammitis suspensa (L.) Proctor	P	
52		1841	PLP	Pteropsis furcata Sw.	Dicranoglossum desvauxii (Klotzsch) Proctor		
53			PLP	Polypodium lycopodioides L.	Microgramma lycopodioides L. Copeland	P	
54		1841	HMP	Hymenophyllum polyanthos (Sw.) Sw.	Hymenophyllum decurrens (Jacq.) Sw.		
55	20		GMM	Grammitis serrulata Sw.			
56		1841	GLC	Mertensia pectinata Willd.	Dicranpteris pectinata (Willd.) Underw.	B BM G	
57			FAB	Crotalaria incana L.	Crotalaria incana L.		
58	18		SCZ	Schizaea			
59	17		SOL	Solanum callicarpifolium Kunth & Bouché	Solanum callicarpifolium Kunth & Bouché		
60			CLU	Clusia camphorata L. ??			
61		1841	RUB	Spermacoce tenuior L.	Borreria latifolia (Aubl.) K. Schum. var. latifolia	BM	
62		1841	MIM	Inga corymbifera Benth.	Inga nobilis Willd.	P K	
63			ADI	Adiantum			
64		1841	RUB	Psychotria cornigera Benth.	Psychotria bahiensis DC. var. cornigera (Benth.) Steyerm.	BM	

Index 3. Robert Schomburgk's second collection series (1841–1844) – continued

65		MLS	Miconia fockeana Miq.	Miconia nervosa (J.E. Sm.) Triana		
66	14	MLS	Clidemia divaricata Naudin	Leandra divaricata (Naudin) Cogn.		
67		CMM	Tradescantia		ht: P	
68		DST	Lindsaea reniformis Dryand.	Lindsaea reniformis Dryand.		
69		LAU	Oreodaphne costulata Nees	Ocotea aciphylla (Nees) Mez	BM F	
70	1841	MIM	Pithecellobium glomeratum Benth. ('Pithecolobium')	Zygia cataractae (Kunth) L. Rico		
71		MLS	Miconia racemosa DC.	Miconia racemosa DC.		
72		MIM	Inga coruscans Humb. & Bonpl.	Inga bourgoni (Aubl.) DC.	P	
73		MLS	Clidemia heteroclita Naudin	Clidemia japurensis DC. var. japurensis		
74	1841	MIM	Inga tenuiflora Salzm.	Inga thibaudiana DC.		
75	1841	CLU	Vismia macrophylla Kunth	Vismia macrophylla Kunth	K P	
76	1841	HMP	Trichomanes heterophyllum Willd.	Trichomanes humboldtii (Bosch) Lellinger	P	
77	1841	HMP	Trichomanes crispum L.	Trichomanes crispum L.	P	
78		OLN	Aspidium nodosum Willd.	Oleandra articulata (Sw.) C. Presl	K P	
79	1841	AST	Spilanthes		BM	
80	1841	CLU	Tovomita myriandra Benth.	Clusia myriandra (Benth.) Planch. & Triana	tc: BM	
81	10	TIL	Triumfetta althaeoides Lam.	Triumfetta althaeoides Lam.		
82	33	RUB	Bertiera guianensis Aubl.	Bertiera guianensis Aubl.	BM	
83	24	MLS	Clidemia dentata D. Don	Clidemia dentata D. Don		
84	32	RHZ	Cassipourea guianensis Aubl.	Cassipourea guianensis Aubl.		
85	23*	ARS	Aristolochia rumicifolia M.R. Schomb. [n.n.]	Aristolochia rugosa Lam.		
86		ASC	Tassadia propinqua Decne.	Tassadia propinqua Decne.	K P	
87	77	MRN	Phrynium pumilum Klotzsch [n.n.]	Calathea micans (Mathieu) Körn.		
88	70	MLS	Clidemia dentata D. Don	Clidemia dentata D. Don	G K P	
89		MRT	Eugenia incanescens Benth.	Eugenia incanescens Benth.		
90		CSL	Brownea capitella Jacq.	Brownea capitella Jacq.		
91	1841	MEL	Guarea aubleti A. Juss.	Guarea guidonia (L.) Sleumer	BM	
92	39	CHB	Chrysobalanus icaco L.	Chrysobalanus icaco L.	BM K OXF P	Roraima
93	20	AMA	Pupalia densiflora (Humb. & Bonpl.) Mart.	Cyathula achyranthoides (Kunth) Moq.		
94	30	SAP	Paullinia podocarpa Klotzsch [n.n.]	Paullinia pinnata L.		
95		MLS	Clidemia rubra (Aubl.) Mart. var. cordifolia Benth.	Clidemia rubra Mart.		
96	1841	MLV	Sida acuta Burm.f. or S. rhombifolia L.	Sida acuta Burm.f. or S. rhombifolia L.	BM	
97	75	PAS	Cieca appendiculata Roem.	Passiflora auriculata Kunth	BM	

No.	Cat.	Year	Family	Name as published	Accepted name	Herbaria
98	19	1841	CNV	Batatas cissoides Choisy	Merremia cissoides (Lam.) Hallier f.	
99	31		RUB	Cephaëlis		
100			CLU	Clusia insignis Mart.	Clusia insignis Mart.	
101	35		CNV	Ipomoea guyanensis Choisy	Jacquemontia guyanensis (Aubl.) Meisn.	
102	175		FAB	Swartzia triphylla Willd.	Swartzia arborescens (Aubl.) Pittier	
103	82	1841	MLP	Heteropterys platyptera DC.	Heteropterys multiflora (DC.) Hochr.	BM K
104	73		MLS	Spennera circaeoides Mart.	Aciotis purpurascens (Aubl.) Triana	
105	53		MIM	Pentaclethra filamentosa Benth.	Pentaclethra macroloba (Willd.) Kuntze	
106	94	1841	AST	Schomburgkia caleoides DC.	Calea caleoides (DC.) H. Rob.	BM
107	55	1841	RUT	Monnieria trifolia L.	Ertela trifolia (L.) Kuntze	BM
108	97		FAB	Machaerium floribundum Benth.	Machaerium floribundum Benth.	BM K
109	38		ACA	Aphelandra pulcherrima (Jacq.) Kunth	Aphelandra pulcherrima (Jacq.) Kunth	BM G K W
110	46	1841	CNV	Maripa scandens Aubl.	Maripa scandens Aubl.	
111	65		APO	Odontadenia speciosa Benth.	Odontadenia macrantha (Roem. & Schult.) Markgr.	
112	47		PHT	Microtea debilis Sw.	Microtea debilis Sw.	P K
113		1841	EUP	Anisophyllum thymifolium Haw.	Euphorbia thymifolia L.	lt: K
114		1841	MIM	Pithecellobium corymbosum Benth. ('Pithecolobium')	Hydrochorea corymbosa (Rich.) Barneby & J. W. Grimes	P
115	98		MLS	Clidemia benthamiana Miq.	Clidemia hirta (L.) D. Don var. hirta	BM
116	80	1841	RUB	Psychotria fastigiata Spreng.	Palicourea croceoides Desv.	
117	23		APO	Malouetia	Tabernaemontana rupicola Benth.	tc: GH
118	216		PLG	Coccoloba marginata Benth.	Coccoloba marginata Benth.	
119	58		CNV	Ipomoea tamnifolia L.	Jacquemontia tamnifolia (L.) Griseb.	BM
120	90	1841	AST	Wulffia stenoglossa (Cass.) DC.	Tilesia baccata (L.) Pruski	BM K
121	40	1841	PGL	Securidaca latifolia Benth.	Securidaca paniculata Rich.	
122	185		MNY	Limnanthemum humboldtianum Griseb.	Nymphoides humboldtianum (Kunth) Kuntze	
122*	101		CMM	Commelina		
123	66		SOL	Solanum pensile Sendtn.	Solanum pensile Sendtn.	ht: K; it: BM NY P
124	50	1841	MIM	Inga pezizifera Benth.	Inga pezizifera Benth.	
125	78		GSN	Episcia mimuloides Benth.	Nautilocalyx mimuloides (Benth.) C. Morton	BM K
126	57		MLP	Byrsonima spicata (Cav.) DC.	Byrsonima spicata (Cav.) DC.	BM
127	93	1841	AST	Mikania racemosa Benth.	Mikania psilostachya DC.	
128	71		ULM	Sponia micrantha (L.) Decne.	Trema micrantha (L.) Blume	BM
129	99	1841	STR	Waltheria americana L.	Waltheria indica L.	
130			MLS	Miconia myriantha Benth.	Miconia myriantha Benth.	
131	81	1841	SML	Smilax schomburgkiana Kunth	Smilax schomburgkiana Kunth	tc: BM
132	84		MLS	Spennera circaeoides Mart.	Aciotis purpurascens (Aubl.) Triana	
133	49		SAP	Serjania micrantha Klotzsch [n.n.]	Serjania oblongifolia Radlk.	syntype

Index 3. Robert Schomburgk's second collection series (1841–1844) – continued

134	87	1841	RUB	Posoqueria trinitatis DC.	Posoqueria trinitatis DC.	BM
135	68		CHB	Hirtella paniculata Sw.	Hirtella paniculata Sw.	BM BR F K OXF P
136	52	1841	MRS	Icacorea guianensis Aubl.	Ardisia guianensis (Aubl.) Mez	BM
137	27	1841	LYT	Crenea repens G. Mey.	Crenea maritima Aubl.	
138	21	1841	APO	Tabernaemontana bicolor Klotzsch [n.n.]	Tabernaemontana lorifera (Miers) Leeuwenb.	tc: BM
139	96		ERO	Paepalanthus fasciculatus (Rottb.) Körn.	Paepalanthus bifidus (Schrad.) Kunth	
140	140		EUP	Phyllanthus microphyllus Kunth	Phyllanthus microphyllus Kunth	
141	74	1841	AST	Clibadium surinamense L.	Clibadium sylvestre (Aubl.) Baill.	BM
142	1155	1841	HMP	Trichomanes pinnatum Hedw.	Trichomanes pinnatum Hedw.	P
143			SCZ	Schizaea trilateralis Schkuhr	Actinostachys pennula (Sw.) Hook.	
144			SCZ	Schizaea flabellum Mart.	Schizaea elegans (Vahl) Sw.	
145			DST	Lindsaea dubia Spreng.	Lindsaea dubia Spreng.	BM K OXF P
146			DST	Lindsaea pendula Klotzsch	Lindsaea pendula Klotzsch	G K OXF P W
147			ADI	Adiantum fovearum Raddi		
148			DST	Lindsaea trapeziformis Salisb.	Lindsaea lancea (L.) Bedd. var. lancea	BM G P
149	29	1841	LYG	Lygodium volubile Sw.	Lygodium volubile Sw.	G P U
150			PLP	Polypodium persicariifolium Schrad. ('persicariaefolium')	Micrograma persicariifolia (Schrad.) C. Presl	
151			PLP	Polypodium lycopodioides L.	Microgramma lycopodioides L. Copeland	
152			CTH	Alsophila		
153			CTH	Alsophila armata (Sw.) J. Presl	Trichypteris armata (Sw.) Tryon	
154			OLN	Nephrolepis exaltata (Sw.) Schott	Nephrolepis rivularis (Vahl) Mett. ex Krug	
155	88		MLS	Clidemia	Clidemia capitellata (Bonpl.) D. Don var. dependens (D. Don) J.F. Macbr.	P
156	132		FAB	Mullera moniliformis L.f.	Lonchocarpus	
157	110		ACA	Thyrsacanthus schomburgkianus Nees	Odontonema schomburgkianum (Nees) Kuntze	
158		1841	MIM	Pithecellobium lasiopus Benth. ('Pithecolobium') Barneby & J.W. Grimes	Zygia latifolia (L.) Fawc. & Rendle var. lasiopus (Benth.)	
159	294		RUB	Malanea angustifolia Bartl. [n.n.]	Malanea obovata Hochr.	
160	204	1841	MLV	Pavonia cancellata (L.) Cav.	Pavonia cancellata (L.) Cav.	U
161			RUB	Malanea		
162			FAB	Alysicarpus nummularifolius DC.	Alysicarpus vaginalis (L.) DC.	
163	205	1841	MLV	Urena americana L.	Urena lobata L.	
164	198		MLS	Mouriria guianensis Aubl.	Mouriri guianensis Aubl.	BM
165	186	1841	MLS	Chaetogastra glomerata (Rottb.) Benth.	Pterolepis glomerata (Rottb.) Miq.	st: NY

							Upper Essequibo R.
166	1841	135	HUM	Humiria obovata Benth. ('Humirium obovatum')	Humiriastrum obovatum (Benth.) Cuatrec.		
167	1841	118	CEC	Coussapoa fagifolia Klotzsch	Coussapoa microcephala Trécul	BM G K W	
168	1841	125	CHB	Parinari guyanense Fritsch	Hirtella guyanensis (Fritsch) Sandw.	ht: W; it: K P	
169		138	MLS	Clidemia surinamensis Miq.	Clidemia pustulata DC.		
170		234	MLS	Diplochita fothergilla DC.	Miconia mirabilis (Aubl.) L.O. Williams		
171	1841	240	LAU	Goeppertia multiflora Miq.	Endlicheria multiflora (Miq.) Mez	tc: G P U US	
172	1841	147	SPT	Sideroxylon cuspidatum A. DC. var. ellipticum Miq.	Pouteria cuspidata (A. DC.) Baehni subsp. cuspidata	tc: BM US	
173		108	VRB	Petrea schomburgkiana Schauer	Petrea bracteata Steud.	L	
174			CSL	Eperua falcata Aubl.	Eperua rubiginosa Miq. var. rubiginosa		
175	1841	131	CLU	Caraipa richardiana Cambess.	Caraipa richardiana Cambess.	st: BM G K P US	
176		141	ACA	Rhytiglossa pectoralis Nees	Justicia pectoralis Jacq.		
177		142	MLS	Miconia alata (Aubl.) DC.	Miconia alta (Aubl.) DC.		
178	1841	117	APO	Tabernaemontana rupicola Benth.	Tabernaemontana rupicola Benth.	BM	
179			ANA	Spondias			
180		153	MLS	Diplochita parviflora Benth.	Miconia pubipetala Miq.		
181			LAM	Hyptis atrorubens Poit.	Hyptis atrorubens Poit.		
182			MLS	Tococa aristata Benth.	Tococa aristata Benth.		
183	1841	146	VRB	Aegiphila cuspidata Mart.	Aegiphila racemosa Vell.		
184	1841	126	AST	Mikania hookeriana DC.	Mikania hookeriana DC.	BM	
185			MLV	Pavonia typhalea (L.) Cav.	Pavonia fruticosa (Mill.) Fawc. & Rendle		
186			RUB	Psychotria			
187		151	LAM	Hyptis atrorubens Poit.	Hyptis atrorubens Poit.		
188	1841	104	MIM	Pithecellobium lasiopus Benth. ('Pithecolobium') Barneby & J.W. Grimes	Zygia latifolia (L.) Fawc. & Rendle var. lasiopus (Benth.)		
189	1841	188	EUP	Anisophyllum thymifolium Haw.	Euphorbia thymifolia L.	P K	
190	1841	134	ACA	Rhytiglossa secunda Nees	Justica secunda Vahl	G K P	
191	1841	157	SCR	Capraria biflora L.	Capraria biflora L.	BM	
192	1841	163	MRT	Eugenia incanescens Benth.		BM	
193		161	AQF	Ilex martiniana D. Don	Ilex martiniana D. Don	G	
194		158	ERO	Tonina fluviatalis Aubl.	Tonina fluviatalis Aubl.		
195	1841	144	AST	Acanthospermum xanthioides DC.	Acanthospermum australe (Loefl.) Kuntze	BM US	
196		163*	SAP	Cupania retusa Klotzsch [n.n.]	Cupania scrobiculata Rich.		
197		168	CSL	Cynometra spruceana Benth.	Cynometra spruceana Benth. var. spruceana	W	
198	1841		CSL	Campsiandra comosa Benth.	Campsiandra comosa Benth. var comosa	BM	
199		167	MLS	Henriettea multiflora Naudin	Henriettea multiflora Naudin	BM	
200	1841	100	OCH	Sauvagesia erecta L.	Sauvagesia erecta L.	BM	

Index 3. Robert Schomburgk's second collection series (1841–1844) – continued

No.	No. 2	Code	Year	Name	Determination	Herb.	Loc.
201	149	LNT		Polypompholyx schomburgkii Klotzsch [n.n.]	Utricularia longeciliata A. DC.		
202	150	BMN		Burmannia bicolor Mart.	Burmannia bicolor Mart.		
203		FAB		Lonchocarpus			
204	170	FLC	1841	Homalium racoubea Sw.	Homalium guianense (Aubl.) Oken	BM	
205	169	APO		Malouetia tamaquarina (Aubl.) A. DC.	Malouetia tamaquarina (Aubl.) A. DC.		
206	119	VRB		Vitex umbrosa Sw.	Vitex compressa Turcz.		
207	123	MLP	1841	Tetrapteris styloptera A. Juss.	Hiraea	BM K	
207*	123*	MLP		Hiraea chrysophylla A. Juss.	Hiraea faginea (Sw.) Nied.	K	
208		SCR		Beyrichia ocymoides Cham. & Schltdl.	Achetaria ocimoides (Cham. & Schltdl.) Wettst.		
209	109	RUB	1841	Coccocypselum tontanea Kunth	Coccocypselum guianense (Aubl.) K. Schum.	BM	
210	133	CSL		Elizabetha princeps M.R. Schomb. ex Benth.	Elizabetha coccinea M.R. Schomb. ex Benth. var. oxyphylla (Harms) R.S. Cowan		Pirara
211	189	CMM		Tradescantia schomburgkiana Kunth	Tripogandra serrulata (Vahl) Handlos	BM	
211*	189*	POA	1841	Panicum trichoides Sw.	Panicum trichoides Sw.		
212	145	AST	1841	Vernonia tricholepis DC.	Lepidaploa remotiflora (Rich.) H. Rob.	BM	
213		CAB	1841	Cabomba aquatica DC.	Cabomba aquatica DC.		
214		MIM	1841	Inga coruscans Humb. & Bonpl.	Inga bourgoni (Aubl.) DC.		
215	239	ORC	1841	Oncidium iridifolium Kunth	Psygmorchis pusilla (L.) Dodson & Dressler	P	
216	200	APO		Echites nitida Vahl	Odontadenia nitida (Vahl) Muell. Arg.	BM P	
217	193	SCR		Vandellia diffusa L.	Lindernia diffusa (L.) Wettst.		
218	128	PLG		Coccoloba excelsa Benth.	Coccoloba excelsa Benth.		
219	112	MLS		Clidemia divaricata Naudin	Leandra divaricata (Naudin) Cogn.		
220	159	SML		Smilax schomburgkiana Kunth	Smilax latipes Gleason		
221		MNS	1841	Cissampelos fasciculata Benth.	Cissampelos fasciculata Benth.		
222	136	FAB		Amphymenium	Pterocarpus amazonum (Benth.) Amshoff		
223	164						
224	315	TRG	1841–42	Trigonia hypoleuca Griseb.	Trigonia hypoleuca Griseb.	BM CGE G K P W	
225	292	SCR		Bacopa aquatica Aubl.	Bacopa aquatica Aubl.		
226	316	MRT	1841–42	Eugenia subobliqua Benth.	Myrcia subobliqua (Benth.) Nied.	BM W	Pirara
227	302	OCH	1841–42	Gagernia essequiboensis Klotzsch & M.R. Schomb. [n.n.]	Elvasia essequibensis Engl.	BM	Pirara
228	321	BOR	1841–42	Heliotropium helophilum Mart.	Heliotropium filiforme Lehm.	B† BM K P / ht: P	
229	303	MLS		Tococa castrata Naudin	Tococa subciliata (DC.) Triana	BM NY P	
230	349	CHB		Licania aperta Benth.	Licania apetala (E. Mey.) Fritsch var. aperta (Benth.) Prance		Pirara

No.	Coll.	Date	Fam.	Name as listed	Accepted name	Herbarium	Locality
231	298	1841–42	RUB	Cephaelis bracteocardia DC.	Psychotria bracteocardia (DC.) Muell. Arg.	BM	Pirara
232		1841–42	TIL	Vasivaea alchorneoides Baill.	Vasivaea alchorneoides Baill.	st: BM K P W	Pirara
233	384	1841–42	AST	Vernonia gracilis Kunth	Lepidaploa remotiflora (Rich.) H. Rob.	BM	
234	327		FAB	Andira retusa Kunth	Andira surinamensis (Bondt) Splitg. ex Amshoff	st: K	
235	313		CSL	Martia excelsa Benth.	Martiodendron excelsum (Benth.) Gleason	BM P W	
236	350	1841–42	DLL	Doliocarpus calinea J.F. Gmel.	Doliocarpus spraguei Cheesm.	BM	Pirara
237	371	1841–42	VIO	Ionidium oppositifolium (L.) Roem. & Schult.	Hybanthus oppositifolius (L.) Taub.	K P	
238	331		OLC	Heisteria cauliflora Sm.	Heisteria cauliflora J.E. Sm.		
239	379		FAB	Leptolobium nitens Vogel	Acosmium nitens (Vogel) Yakovlev		
240		1841–42	CLU	Hypericum cayennense Jacq.	Vismia cayennensis (Jacq.) Pers.	BM FI U	Pirara
241	354		CSL	Cynometra bauhiniifolia Benth. ('bauhiniaefolia')	Cynometra bauhiniifolia Benth. var. bauhiniifolia	K US W	Pirara
242	374		OCH	Gomphia dura Klotzsch [n.n.]	Ouratea rigida Engl.	st: BM	Pirara
243	380	1841–42	OCH	Gomphia rupununiensis Klotzsch [n.n.]	Ouratea rupununiensis Engl.	BM K OXF	Upper Takutu R.
244	356	1841	CHB	Moquilea comosa Benth.	Couepia comosa Benth.		
245	363		FAB	Aeschynomene sensitiva Sw.	Aeschynomene sensitiva Sw.		
246	364	1841–42	AST	Eupatorium ivaefolium L.	Chromolaena ivaefolium (L.) R.M. King & H. Rob.	BM	Pirara
247	367	1841–42	AST	Trichospira menthoides Kunth	Trichospira verticillata (L.) S.F. Blake	BM	
248	326		RUB	Palicourea riparia Benth.	Palicourea croceoides Desv.		
249	373		TRG	Trigonia subcymosa Benth.	Trigonia subcymosa Benth.	F W	
250	366	1841–42	AST	Gnaphalium americanum Mill.	Gamochaeta americana (Mill.) Wedd.	BM	Pirara
251	332		CSL	Elizabetha princeps M.R. Schomb. ex Benth.	Elizabetha coccinea M.R. Schomb. ex Benth. var. oxyphylla (Harms) R.S. Cowan		
252	402		CHB	Licania aperta Benth.	Licania apetala (E. Mey.) Fritsch var. aperta (Benth.) Prance	P	
253	445	1841–42	FAB	Eriosema lanceolatum Benth.	Eriosema simplicifolium (Kunth) G. Don var. simplicifolium	BM G NY	Pirara
254	385		MIM	Machaerium schomburgkii Benth.	Mimosa annularis Benth. var. odora Barneby	tc: BM NY W	
255	487		AMA	Serturnera guianensis Klotzsch [n.n.]	Pfaffia glomerata (Spreng.) Pedersen		
256	454	1841–42	FLC	Casearia carpinifolia Benth.	Casearia sylvestris Sw. var. lingua (Cambess.) Eichl.	BM F G K P W	Pirara
257	453	1841–42	RUB	Diodia setigera DC.	Diodia apiculata (Willd. ex Roem. & Schult.) K. Schum.	BM	
258	449	1841–42	FAB	Eriosema violaceum (Aubl.) G. Don	Eriosema violaceum (Aubl.) G. Don	BM G NY US	Pirara
259	447		MLP	Byrsonima verbascifolia (L.) DC.	Byrsonima verbascifolia (L.) DC.	st: BM K	Pirara
260	361		FAB	Dioclea guianensis Benth.	Dioclea guianensis Benth.		Pirara
261	410	1841–42	CSL	Cassia lotoides Kunth	Chamaecrista hispidula (Vahl) H.S. Irwin & Barneby	BM	
262	399	1841–42	PGL	Polygala variabilis Kunth	Polygala trichosperma L.	G K W	Pirara

108

Index 3. Robert Schomburgk's second collection series (1841–1844) – continued

263	450	1841–42	MLV	Abutilon periplocifolium (L.) Sweet var. caribaeum G. Don	Wissadula periplocifolia (L.) C. Presl ex Thwaites	BM	Pirara
264	456	1841–42	CNN	Omphalobium opacum Klotsch [n.n.]	Connarus coriaceus Schellenb.	BM	Pirara
264*	355		SAP	Cupania			
265	463		HYD	Hydrolea spinosa L.	Hydrolea spinosa L.		
266	389		MLP	Byrsonima crassifolia (L.) Kunth	Byrsonima crassifolia (L.) Kunth	st: BM K	
267	483		FAB	Tephrosia toxicaria (Sw.) Pers.	Tephrosia sinapou (Bucholz) A. Chev.		
268	458	1841–42	AST	Elephantopus angustifolius Sw.	Orthopappus angustifolius (Sw.) Gleason	BM	Pirara
269	462	1841–42	STR	Helicteres guazumaefolia Kunth	Helicteres guazumaefolia Kunth	BM U	Pirara
270		1841–42	HPC	Hippocratea schomburgkii Klotzsch [n.n.]	Prionostemma aspera (Lam.) Miers	BM P	Pirara
271	466	1841–42	AST	Wulffia platyglossa DC.	Tilesia baccata (L.) Pruski	BM	
272	470	1841–42	PGL	Polygala adenophora DC.	Polygala adenophora DC.	BM K P W	Pirara
273	472	1841–42	TNR	Turnera opifera Mart.	Turnera caerulea Moç. & Sessé ex DC. var. surinamensis (Urb.) Arbo	BM P	Pirara
273*			TNR	Wormskioldia tanacetifolia Klotzsch	Tricliceras tanacetifolium (Klotzsch) R. Fern.	G	Pirara
274	476	1841–42	MLS	Microlicia recurva (Rich.) DC.	Acisanthera uniflora (Vahl) Gleason		
275	544	1841–42	MLP	Bunchosia mollis Benth.	Bunchosia mollis Benth.	BM	Pirara
276	521		FAB	Tephrosia penicillata Benth.	Tephrosia adunca Benth.		
277	543	1841–42	APO	Thevetia humboldtii M.R. Schomb. [n.n.]	Aspidosperma macrophyllum Muell. Arg.	P	Pirara
278	543	1841–42	ERX	Erythroxylum campestre A. St.-Hil. ('Erythroxylon')	Erythroxylum campestre A. St.-Hil.	BM P	Pirara
279	424		PAS	Astrophea glaberrima Klotzsch [n.n.]	Passiflora leptopoda Harms	st: P W	
280	523	1841–42	SAP	Schmidelia mollis Klotzsch [n.n.]	Allophylus racemosus Sw.	BM K P	
281	507*	1841–42	EUP	Adenogyne guyanensis Klotzsch [n.n.]	Gymnanthes guyanensis Klotzsch ex Muell. Arg.	BM	Pirara
282	576	1841–42	RUB	Psychotria cordifolia Kunth	Psychotria cordifolia Kunth subsp. perpusilla Steyerm.	BM	Pirara
283	573	1841–42	BOR	Heliophytum passerinoides Klotzsch [n.n.]	Heliotropium ternatum Vahl	BM P	Pirara
284	572		CPP	Physostemon intermedium Moric.	Cleome guianensis Aubl.		
285	448		ERO	Eriocaulon tenuifolium Klotzsch ex Körn.	Eriocaulon tenuifolium Klotzsch ex Körn.	st: GH	
286	459		FAB	Tephrosia brevipes Benth.	Tephrosia brevipes Benth.		
287	442	1841–42	PON	Pontederia schomburgkiana Klotzsch [n.n.]	Pontederia subovata (Seub.) Lowden	BM G K P	Pirrara
288	421		VRB	Vitex schomburgkiana Schauer	Vitex schomburgkiana Schauer		
289	558		PAS	Dysosmia foetida (L.) Roem.	Passiflora foetida L. var. foetida		
290	543*		PGL	Polygala mollis Kunth	Polygala hebeclada DC.	BM K P	Pirara

No.	No.	Fam.	Original name	Date	Accepted name	Herbarium	Locality
291	477	ACA	Dipteracanthus canescens Nees		Ruellia geminiflora Kunth var. angustifolia (Nees) Griseb.	ht: BM; it: K P	
292	583	RUB	Coccocypselum canescens Willd.	1841–42	Coccocypselum aureum (Spreng.) Cham. & Schltdl. var. capitatum (Benth.) Steyerm.	BM	Pirara
293	585	FAB	Swartzia alterna Benth.		Bocoa alterna (Benth.) R.S. Cowan		
294	591	MRT	Eugenia perforata O. Berg	1841–42	Eugenia egensis DC.	BM	Pirara
295	500	MLP	Heteropterys candolleana A. Juss.	1841–42	Heteropterys macradena (DC.) W.R. Anderson	BM K	
296	557	CNV	Evolvulus glomeratus Nees & Mart.		Evolvulus glomeratus Nees & Mart.		
297	561	FAB	Bowdichia			P	Pirara
298	567	MRT	Eugenia flavescens DC.		Eugenia flavescens DC.		
299	586	MIM	Acacia paniculata Willd.		Acacia tenuifolia (L.) Willd.		
300	398	SCR	Buchnera palustris (Aubl.) Spreng.		Buchnera palustris (Aubl.) Spreng.	BM G K NY P W	Pirara
301	432	PGL	Catocoma lucida Benth.	1841–42	Bredemeyera lucida (Benth.) Klotzsch ex Hassk.	st: BM US	Pirara
302	289	MRT	Campomanesia glabra Benth.	1841–42	Calycolpus goetheanus (Mart. ex. DC.) O. Berg	BM	Pirara
303	295	SCR	Conobea aquatica Aubl.	1841–42	Conobea aquatica Aubl.	BM	Pirara
304	290	LNT	Utricularia calycifida Benj.	1841–42	Utricularia calycifida Benj.	BM	Pirara
305	293	XYR	Xyris surinamensis Miq.	1841–42	Xyris jupicai Rich.	tc: BM G U US W	Pirara
306	296	MRT	Aulomyrcia pirarensis O. Berg	1841–42	Myrcia inaequiloba (DC.) D. Legrand	BM	Pirara
307	325	APO	Malouetia gracilis (Benth.) A. DC	1841–42	Malouetia gracilis (Benth.) A. DC	BM	
308	308	EUP	Phyllanthus guianensis Klotzsch		Phyllanthus caroliniensis Walter subsp. guianensis (Klotzsch) Webster	BM	Pirara
309	327*	RUB	Psychotria rubra (Willd.) Muell. Arg.	1841–42	Psychotria hoffmannseggiana (Willd. ex Roem. & Schult.) Muell. Arg.	BM	Pirara
310	305	OXL	Oxalis schomburgkiana Prog.		Oxalis frutescens L.	BM G K MO OXF P W	Pirara
311	528	MRT	Campomanesia coaetanea O. Berg	1841–42	Campomanesia aromatica (Aubl.) Griseb.	lt: W; ilt: BM F K US	Pirara
312	574	RUB	Brignolia pubigera Benth.	1841–42	Isertia parviflora Vahl	BM	
313	497	MIM	Mimosa schomburgkii Benth.		Mimosa schomburgkii Benth.		
314	590	FAB	Centrolobium robustum Mart. ex Benth.		Centrolobium paraense Tul.	BM	Pirara
315	610	MEL	Trichilia brachystachya Klotzsch ex C. DC.	1841–42	Trichilia pallida Sw.	BM	Pirara
316	382	EUP	Croton essequeboensis Klotzsch	1841–42	Barhamia macrostachya Klotzsch	BM	Pirara
317	300	RUB	Psychotria inundata Benth.	1841–42	Psychotria capitata Ruiz & Pav. subsp. inundata (Benth.) Steyerm.		
318	311	CHB	Licania incana Aubl.		Licania incana Aubl.		
319	336	FAB	Etaballia guianensis Benth.		Etaballia dubia (Kunth) Rudd	st: BM US	Pirara
320	342	MRT	Calycampe angustifolia O. Berg	1841–42	Myrcia calycampa Amshoff	BM CGE F K NY OXF	
321	309	CHB	Moquilea multiflora Benth.	1841–42	Couepia multiflora Benth.	BM	
322	301	MLS	Aciotis aequatorialis Cogn.		Aciotis aequatorialis Cogn.		
323		LOG	Spigelia humilis Benth.		Spigelia humilis Benth.		

Index 3. Robert Schomburgk's second collection series (1841–1844) – continued

				Name as written	Current name	Herbaria	Locality
324	397		GEN	Coutoubea ramosa Aubl.	Coutoubea ramosa Aubl.		
325	387		PHT	Microtea maypurensis (Kunth) G. Don	Microtea maypurensis (Kunth) G. Don		
326	404		VRB	Lippia schomburgkiana Schauer	Lippia origanoides Kunth		
327	388	1841–42	RUB	Palicourea rigida Kunth	Palicourea rigida Kunth	BM	
328			LAU				
329	323	1841–42	AQF	Ilex laurina Klotzsch [n.n.]	Ilex daphnogenea Reissek	BM P	
330	335		LAM	Hyptis parkeri Benth.	Hyptis parkeri Benth.		Pirara
331	291		ACA	Hygrophila guianensis Nees ex Benth.	Hygrophila costata Nees		
332	322		CSL	Tachigali pubiflora Benth. ('Tachigalia')	Tachigali pubiflora Benth. ('Tachigalia')	ht: G; it: BM K P W	Pirara
333	394	1841–42	RUB	Diodia articulata DC.	Diodia hyssopifolia (Willd. ex. Roem. & Schult.) Cham. & Schltdl. var. articulata (Pohl ex DC.) Steyerm.	BM	Pirara
334	542		LYT	Cuphea antisyphilitica Kunth	Cuphea antisyphilitica Kunth	BM	Pirara
335	513	1841–42	ERX	Erythroxylum rufum Cav. ('Erythroxylon')	Erythroxylum rufum Cav.	BM	Pirara
336	505		SAP	Schmidelia edulis A. St.-Hil.	Allophylus edulis (A. St.-Hil.) Radlk.	BM	
337	514	1841–42	RUB	Randia spinosa (Jacq.) K. Schum. var. nitida K. Schum.	Randia armata (Sw.) DC.	BM	Pirara
338	526		CNV	Evolvulus sericeus Sw.	Evolvulus sericeus Sw.	BM	Pirara
339	509		AST	Eupatorium subobtusum DC.	Ayapana amygdalina (Lam.) R.M. King & H. Rob.	BM	
340	538	1841–42	MRT	Eugenia incanescens Benth.	Eugenia incanescens Benth.	BM	Pirara
341	530	1841–42	AST	Vernonia odoratissima Kunth	Vernonia scabra Pers.	BM	Pirara
342	588		NYC	Pisonia schomburgkiana Heimerl	Guapira cuspidata (Heimerl) Lundell	BM	
343	536	1841–42	SMR	Simaba guianensis Aubl.	Simaba guianensis Aubl.	BM	Pirara
344	508		FAB	Swartzia microstylis Benth.	Swartzia dipetala Willd. ex Vogel	BM	
345	541		PLG	Ruprechtia brachystachya Benth.	Ruprechtia brachystachya Benth.	tc: NY W	Pirara
346	560	1841–42	HUM	Humiria laurina Klotzsch ex Urb. ('Humirium laurinum')	Humiria balsamifera (Aubl.) J. St.-Hil. var. laurina (Urb.) Cuatrec.	P K	Pirara
347	496	1841–42	HPC	Salacia guianensis Klotzsch ex Peyr.	Salacia elliptica (Mart. ex Roem. & Schult.) G. Don	ht: K; it: BM G L	Pirara
348	494		HPC	Raddia pachyphylla Miers	Salacia pachyphylla Peyr.	lt: K; ilt: BM L	
348*	494*		HPC	Tontelea	Tontelea sandwithii A.C. Sm.		
349	599		MOR	Brosimum aubletti Poepp. & Endl.	Brosimum guianense (Aubl.) Huber	G K P U	
350	524	1841–42	BRS	Protium	Protium	BM	
351						BM	
352	439	1841–42	ACA	Mendoncia puberula Mart.	Pulchranthus variegatus (Aubl.) Baum, Reveal & Nowicke	P	Pirara

353	495	1841–42	PON	Eichhornia azurea Kunth	Eichhornia azurea (Sw.) Kunth	BM G P	Pirara
353*	488		AST	Ageratum scorpioideum Baker	Ageratum scorpioideum Baker	ht: K	Pirara
354	489	1841–42	OLC	Jussiaea repens L. ('Jussieua')	Ludwigia inclinata (L.f.) Gómez	BM	Pirara
355			BIG	Bignonia robusta Klotzsch [n.n.]	Anemopaegma robustum Bureau & K. Schum.	tc: K	Essequibo R.
356	411		RUB	Calycophyllum sp. nov.	Calycophyllum stanleyanum R.H. Schomb.	ht: K; it: BM NY	
357	602		ERX	Erythroxylum amazonicum Peyr.	Erythroxylum amazonicum Peyr.		Pirara
358	535		EUP	Mabea schomburgkii Benth.	Mabea taquari Aubl.	st: G K L P W	Pirara
359	571		ASC	Ditassa pauciflora Decne.	Ditassa pauciflora Decne.		
360	549		MLS	Rhynchanthera grandiflora (Aubl.) DC.	Rhynchanthera grandiflora (Aubl.) DC.		
361	578	1841–42	STR	Helicteres brevispica A. St.-Hil. or H. urbani K. Schum.	Helicteres brevispica A. St.-Hil. or H. urbani K. Schum.	BM	Pirara
362	525		FAB	Zornia	Zornia		
363	532		MLS	Miconia	Miconia prasina (Sw.) DC.		
364	414	1841–42	MRT	Eugenia pyrroclada O. Berg	Eugenia punicifolia (Kunth) DC.	BM	Pirara
365	498	1841–42	MRT	Psidium ciliatum Benth.	Psidium salutare (Kunth) O. Berg	BM	Pirara
366	533		MLS	Miconia hypargyrea Miq.	Miconia stenostachya DC.		
367	441	1841–42	DLL	Curatella americana L.	Curatella americana L.	BM	Pirara
368	457		MLS	Clidemia pustulata DC.	Clidemia pustulata DC.	K	
369	480		EUP	Tragia			
370	654	1841–42	CNV	Jacquemontia hirsuta Choisy	Jacquemontia sphaerostigma (Cav.) Rusby	BM	Pirara
371	641*		CMM	Tradescantia			
372	638	1841–42	ASC	Ditassa glaucescens Decne.	Blepharodon glaucescens (Decne.) Fontella	ht: P; it: F NY	Pirara
372*	664	1841–42	VIO	Ionidium itoubou Kunth	Hybanthus calceolaria (L.) G.K. Schulze	BM	
373	663	1841–42	RUB	Declieuxia chiocococoides Kunth	Declieuxia fruticosa (Willd. ex Roem. & Schult.) Kuntze	BM	
374	649		APO	Prestonia latifolia Benth.	Prestonia tomentosa R. Br.		
375	660		FAB	Zornia			
376	656	1841–42	AST	Synedrella nodiflora (L.) Gaertn.	Synedrella nodiflora (L.) Gaertn.	BM	Pirara
377	658	1841–42	EUP	Euphorbia pilulifera L.	Euphorbia hirta L.	BM K	Pirara
378							
379	639	1841–42	ERX	Erythroxylum ('Erythroxylon')	Erythroxylum vernicosum O.E. Schulz	st: BM P	Pirara
380	645		MIM	Schrankia leptocarpa DC.	Mimosa quadrivalvis L. var. leptocarpa (DC.) Barneby		
381	646		FAB	Trifolium guianense Aubl.	Stylosanthes guianensis (Aubl.) Sw.		
382	655		SAP	Cardiospermum halicacabum L.	Cardiospermum halicacabum L.		
383	641	1841–42	PAS	Passiflora sp. nov.	Passiflora tuberosa Jacq.	P	
384	661		IRI	Cipura palludosa Aubl.	Cipura palludosa Aubl.		
385	643	1841–42	EUP	Chamaesyce hyssopifolia (L.) Small	Euphorbia hyssopifolia L. subsp. hyssopifolia	BM P	Pirara

Index 3. Robert Schomburgk's second collection series (1841–1844) – continued

No.		Date	Fam.	Name	Current name	Herbaria	Locality
386	207	1841–42	MRT	Eugenia subobliqua Benth.	Myrcia subobliqua (Benth.) Nied.	BM U	
387	353		MIM	Pithecellobium adiantifolium Benth. var. multipinnum Benth	Macrosamanea pubiramea (Steud.) Barneby & J.W. Grimes var. pubiramea		Pirara
388	604	1841–42	MRT	Psidium ciliatum Benth.	Psidium sartorianum (O. Berg) Nied.	pt: BM G NY US	
389	589		FAB	Desmodium axillare (Sw.) DC.	Desmodium axillare (Sw.) DC.	BM	Pirara
390	555		MLS	Comolia hirtella Naudin	Comolia villosa (Aubl.) Triana var. villosa		
391	552		RUB	Perama hirsuta Aubl.	Perama hirsuta Aubl.	ht: P	
392	307		MLS	Comolia veronicaefolia Benth.	Comolia villosa (Aubl.) Triana var. C.		
393	603		FAB	Rhynchosia schomburgkii Benth.	Rhynchosia schomburgkii Benth.		Pirara
394	490		CMM	Commelina schomburgkiana Klotzsch var. latifolia Klotzsch [n.n.] ('Commelyna')	Commelina schomburgkiana Klotzsch ('Commelyna')		
395	478	1841–42	PGL	Polygala timoutou Aubl.	Polygala timoutou Aubl.	BM G K NY P	Pirara
396	493		MIM	Piptadenia peregrina (L.) Benth.	Adenanthera peregrina (L.) Speg.		
397	428	1841–42	OLC	Jussiaea nervosa Poir. ('Jussieua')	Ludwigia nervosa (Poir.) Hara	BM	Pirara
398	607		CSL	Cassia disadena Steud.	Chamaecrista nictitans (L.) Moench var. disadena (Steud.) H.S. Irwin & Barneby		
399	507	1841–42	SML	Smilax pirarensis Kunth & M.R. Schomb. ex Kunth	Smilax cumanensis Willd.	BM	Pirara
400	609		TIL	Apeiba tibourbou Aubl.	Apeiba schomburgkii Szyszyl.		
401	592		VRB	Aegiphila laevis Willd.	Aegiphila laxiflora Benth.		
402	593	1841–42	STR	Helicteres althaeifolia Benth.	Helicteres baruensis Jacq.	BM	Pirara
403	377		PAS	Tacsonia spinescens Klotzsch [n.n.]	Passiflora securiclata Mast.		
404	314	1841	MNM	Citriosma guianensis (Aubl.) Tul.	Siparuna guianensis Aubl.		
405	648	1841–42	MLV	Sida ciliaris var. guianensis K. Schum.	Sida ciliaris L.	tc: BM US	Pirara
406	652		EUP	Brachystachys hirta Klotzsch	Croton glandulosus L.		Pirara
407	651	1841–42	EUP	Traganthus sidioides Klotzsch	Bernardia sidoides (Klotzsch) Muell. Arg.	BM	Pirara
408	647		ASC	Oxypetalum capitatum Mart.	Oxypetalum capitatum Mart.		
409	501		BIG	Tecoma nigricans Klotzsch [n.n.]	Tabebuia serratifolia (Vahl) G. Nichols.		
410	550	1841–42	MLV	Pavonia bracteosa Benth.	Peltaea trinervis (C. Presl) Krapov. & Cristóbal	P	
411	348	1841–42	SAP	Lamprospermum schomburgkii Klotzsch [n.n.]	Matayba camptoneura Radlk.		
412	417		FAB	Neurocarpum angustifolium Kunth	Clitoria guianensis (Aubl.) Benth.		
413	610*		SOL	Cestrum vespertinum L.			
414	611		ARS	Aristolochia rumicifolia M.R. Schomb. [n.n.]	Aristolochia rugosa Lam.	ht: P; it: BM K	
415	606		VRB	Lantana canescens Kunth	Lantana canescens Kunth		Pirara
416	518	1841–42	PGL	Polygala hygrophila Kunth	Polygala hygrophila Kunth	BM K P	Pirara

No.	Coll.	Date	Fam.	Name as written	Current name	Herbaria	Locality
417	386		MLV	Sida angustifolia Lam.	Sida serrata Willd. ex Spreng.	BM CGE F G OXF P U W	Pirara
418	444		SCR	Herpestis gratioloides Benth.		BM	Pirara
419	659	1841–42	TEO	Clavija ornata D. Don	Clavija macrophylla (Link ex Roem. & Schult.) Radlk.		
420	680	1841–42	SPT	Chrysophyllum sparsiflorum Klotzsch ex Miq.	Chrysophyllum sparsiflorum Klotzsch ex Miq.		Pirara
421	681		APO	Secondatia densiflora A. DC.	Secondatia densiflora A. DC.		
422	682		RUB	Amaioua corymbosa Kunth	Amaioua corymbosa Kunth	BM P	Pirara
423	691	1841–42	DSC	Dioscorea syringifolia Kunth & M.R. Schomb. [n.n.] ('syringaefolia')	Helmia syringifolia Kunth & M.R. Schomb. ('syringaefolia')	BM F Fl G K P U W	Pirara
424	650	1841–42	FLC	Casearia densiflora Benth.	Casearia commersoniana Cambess.		
425	690		FAB	Aeschynomene hystrix Poir.	Aeschynomene hystrix Poir.	F K OXF P W	Pirara
426	694	1841–42	EUP	Manihot guianensis Klotzsch [n.n.]	Manihot esculenta Crantz	BM G K P	
427	669	1841–42	BOR	Tournefortia spigeliiflora A. DC. ('spigeliaeflora')	Tournefortia paniculata Cham. var. spigeliiflora (A. DC.) I.M. Johnst.		Pirara
428	672		SOL	Solanum radula Vahl	Solanum asperum Rich.		
429	678		RUB	Cephaëlis			
430	688	1841–42	LIL	Bomarea fuscata Klotzsch [n.n.]	Bomarea edulis (Tussac) Herbert	P	
431	683		MLS	Miconia rufescens (Aubl.) DC.	Miconia rufescens (Aubl.) DC.	ht: G	Pirara
432	667		PGL	Polygala longicaulis Kunth	Polygala longicaulis Kunth	BM G K P W	
433	675		LAU	Oreodaphne glomerata Nees	Ocotea glomerata (Nees) Mez		
434	674		TIL	Corchorus argutus Kunth	Corchorus orinocensis Kunth		
435	653	1841–42	RUB	Richardsonia divergens (Pohl) DC.	Richardia	BM	Pirara
436	671	1841–42	BIG	Amphilophium paniculatum (L.) Kunth	Amphilophium paniculatum (L.) Kunth	BM	Pirara
437	676		BIG	Arrabidaea cordifolia Klotzsch [n.n.]	Arrabidaea pubescens (L.) A. Gentry		
438		1841–42	RUB	Guettarda macrantha Benth.	Guettarda macrantha Benth.	BM	
439	1686	1841–42	CNN	Connarus incomptus Baker var. subcordata Baker	Connarus incomptus Planch.	lt: K; ilt: F G GH W	Pirara
440	696	1842–43	MNS	Trichoa guianensis Klotzsch [n.n.]	Abuta grandifolia (Mart.) Sandw.	ht: BM	savanna at Pirara
440*	372		CLU	Vismia cayennensis (Jacq.) Pers.	Vismia cayennensis (Jacq.) Pers.	NY	
441	689	1841–42	CSL	Cassia patellaria DC.	Chamaecrista nictitans (L.) Moench subsp. patellaria (Collad.) H.S. Irwin & Barneby	ht: K; it: BM G	Pirara
442	698	1841–42	AST	Porophyllum latifolium Benth.	Porophyllum ruderale (Jacq.) Cass. var. ruderale		Pirara
443			RUB	Chiococca	Chiococca	BM	
444	693	1841–42	RUB	Chiococca			Pirara
445	695	1841–42	OXL	Oxalis barrelieri L.	Oxalis barrelieri L.	K P	Pirara
446			CSL	Cassia			

Index 3. Robert Schomburgk's second collection series (1841–1844) – continued

No.	Coll.	Fam.	Name	Determination	Herbaria	Locality
447		CSL	Cassia brevipes DC.	Chamaecrista desvauxii (Collad.) Killip var. brevipes (Benth.) H.S. Irwin & Barneby	BM K NY	Pirara
448	1841–42	RUB	Diodia setigera DC.	Diodia apiculata (Willd. ex Roem. & Schult.) K. Schum.	BM	Pirara
449	726	MLS	Tibouchina aspera Aubl.	Tibouchina aspera Aubl. var. aspera		
450	723	MLV	Sida althaeifolia Sw. var. aristosa DC.	Sida cordifolia L.	BM	
451	724	MLV	Abutilon lucianum Sweet	Wissadula contracta (Link) R.E. Fr.	BM	
452						
453	722	AST	Pectis elongata Kunth	Pectis elongata Kunth var. elongata	BM US	Roraima
454	725	CSL	Cassia hispida Collad.	Chamaecrista hispidula (Vahl) H.S. Irwin & Barneby		
455	728	MLV	Fugosia guianensis Klotzsch [n.n.]	Cienfuegosia phlomidifolia (A. St.-Hil.) Garcke		
456	737	CSL	Outea acaciifolia Benth. ('acaciaefolia')	Macrolobium acaciifolium (Benth.) Benth.		
457	730	FAB	Leptolobium nitens Vogel	Acosmium nitens (Vogel) Yakovlev		
458	729	LCY	Lecythis schomburgkii O. Berg	Lecythis schomburgkii O. Berg	BM P	
459	733	FAB	Amphymenium rohrii (Vahl) Kunth	Pterocarpus rohrii Vahl		
460	736	CSL	Outea multijuga DC.	Macrolobium multijugum (DC.) Benth. var. multijugum		
461	734	CSL	Outea multijuga DC.	Macrolobium multijugum (DC.) Benth. var. multijugum		
462	735	RUB	Cordiera latifolia Benth.	Alibertia latifolia Benth. var. latifolia		
463	738	EUP	Amanoa guianensis Aubl.	Amanoa guianensis Aubl.	BM G K W	Roraima
464	744	RUB	Sipanea sp. nov.	Limnosipanea schomburgkii Hook.f.	tc: BM G US	Roraima
465	773	MLS	Noterophila pusilla Naudin	Acisanthera limnobios (DC.) Triana	ht: P	near Mt. Roraima
466	740	CPP	Cleome stenophylla Klotzsch ex Urb.	Cleome stenophylla Klotzsch ex Urb.	BM	Roraima
467	750	SPT	Lucuma glomerata Miq.	Pouteria glomerata (Miq.) Radlk. subsp. glomerata	BM	Roraima
468	764	MLV	Pavonia sp. nov.	Pavonia geminiflora Moric.	BM OXF	Roraima
469	758	MIM	Pithecellobium glomeratum Benth. ('Pithecolobium')	Zygia cataractae (Kunth) L. Rico		
470	765	MIM	Mimosa schrankioides Benth.	Mimosa schrankioides Benth. var. schrankioides	ht: K; it: G W	
471	776	CSL	Cassia arowanna R.H. Schomb. [n.n.]	Senna velutina (Vogel) H.S. Irwin & Barneby	US	
472	766	CSL	Peltogyne pubescens Benth.	Peltogyne paniculata Benth. subsp. pubescens (Benth.) M.F. Silva		
473	775	CSL	Cassia prostrata L.	Chamaecrista serpens (L.) Greene var. serpens		

No.	Coll.	Date	Fam.	Name as cited	Accepted name	Vouchers	Locality
474	759		TIL	Carpodiptera schomburgkii Baill.	Christiana africana DC.	BM K P	Roraima
475	801	1842–43	LAU	Goeppertia reflectens Nees	Endlicheria reflectens (Nees) Mez	F G MO NY US	Roraima
476	760	1842–43	STR	Bytneria divaricata Benth. ('Büttneria')	Byttneria divaricata Benth.	BM	Roraima
477	739	1842–43	RUB	Randia mussaendae (Thunb.) DC.	Randia formosa (Jacq.) K. Schum. var. densiflora Bartl. ex K. Schum.	lt: BM; ilt: BM F G GH K P	Roraima
478	772	1842–43	RUB	Tocoyena neglecta N.E. Br.	Tocoyena neglecta N.E. Br.	BM	Roraima
479	785	1842–43	EUP	Euphorbia amoena Klotzsch [n.n.]	Euphorbia dioeca Kunth	BM P K	Roraima
480	770		SOL	Schwenckia hirta Klotzsch var. angustifolia Benth. ('Schwenkia')	Schwenckia hirta Klotzsch ex Schltdl. ('Schwenkia')	tc: W	Mt. Roraima
481	793		GEN	Schultesia subcrenata Klotzsch ex Griseb.	Schultesia subcrenata Klotzsch ex Griseb.		Mt. Roraima
482	792	1842–43	LOG	Strychnos schomburgkiana Klotzsch [n.n.]	Strychnos bredemeyeri (Schult.) Sprague & Sandw.	tc: F GH NY US	Roraima
483	741	1842–43	CNV	Evolvulus linifolius L.	Evolvulus filipes Mart.	BM	Roraima
484	749		FAB	Desmodium pachyrhizum Vogel	Desmodium pachyrhiza Vogel		
485	746	1842	SCR	Gerardia hispidula Mart.	Agalinis hispidula (Mart.) D'Arcy	BM	Roraima
486	742	1843	LYT	Maja hypericoides Klotzsch [n.n.]	Cuphea anagalloidea A. St.-Hil.	ht: P; it: K P	Roraima
487	773*	1842	GEN	Schultesia benthamiana Klotzsch [n.n.]	Schultesia benthamiana Klotzsch ex Griseb.		Mt. Roraima
488	754	1842–43	MLP	Banisteria corymbosa Griseb.	Banisteriopsis cinerascens (Benth.) B. Gates	ht: K; it: BM F G P W	Roraima
489	756		CSL	Cassia obtusifolia L.	Senna obtusifolia (L.) H.S. Irwin & Barneby		
490			MIM	Pithecellobium multiflorum (Kunth) Benth. ('Pithecolobium')	Albizia subdimidiata (Splitg.) Barneby & J.W. Grimes var. subdimidiata		
491	794	1842–43	SCR	Ilysanthes gratioloides (L.) Benth.	Lindernia dubia (L.) Pennell	BM	Roraima
492	791		MLS	Microlicia brevifolia (Rich.) DC.	Acisanthera bivalvis (Aubl.) Cogn.		
493	763		MIM	Acacia polyphylla DC.	Acacia polyphylla DC.		
494	767		MIM	Entada polystachya (L.) DC.	Entada polystachya (L.) DC.		
495	768		MIM	Acacia westiniana DC.	Acacia riparia Kunth		
496	790	1842–43	KRM	Krameria spartioides Klotzsch ex O. Berg	Krameria spartioides Klotzsch ex O. Berg	BM G K P	Roraima
497	761	1842–43	STR	Waltheria americana L.	Waltheria indica L.	BM	Roraima
498	771		CSL	Cassia filipes Benth.	Chamaecrista rotundifolia (Pers.) Greene var. grandiflora (Benth.) H.S. Irwin & Barneby		
499	769		VRB	Stachytarpheta roraimensis Moldenke ('Stachytarpha')	Stachytarpheta sprucei Moldenke	tc: BM K P US	Roraima
500	774	1842–43	STR	Waltheria paniculata Benth.	Waltheria paniculata Benth.	BM	Roraima
501	753	1842–43	STR	Waltheria involucrata Benth.	Waltheria involucrata Benth.	BM	Roraima
502	784		FAB	Crotalaria stipularis Desv.	Crotalaria stipularis Desv.		
503	785*		SCR	Buchnera palustris (Aubl.) Spreng.	Buchnera palustris (Aubl.) Spreng.	st: K NY P	
504	803	1842–43	PGL	Catocoma cuneata Klotzsch	Bredemeyera cuneata Klotzsch ex Hassk.		Roraima
505	797	1842–43	CNV	Ipomoea evolvuloides Moric.	Jacquemontia evolvuloides (Moric.) Meisn.		Roraima
506	802		FAB	Crotalaria anagyroides Kunth	Crotalaria micans Link		Roraima

Index 3. Robert Schomburgk's second collection series (1841–1844) – continued

No.	Year	Coll.	Family	Name	Determination	Herbarium	
507	1842–43	777	MLP	Byrsonima schomburgkiana Benth.	Byrsonima schomburgkiana Benth.	st: BM CGE G K	
508		788	MIM	Mimosa pudica L.	Mimosa pudica L.		Roraima
509		778	CMB	Combretum laxum Jacq.	Combretum laxum Jacq.	BM CGE G K P OXF TCD W	Roraima
510	1842–43	795	FAB	Copaifera	Andira surinamensis (Bondt) Splitg. ex Amshoff		Roraima
511	1842–43	782	CNV	Quamoclit coccinea Choisy	Ipomoea angulata Mart. ex Choisy	BM	Roraima
512		780	FAB	Andira retusa Kunth	Andira surinamensis (Bondt) Splitg. ex Amshoff		Roraima
513	1842–43	787	SCR	Stemodia foliosa Benth.	Stemodia pratensis (Aubl.) C.P. Cowan	st: BM	Roraima
513*		811	SCR	Capraria biflora L.	Capraria biflora L.	BM	Roraima
514			CSL	Cassia polystachya Benth.	Chamaecrista polystachya (Benth.) H.S. Irwin & Barneby		
515		827	MLS	Dicrananthera hedyotidea C. Presl	Acisanthera hedyotidea (C. Presl) Triana	K	Roraima
515*			MLS	Osbeckia pumila DC.	Pterolepis pumila (Bonpl.) Cogn.	ht: G; it: BR G P	Roraima
516	1842–43	829	EUP	Caperonia angustissima Klotzsch [n.n.]	Caperonia stenophylla Muell. Arg.	BM	Roraima
517	1842–43	796	ERX	Erythroxylum orinocense Klotzsch ('Erythroxylon')	Erythroxylum schomburgkii Peyr.	lt: ?; ilt: BM F K P U	Roraima
518*	1842–43	805	CLU	Mahurea linguiformis Tul.	Mahurea exstipulata Benth.	tc: BM	Roraima
519		812	CHB	Licania flavicans Klotzsch [n.n.]	Licania compacta Fritsch	ht: W; it: BR CGE F G K NY OXF P U	Roraima
520		822	LOG	Antonia pilosa Hook.	Antonia ovata Pohl		
521	1842–43	823	MLS	Macairea calvescens Naudin	Macairea pachyphylla Benth.	ht: P; it: NY	Roraima
522		825	FAB	Dioclea guianensis Benth.	Dioclea guianensis Benth.	BM	
523	1842–43	817	MLS	Rhynchanthera serrulata (Rich.) DC.	Rhynchanthera serrulata (Rich.) DC.	ht: P; it: NY U	Roraima
524		814	MIM	Mimosa microcephala Humb. & Bonpl. ex Wild.	Mimosa microcephala Humb. & Bonpl. ex Wild.	type	
525		824	MIM	Pithecellobium polycephalum Benth. ('Pithecolobium')	Albizia glabripetala (H.S. Irwin) G.P. Lewis & P.E. Owen		
526		813	RUB	Coutarea	Luehea speciosa Willd.		
527		821	TIL	Luehea rufescens A. St.-Hil.			
528		839	CSL	Cassia pulchra Kunth	Cassia tetraphylla Desv. var. tetraphylla	BM K NY	Roraima
529		834	FAB	Tephrosia leptostachya DC.	Tephrosia purpurea (L.) Pers.		
530		833	FAB	Aeschynomene trisperma Klotzsch [n.n.]	Aeschynomene brasiliana (Poir.) DC.		
531		836	CSL	Cassia prostrata L.	Chamaecrista serpens (L.) Greene var. serpens	US	
532	1842–43	835	AST	Ooclinium villosum DC.	Praxelis pauciflora (Kunth) R.M. King & H. Rob.	BM	
533	1842–43	837	AST	Pectis elongata Kunth	Pectis elongata Kunth var. elongata	BM	
534		832	FAB	Swartzia latifolia Benth.	Swartzia latifolia Benth. var. sylvestris R.S. Cowan	ht: G; it: BM CGE NY P W	Roraima

No.	Coll.	Year	Fam.	Name	Accepted name	Herbarium	Roraima
535	831		NYC	Boerhavia surinamensis Miq.	Boerhavia diffusa L.		
536	816		FAB	Stylosanthes angustifolia Vogel	Stylosanthes angustifolia Vogel		
537	810		MRT	Eugenia ochra O. Berg	Eugenia tapacumensis O. Berg		
538	819	1842–43	MLV	Sida pitifera Klotzsch [n.n.]	Sida aggregata C. Presl	BM	Roraima
539	820	1842–43	STR	Waltheria viscosissima A. St.-Hil.	Waltheria viscosissima A. St.-Hil.	BM	
540	818		LNT	Polypompholyx bicolor Klotzsch [n.n.]	Utricularia simulans Pilg.		
541	828		FAB	Nicolsonia cayennensis DC.	Desmodium barbatum (L.) Benth. & Oerst.	BM	Roraima
542	804	1842–43	MLV	Pavonia speciosa Kunth	Peltaea speciosa (Kunth) Standley	BM	Roraima
543	889	1842–43	AST	Stiftia condensata Baker	Stomatochaeta condensata (Baker) Maguire & Wurdack		
544	855	1842–43	PRT	Roupala suaveolens Klotzsch	Roupala suaveolens Klotzsch	BM NY	Roraima
545	863	1842–43	MLV	Malva sp. nov.	Malva sp. nov.	BM	Roraima
546	853		LAM	Hyptis arborea Benth.	Hyptidendron arboreum (Benth.) Harley	st: US	
547	888	1842–43	RUB	Cascarilla schomburgkii Klotzsch	Ladenbergia lambertiana (A. Br. ex Mart.) Klotzsch	BM	Roraima
548	852		FAB	Swartzia oblonga Benth.	Swartzia oblonga Benth.	ht: K; it: U	Roraima
549	858		LAU	Nectandra salicifolia (Kunth) Nees	Nectandra sanguinea Rol. ex Rottb.	st: US	Roraima
550	856	1842–43	SPT	Chrysophyllum guyanense Klotzsch ex Miq. [n.n.]	Chrysophyllum argenteum Jacq. subsp. auratum (Miq.) T.D. Penn.	BM	
551	879	1842–43	PGL	Polygala paniculata L.	Polygala paniculata L.	BM K P W	Roraima
551*	881		CNV	Quamoclit coccinea Choisy	Ipomoea angulata Mart. ex Choisy		
552	880	1842–43	CLU	Vismia guianensis (Aubl.) Choisy	Vismia guianensis (Aubl.) Choisy	BM	Roraima
553	890	1842–43	CSL	Cassia uniflora Spreng. var. ramosa (Vogel) Benth.	Chamaecrista ramosa (Vogel) H.S. Irwin & Barneby var. ramosa	BM K	Roraima
554	887		MRS	Badula schomburgkiana A. DC.	Stylogyne schomburgkiana (DC.) Mez	st: B BM G K P W	Roraima
555	882		AQF	Ilex laureola Triana & Planch.	Ilex laureola Triana & Planch.	K	Roraima
556	876		BNN	Archytaea multiflora Benth.	Archytaea triflora Mart.		
557	857		MLS	Melastoma dodecandra Desv.	Miconia dodecandra (Desv.) Cogn.		
558	877		MRT	Myrcia subcordata DC.	Marlierea lituatinervia (O. Berg) McVaugh		
559	868	1842–43	ERI	Vaccinium puberulum Klotzsch [n.n.]	Vaccinium puberulum Klotzsch ex Meisn.	st: BM NY	Roraima
560	887*		MLS	Rhexia taxifolia A. St.-Hil.	Marcetia taxifolia (A. St.-Hil.) DC.	G P US	Roraima
561	847	1842–43	EUP	Croton	Croton	st: BM US	Roraima
562	850	1842–43	MRT	Myrciaria uliginosa O. Berg	Myrciaria uliginosa O. Berg	BM	Roraima
563	848	1842–43	AST	Clibadium surinamense L.	Clibadium surinamense L.	OXF P	
564			CHB	Parinari campestris Aubl. ('Parinarium campestre')	Parinari campestris Aubl. ('Parinarium campestre')		
565	849	1842–43	AST	Eupatorium martiusii DC.	Chromolaena squalida (DC.) R.M. King & H. Rob.	BM	Roraima
566	867	1842–43	ERI	Thibaudia nutans Klotzsch [n.n.]	Thibaudia nutans Klotzsch ex Mansf.	BM	Roraima
566*		1842–43	ERI	Notopora schomburgkii Hook.f.	Notopora schomburgkii Hook.f.	BM	Roraima

Index 3. Robert Schomburgk's second collection series (1841–1844) – continued

No.	Coll.	Year	Fam.	Name as collected	Accepted name	Herbarium	
567	873	1842–43	ERI	Thibaudia nutans Klotzsch [n.n.]	Thibaudia nutans Klotzsch ex Mansf.	BM	Roraima
568	885	1842–43	MLS	Graffenrieda ovalifolia Naudin	Graffenrieda weddelii Naudin	ht: P	Roraima
569	872		OCH	Poecilandra retusa Tul.	Poecilandra retusa Tul.	tc: BM	Roraima
570	865		CHB	Hirtella scabra Benth.	Hirtella scabra Benth.	F K NY OXF P	
571	806		MIM	Pithecellobium aff. polycephalum Benth. ('Pithecolobium')	Pithecellobium aff. polycephalum Benth. ('Pithecolobium')		
572	846	1842–43	PRT	Roupala complicata Kunth	Roupala montana Aubl. var. montana	BM	
573	844	1842–43	TEA	Ternstroemia roraimae Klotzsch [n.n.]	Ternstroemia laevigata Wawra	tc: NY	Roraima
574	842		HUM	Sacoglottis guianensis Benth.	Sacoglottis guianensis Benth. var. guianensis	ht: K; it: P	Roraima
575	900		AST	Vernonia opaca Benth.	Piptocarpha opaca (Benth.) Baker		
576	845		HUM	Humiria elliptica Klotzsch ex Urb. ('Humirium ellipticum')	Humiria balsamifera (Aubl.) J. St.-Hil. var. savannarum (Gleason) Cuatrec.	P K	Roraima
577	871		ERI	Befaria schomburgkiana Klotzsch [n.n.]	Bejaria sprucei Meisn.	P	
578	870	1842–43	RUB	Pagamea capitata Benth.	Pagamea capitata Benth. subsp. capitata	ht: K; it: BM NY	Roraima
579	883		RUB	Cephaëlis tomentosa Willd.	Psychotria poeppigiana Muell. Arg.	st: BM G P U	Roraima
580	905	1842–43	EUP	Peridium schomburgkianum Benth.	Pera bicolor (Klotzsch) Muell. Arg.	st: BM G P U	Roraima
581	892	1842–43	RUB	Psychotria inundata Benth.	Psychotria capitata Ruiz & Pav. var. roraimensis (Wernham) Steyerm.	tc: BM	Roraima
582	840		CSL	Cassia roraimae Benth.	Chamaecrista roraimae (Benth.) Gleason		
583	894	1842–43	MEL	Guarea aubletii A. Juss.	Guarea guidonia (L.) Sleumer	lt: F FI G K NY	Roraima
584	893		VOC	Qualea schomburgkiana Warm.	Qualea schomburgkiana Warm.	BM	Roraima
585	964	1842–43	VOC	Vochisia curvata Klotzsch [n.n.]	Vochysia crassifolia Warm.		Roraima
586	959	1842	MIM	Inga setifera DC.	Inga pilosula (Rich.) J.F. Macbr.	ht: K; it: BM G U	
587	912	1842–43	MLP	Byrsonima concinna Benth.	Byrsonima concinna Benth.	ht: K; it: BM CGE F G MICH P W	Roraima
588	906	1842–43	MYS	Myristica sebifera Sw.	Virola sebifera Aubl.	G NY	Roraima
589	911		STY	Styrax subleprosum Klotzsch [n.n.]	Styrax roraimae Perkins	tc: US	Roraima
590	913		LAU	Aioua schomburgkii Meisn.	Aiouea guianensis Aubl.	ht: G; it: BM F K NY US	Roraima
591	909		TEA	Lettsomia guianensis Klotzsch [n.n.]	Freziera roraimensis Tul.	st: GH	Roraima
592	910		LAU	Oreodaphne guianensis (Aubl.) Nees	Ocotea guianensis Aubl.	st: BM NY	Roraima
593	948	1842–43	EUP	Tragia grandifolia Klotzsch	Adenophaedra grandifolia (Klotzsch) Muell. Arg.	st: BM K NY	Roraima
594	901	1842–43	EUP	Peridium schomburgkianum Benth.	Pera bicolor (Klotzsch) Muell. Arg.	BM	Roraima
595	903	1842–43	BRS	Trattinickia guianensis Klotzsch [n.n.]	Trattinickia burserifolia Mart.		Roraima
596	843		CYR	Cyrilla antillana Michx.	Cyrilla racemiflora L.		Roraima
597	878		ASC	Ditassa taxifolia Decne.	Ditassa taxifolia Decne.		Roraima

No.	No.	Year	Fam.	Original name	Accepted name	Herbarium	Locality
598	931	1842–43	CSL	Amorphocalyx roraimae Klotzsch [n.n.]	Sclerolobium guianense Benth.	tc: NY	Roraima
599	936		LAU	Oreodaphne guianensis (Aubl.) Nees	Ocotea guianensis Aubl.	st: GH	Roraima
600	937		TEA	Ternstroemia punctata (Aubl.) Sw.	Ternstroemia punctata (Aubl.) Sw.	FM	Roraima
601	935		CHB	Licania rufescens Klotzsch [n.n.]	Licania rufescens Klotzsch ex Fritsch	ht: W; it: BR CGE K NY P	
602	942	1842–43	TEA	Ternstroemia suborbicularis Klotzsch [n.n.]	Ternstroemia crassifolia Benth.	tc: FM G NY	Roraima
603	934		LAU	Oreodaphne crassifolia Nees	Ocotea crassifolia (Nees) Mez		Roraima
604	945	1842–43	AQF	Ilex thyrsifolia Klotzsch [n.n.]	Ilex thyrsifolia Klotzsch ex Reissek var. thyrsifolia	st: BM G W	Roraima
605	943		AQF	Ilex schomburgkii Klotzsch [n.n.]	Ilex thyrsifolia Klotzsch ex Reissek var. schomburgkii (Klotzsch ex Reissek) Loes.	st: BM G P W syntype	
606	951		FAB	Dipteryx reticulata Benth.	Dipteryx reticulata Benth.	st: BM G GH P U	Roraima
607	946	1842–43	AQF	Ilex retusa Klotzsch [n.n.]	Ilex retusa Klotzsch ex Reissek	BM G	Roraima
608	957	1842–43	MRT	Myrciaria cordata O. Berg	Myrciaria cordata O. Berg	st: BM	Roraima
609	923	1842–43	SYM	Symplocos schomburgkii Klotzsch [n.n.]	Symplocos guianensis (Aubl.) Guerke	BM	Roraima
610	920	1842–43	SYM	Symplocos schomburgkii Klotzsch [n.n.]	Symplocos guianensis (Aubl.) Guerke		Roraima
611	930		FAB	Aeschynomene hystrix Poir.	Aeschynomene hystrix Poir.	BM W	Roraima
612	947	1842–43	CLU	Calophyllum lucidum Benth.	Calophyllum lucidum Benth.	tc: BM U	Roraima
613	907		MRT	Myrcia ferruginea (Poir.) DC.	Marlierea ferruginea (Poir.) McVaugh	tc: P U	Roraima
614	939	1842–43	LOG	Bonyunia superba M.R. Schomb. ex Progel	Bonyunia superba M.R. Schomb. ex Progel		Roraima
615	929		LAU	Ocotea costulata Nees	Ocotea aciphylla (Nees) Mez	BM Fl K P	Roraima
616	932	1842–43	FLC	Patrisia bicolor A. DC.	Ryania speciosa Vahl var. tomentella Sleumer	BM	Roraima
617	940	1842–43	ANN	Guatteria ouregou (Aubl.) Dunal	Guatteria	st: BM K U	Roraima
618	952	1842–43	ERX	Erythroxylum roraimae Klotzsch ex O.E. Schulz	Erythroxylum roraimae Klotzsch ex O.E. Schulz		Roraima
619	944		TEA	Ternstroemia punctata (Aubl.) Sw.	Ternstroemia punctata (Aubl.) Sw.	BM	Roraima
620		1842–43	BOR	Cordia	Cordia	st: BM NY	
621	955	1842–43	RUB	Aspidanthera klotzschiana M.R. Schomb. [n.n.]	Ferdinandusa goudotiana K. Schum.		
622	953		SAP	Cupania	Miconia lepidota DC.		Roraima
623	926		MLS	Miconia schomburgkii Benth.	Miconia schomburgkiana Urb.	st: BM G	Roraima
624	922	1842–43	TNR	Turnera schomburgkiana Urb.	Turnera brasiliensis Willd. ex Roem. & Schult.	K	Roraima
624*			TNR	Turnera brasiliensis Willd. ex Roem. & Schult.	Hypolytrum pulchrum (Rudge) H. Pfeiff.	P	Roraima
625	924	1842–43	CYP	Hypolytrum pungens (Vahl) Kunth	Piptocoma schomburgkii (Sch. Bip.) Pruski		Roraima
626	921		AST	Oliganthes schomburgkii Sch. Bip.	Ditassa angustifolia Decne.	ht: P; it: MO NY W	Roraima
627	915	1842–43	ASC	Ditassa angustifolia Decne.	Humiria balsamifera (Aubl.) J. St.-Hil. var. laurina (Urb.) Cuatrec.	BM P K U	Roraima
628	968		HUM	Humiria laurina Klotzsch ex Urb. ('Humirium laurinum')	Ternstroemia schomburgkiana Benth.		Roraima
629	967		TEA	Ternstroemia schomburgkiana Benth.	Clethra guianensis Klotzsch ex Meisn.	tc: BM F NY	Roraima
630	956	1842–43	CLE	Clethra guianensis Klotzsch ex Meisn.	Ouratea roraimae Engl.	tc: BM	Roraima
631	965	1842–43	OCH	Ouratea roraimae Engl.			Roraima

Index 3. Robert Schomburgk's second collection series (1841–1844) – continued

				Collection name	Current name	Herbarium	
632	958	1842–43	MLP	Heteropterys carinata Benth.	Heteropterys cristata Benth.	ht: K; it: BM G P	Roraima
633	961	1842–43	CLU	Quapoya robusta Klotzsch [n.n.]	Clusia schomburgkiana (Planch. & Triana) Benth. ex Engl.	tc: BM NY	Roraima
634	950		PRT	Andripetalum sessilifolium (Rich.) Klotzsch	Panopsis sessilifolia (Rich.) Sandw.	BM	Roraima
635	962		CSL	Dimorphandra macrostachya Benth.	Dimorphandra macrostachya Benth. subsp. macrostachya	G	Roraima
636	866	1842–43	BNN	Bonnetia sessilis Benth.	Bonnetia sessilis Benth.	BM	
637	977		CPR	Viburnum		BM BR G L K P U	Roraima
638	1059	1842–43	EUP	Maprounea guianensis Aubl.	Maprounea guianensis Aubl.	BM	Roraima
639	979	1842–43	ERI	Gaultheria roraimae Klotzsch ex Meisn.	Gaultheria erecta Vent.	lt: W; it: F NY FI G K P W	Roraima
640	981	1842–43	PLG	Coccoloba schomburgkii Meisn.	Coccoloba schomburgkii Meisn.	tc: G GH P NY	Roraima
641	980	1842	BIG	Tabebuia triphylla (L.) DC.	Tabebuia insignis (Miq.) Sandw. var. insignis	BM G P W	Roraima
642	841		VOC	Vochisia lucida Klotzsch [n.n.]	Vochysia glaberrima Warm.	ht: K; it: BM G GH L OXF P W	Roraima
643	982	1842–43	AST	Baccharis roraimae M.R. Schomb. [n.n.]	Baccharis brachylaenoides DC. var. ligustrina (DC.) Maguire & Wurdack	BM	
644	978	1842–43	MRT	Aulomyrcia triflora O. Berg	Myrcia citrifolia (Aubl.) Urb.	BM	Roraima
645	983		SAR	Heliamphora nutans Benth.	Heliamphora nutans Benth.		
646	984		RUB				Roraima
647	992	1843	MRS	Grammadenia lineata Benth.	Cybianthus lineatus (Benth.) Pipoly	ht: K; it: B BM G K P U W	Roraima
648	996		MLS	Clidemia lutescens Naudin var. lindeniana Naudin	Leandra lindeniana (Naudin) Cogn.	K	
649	985		LNT	Utricularia humboldtii R.H. Schomb.	Utricularia humboldtii R.H. Schomb.		
650	998	1842–43	MLP	Hiraea oleifolia Benth.	Tetrapterys oleifolia (Benth.) Nied.	ht: K; it: BM CGE G K P	
651	999		MLP	Banisteria leptocarpa Benth.	Banisteria martiniana (A. Juss.) Cuatrec. var. martiniana	st: BM C F G K	Roraima
652	949		AST	Vernonia schomburgkiana Sch. Bip. var. lanceolata Sch. Bip.	Vernonia ehretifolia Benth.		Roraima
653	919	1842–43	AST	Vernonia tricholepis DC.	Lepidaploa remotiflora (Rich.) H. Rob.	BM	
654	993	1842–43	AST	Verbesina schomburgkii Sch. Bip. [n.n.]	Verbesina schomburgkii Sch. Bip. ex Klatt	tc: BM GH	Roraima
655	941		MLS	Diplochita fothergilla DC.	Miconia mirabilis (Aubl.) L.O. Williams		Roraima
656	973	1842–43	AST	Vernonia schomburgkiana Sch. Bip. var. elliptica Sch. Bip.	Vernonia ehretifolia Benth.	tc: BM G	
657	914	1842–43	CLU	Clusia nemorosa G. Mey.	Clusia nemorosa G. Mey.	BM	Roraima
658	928		BIG	Tecoma dura Bureau & K. Schum.	Tabebuia insignis (Miq.) Sandw. var. insignis	ht: B †; it: K	Roraima
659	1006	1842–43	EUP	Phyllanthus pycnophyllus Muell. Arg.	Phyllanthus pycnophyllus Muell. Arg.	BM	Roraima

No.	Coll. no.	Date	Fam.	Name (Klotzsch)	Accepted name	Herbarium	Locality
660	991		MLS	Ossaea coriacea Naudin	Clidemia tepuiensis Wurdack	ht: P	near Mt. Roraima
661	1005	1842–43	CUN	Weinmannia guyanensis Klotzsch ex Engl.	Weinmannia guyanensis Klotzsch ex Engl.		Roraima
662	971	1842–43	CUN	Weinmannia ovalis Ruiz & Pav.	Weinmannia balbisiana Kunth var. roraimensis (Pampan.) Bernardi		Roraima
663	1013		MLS	Chaetolepis anisandra Naud.	Chaetolepis anisandra Naudin	ht: P	near Mt. Roraima
663*	994	1843	MIM	Pithecellobium ferrugineum Benth. ('Pithecolobium')	Abarema ferruginea (Benth.) Pittier	ht: K; it: U	Mt. Roraima
664	990		AST	Achyrocline			Roraima
665	1029	1842–43	EUP	Croton subincanus Muell. Arg.	Croton subincatus Mull. Arg.	st: BM G P	Roraima
666	1000	1842–43	AST	Mikania	Mikania pannosa Baker	BM	Roraima
667	1009	1842–43	LIL	Nietneria corymbosa Klotzsch & M.R. Schomb. [n.n.]	Nietneria corymbosa Klotzsch & M.R. Schomb. ex Jackson	st: U	Mt. Roraima
668	1039		LOR	Gaiadendron tagua (Kunth) G. Don	Gaiadendron punctatum (Ruiz & Pav.) G. Don	BM	Roraima
669	972	1842–43	ERI	Thibaudia guianensis Klotzsch [n.n.]	Psammisia guianensis Klotzsch	ht: P	near Mt. Roraima
670	974		MLS	Davya crassiramis Naudin	Meriania crassiramis (Naudin) Wurdack		Roraima
671	1004	1842–43	RPT	Stegolepis guianensis Klotzsch ex Körn.	Stegolepis guianensis Klotzsch ex Körn.	tc: G K NY W	Roraima
672	987	1842–43	ERI	Befaria guyanensis Klotzsch [n.n.]	Bejaria sprucei Meisn.	BM F G OXF W	Roraima
673	1041	1842–43	RUB	Isertia hypoleuca Benth.	Isertia hypoleuca Benth.	BM	Roraima
674	1044		MLS	Meisneria microlicioides Naudin	Siphanthera cordifolia (Benth.) Gleason	ht: P	
675	1003	1842–43	ERO	Paepalanthus dichotomus Klotzsch [n.n.]	Paepalanthus dichotomus Klotzsch ex Körn.		
676	849		MLP	Coleostachys hypoleuca Benth.	Blephandra hypoleuca (Benth.) Griseb.	ht: K; it: BM CGE K P	Roraima
677	1043		MRT	Myrcia			Roraima
678	1055	1842–43	BOR	Cordia dichotoma Klotzsch [n.n.]	Cordia bicolor A. DC.	BM K P	Roraima
678*	1032	1842–43	CNN	Rourea subtriplinervis Radlk.	Pseudoconnarus subtriplinervis (Radlk.) Schellenb.	lt: G; ilt: BM K M P US W	Roraima
679	1061	1842–43	SOL			BM	Roraima
680	916		MRS	Conomorpha crotonoides Mez	Cybianthus crotonoides (M.R. Schomb. ex Mez) G. Agostini	BM	Roraima
681	1027	1842–43	CLU	Caraipa tereticaulis Tul.	Caraipa tereticaulis Tul.	BM	
682	925	1842–43	CSL	Cassia calliantha G. Mey.	Senna multijuga (Rich.) H.S. Irwin & Barneby var. multijuga		
683	1053		ANN	Xylopia sericea A. St.-Hil.	Xylopia sericea A. St.-Hil.	BM K	
684	1072	1842–43	EUP	Peridium bicolor Klotzsch var. nitidum Benth.	Pera decipiens (Muell. Arg.) Muell. Arg.	st: BM	Roraima
685	1071	1842–43	EUP	Peridium bicolor Klotzsch var. nitidum Benth.	Pera decipiens (Muell. Arg.) Muell. Arg.	st: BM U	Roraima
686	1070	1842–43	BML	Encholirium augustae M.R. Schomb.	Connellia augustae (M.R. Schomb.) N.E. Br.	ht: K; it: BM GH	Roraima
687	1021		ROS	Rubus guyanensis Focke	Rubus guyanensis Focke		Roraima
688	1038		MLS				Roraima
689	1056		MLS	Miconia sp. nov.	Miconia macrothyrsa Benth.		Roraima
690	1066		MLS	Miconia fallax Benth.	Miconia fallax Benth.		Roraima

Index 3. Robert Schomburgk's second collection series (1841–1844) – continued

No.	Coll.	Year	Fam.	Name (Schomburgk)	Determination	Herbaria	Locality
691	1069		MLS	Miconia heterochroa Miq.	Miconia albicans (Sw.) Triana		
692	1042		CYP	Rhynchospora cephalotes (L.) Vahl	Rhynchospora cephalotes (L.) Vahl		
693	904	1842–43	CLU	Clusia insignis Mart.	Clusia grandiflora Splitg.		
694	1047		RUB	Cordiera	Alibertia myrciifolia K. Schum.	BM	
695	1080	1842–43	MIM	Inga brachyptera Benth.	Inga bracteosa Benth.		
696	1080	1842–43	EUP	Croton gossypiifolius Vahl	Croton gossypiifolius Vahl	ht: K; it: BM G NY P US	Roraima
697	1079		FAB	Machaerium ferrugineum Pers.	Machaerium quinata (Aubl.) Sandw. var. parviflorum (Benth.) Rudd	BM P; BM GH K NY P W	Roraima
698	1054	1842–43	MRT	Myrcia hostmanniana Kiaersk.	Myrcia amazonica DC.		
699	1089	1842–43	MLS	Macairea parvifolia Benth.	Macairea parvifolia Benth.	BM	Roraima
700	1092	1842–43	MRT	Myrciaria ehrenbergiana O. Berg	Myrcia ehrenbergiana (O. Berg) McVaugh	ht: K; it: NY	Mt. Roraima
701	1086	1842–43	MRT	Myrcia schomburgkiana O. Berg	Myrcia servata McVaugh	st: BM; ilt: U	Roraima
702	1088		AST	Eupatorium ixodes Benth.	Ayapana amygdalina (Lam.) R.M. King & H. Rob.	BM	Roraima
702*		1842–43	MRT	Myrcia schomburgkiana O. Berg	Myrcia servata McVaugh	BM	Roraima
703	1091	1842–43	ANN	Xylopia grandiflora A. St.-Hil.	Xylopia aromatica (Lam.) Mart.	BM	Roraima
704	1085	1842–43	MRT	Myrcia guianensis (Aubl.) DC.	Myrcia guianensis (Aubl.) DC. var. guianensis	BM K P	Roraima
705	1090		RUB	Gonzalea spicata DC.	Gonzalagunia spicata (Lamb.) Gómez	BM	Roraima
706	1073		MLS	Macairea multinerva Benth.	Macairea multinerva Benth.	ht: K	Roraima
707	1026	1842–43	AST	Baccharis schomburgkii Baker	Baccharis schomburgkii Baker	tc: BM NY US W	Roraima
707*		1842–43	APO	Thyrsanthus adenobasis (Muell. Arg.) Miers	Forsteronia adenobasis Muell. Arg.	tc: BM G	Roraima
708	1067		TEA	Ternstroemia suborbicularis Klotzsch [n.n.]	Ternstroemia crassifolia Benth.		
709	1036	1842–43	CLU	Clusia crassifolia Planch. & Triana	Clusia crassifolia Planch. & Triana	BM	Roraima
710	1083	1842–43	AST	Elephantopus angustifolius Sw.	Orthopappus angustifolius (Sw.) Gleason	BM	Roraima
711	1081	1842–43	RHM	Gouania virgata Reissek	Gouania virgata Reissek	tc: G	Roraima
712	1087		LAU	Oreodaphne uruphylla Nees	Ocotea cernua (Nees) Mez		
713	1026*		ERO	Paepalanthus schomburgkii Klotzsch [n.n.]	Paepalanthus schomburgkii Klotzsch ex Körn.		
713*		1842–43	MEL				
714	1078		MEL	Moschoxylon propinquum Miq.	Trichilia quadrijuga Kunth subsp. quadrijuga	BM	
715	1077	1842–43	AST	Gnaphalium simplicicaule Willd. ex Spreng.	Gamochaeta simplicicaulis (Willd. ex Spreng.) Cabrera	BM	Roraima
716	1076	1842–43	MLS	Clidemia capitata Benth.	Clidemia capitata Benth.	BM	
717	933	1842–43	CLU	Clusia schomburgkii Vesque	Clusia schomburgkii Vesque	NY	Roraima
718	1065	1842–43	ORC	Cypripedium klotzscheanum Rchb.f.	Phragmipedium klotzscheanum (Rchb.f.) Rolfe	tc: BM	Roraima
719	896	1842–43	RUB	Chiococca nitida Mart. & Eichl.	Chiococca nitida Benth. var. amazonica Muell. Arg	tc: BM P	Roraima
720	976	1842–43	SPT	Lucuma rigida Mart. & Eichl.	Pouteria rigida (Mart. & Eichl.) Radlk. subsp. rigida	BM	Roraima

No.	Coll. no.	Year	Fam.	Name (as listed)	Current name	Herbarium	Locality
721	1052		ANA	Tapirira guianensis Aubl.	Tapirira guianensis Aubl.		near Mt. Roraima
722	752	1843	GEN	Schultesia brachyptera Cham.	Schultesia brachyptera Cham.		Roraima
723	1095		LOG	Strychnos rhexioides Klotzsch [n.n.]	Strychnos tomentosa Benth.	syntype	Roraima
724	815	1842–43	RUB	Patima laxiflora Benth.	Retiniphyllum laxiflorum (Benth.) N.E. Br. var. laxiflorum	tc: BM; nt: BM	
725	854	1842–43	APO	Forsteronia diospyrifolia Muell. Arg.	Forsteronia diospyrifolia Muell. Arg.	BM K	Roraima
726	1099	1842–43	MLP	Heteropterys lessertiana A. Juss.	Heteropterys macradena (DC.) W.R. Anderson	BM P	Roraima
727	1097	1842–43	ORC	Cypripedium lindleyanum R.H. Schomb.	Phragmipedium lindleyanum (R.H. Schomb.) Rolfe	BM	Roraima
728	1107	1842	MNM	Siparuna guianensis Aubl.	Siparuna guianensis Aubl.	BM G NY P U W	Roraima
729	1106	1842–43	PGL	Catocoma lucida Benth.	Bredemeyera lucida (Benth.) Klotzsch ex Hassk.	BM K W	Roraima
729*	1106	1842–43	FLC	Myroxylon benthamii Tul.	Xylosoma benthamii (Tul.) Triana & Planch.		
730	1105		SAP	Cupania subsinuata Klotzsch [n.n.]	Cupania rubiginosa (Poir.) Radlk.	tc: G K P W	Roraima
731	1109	1843	EUP	Mabea piriri Aubl. var. laevigata Muell. Arg.	Mabea biglandulosa Muell. Arg.	BM K P	Roraima
732	1110	1842–43	BOR	Tournefortia volubilis L.	Tournefortia volubilis L.	st: NY	Roraima
733	1111	1842–43	FAB	Lonchocarpus nitidulus Benth.	Lonchocarpus floribundus Benth.	BM K W	Roraima
734	1117		MRT	Aulomyrcia roraimensis O. Berg	Aulomyrcia roraimensis O. Berg	BM	Roraima
734*			APO	Tabernaemontana undulata Vahl	Tabernaemontana undulata Vahl	BM F G K M P W	Roraima
735	1116	1842–43	CNN	Omphalobium opacum Klotsch [n.n.]	Connarus coriaceus Schellenb.	P	Roraima
736		1842–43	CNN	Omphalobium opacum Klotsch [n.n.]	Connarus coriaceus Schellenb.	ht: K	Roraima
737	1119	1842–43	MLP	Lasianthemum unijugum Klotzsch [n.n.]	Tetrapterys maranhamensis A. Juss.	tc ?: GH P	Roraima
738	1351		SAP	Hiraea gracilis Benth.	Talisia squarrosa Radlk.	st: BM	Roraima
739	1338	1842–43	CLU	Clusia alba L.	Clusia palmicida Rich.	GH	Roraima
740	1349		LAU	Persea gratissima Gaertn.	Persea americana P. Mill.	BM K	Roraima
741	1332	1842–43	BIG	Arrabidaea schomburgkii Klotzsch [n.n.]	Arrabidaea candicans (Rich.) DC.	ht: K; it: G NY P	Roraima
742	1322		PED	Sesamum			
743	1335		MLP	Byrsonima propinqua Benth.	Byrsonima spicata (Cav.) DC.	BM	Roraima
744	1343		XYR	Xyris			
745	1380		LAU	Acrodiclidium guianense Nees	Licaria polyphylla (Nees) Kosterm.	BM	Roraima
746	1339	1842–43	BRS	Trattinickia schomburgkii Klotzsch [n.n.]	Trattinickia burserifolia Mart.	st: BM F G NY P	
747	1353		AQF	Ilex umbellata Klotzsch [n.n.]	Ilex umbellata Klotzsch ex Reissek var. umbellata		
748	1324		MEL	Melia sempervirens Sw.	Melia azedarach L.		
749	1336		LAU	Oreodaphne schomburgkiana Nees	Ocotea schomburgkiana (Nees) Mez		
750	1342		ANA	Mangifera indica L.	Mangifera indica L.		
751	1400	1842	MIM	Inga myriantha Poepp.	Inga umbellifera (Vahl) Steud. ex DC.	K MEL P	
752	1346	1842–43	MEL	Trichilia schomburgkii C. DC.	Trichilia schomburgkii C. DC. subsp schomburgkii	lt: G; ilt: BM G K NY P	Roraima
753	1348	1842–43	CLU	Tovomita schomburgkiana Klotzsch [n.n.]	Tovomita schomburgkii Planch. & Triana	tc: BM P	Roraima
754	1375	1842–43	APO	Tabernaemontana undulata Vahl	Tabernaemontana undulata Vahl	BM	Roraima
755	1458		CNV	Operculina pterodes Choisy	Operculina hamiltonii (G. Don) Austin & Staples		Roraima

Index 3. Robert Schomburgk's second collection series (1841–1844) – continued

756	1396	1843	MIM	Inga graciliflora Benth.	Inga graciliflora Benth.	ht: K; it: BM NY P US	Roraima
757	1478		CLU	Mammea americana L.			
758	1472		CMB	Cacoucia coccinea Aubl.	Combretum cacoucia (Baill.) Exell ex Sandw.	K P W	Roraima
759	1345		OLC	Olax schomburgkii Klotzsch [n.n.]	Dulacia guianensis (Engl.) Kuntze	ht: B; it: F K	
760	1451		LCS	Lacistema macrophylla Klotzsch [n.n.]	Lacistema aggregatum (Bergius) Rusby	BM G K NY	
761	1511		MLP	Tetrapterys crispa A. Juss.	Tetrapterys crispa A. Juss.		
762	1531		CMM	Commelina platyphylla Klotzsch [n.n.]	Commelina platyphylla Klotzsch ex C.B. Clarke ('Commelyna')		
763	1532		CNV	Lysiostyles scandens Benth.	Lysiostyles scandens Benth.	tc: BM F K P U W	
764	1497		LOR				
765	1508	1842–43	LAU	Acrodiclidium oppositifolium Nees	Licaria oppositifolia (Nees) Kosterm.	G	
765*	1394						
766	1495		FAB	Machaerium leiophyllum (DC.) Benth.	Machaerium leiophyllum (DC.) Benth.	GH NY P	
767							
768	1388	1841	ELC	Sloanea dentata L.	Sloanea grandiflora Sm.	BM	
768	1428		BIG	Lundia schomburgkii Klotzsch [n.n.]	Callichlamys latifolia (Rich.) K. Schum.		
769	1448		LOG	Strychnos toxifera R.H. Schomb. ex Benth. var. latifolia Klotzsch ex Prog.	Strychnos toxifera R.H. Schomb. ex Benth.		
770	1465		LOG	Strychnos toxifera R.H. Schomb. ex Benth. var. obliqua Klotzsch ex Prog.	Strychnos toxifera R.H. Schomb. ex Benth.		
771	1447	1842–43	MEL	Guarea aubletii A. Juss.	Guarea guidonia (L.) Sleumer	BM	Roraima
772	1518	1841	SPT	Sideroxylon elegans A. DC. var. micranthum Miq.	Pouteria cuspidata (A. DC.) Baehni subsp. cuspidata	tc: BM G K NY P W	
773	1440	1841	ELC	Sloanea schomburgkii Benth.	Sloanea schomburgkii Benth.	tc: G	
773*	1440*	1841	ELC	Sloanea sinemariensis Aubl.	Sloanea sinemariensis Aubl.	BM	
774	1520	1842–43	VIO	Alsodeia pubiflora Benth.	Rinorea pubiflora (Benth.) Sprague & Sandw. var. pubiflora	CGE F G K P	Roraima
774*	1520*		VIO	Alsodeia flavescens Spreng.	Rinorea pubiflora (Benth.) Sprague & Sandw. var. grandifolia (Eichl.) Hekking	CGE G K	Roraima
775	1516	1842–43	LOG	Strychnos smilacina Benth.	Strychnos mitscherlichii R.H. Schomb.	st: F G	
776	1519	1842–43	EUP	Gaedawakka schomburgkiana Kuntze	Chaetocarpus schomburgkianus (Kuntze) Pax & K. Hoffm.	tc: BM K	Roraima
777	1533	1841	CSL	Cynometra marginata Benth. var. guianensis Dwyer	Cynometra marginata Benth. var. guianensis Dwyer	ht: NY; it: BM K V W	
778	1341		ORC	Cypripedium palmifolium Lindl.	Selenipedium palmifolium (Lindl.) Rchb.f.		
779	1434	1842–43	MEL	Guarea schomburgkii C. DC.	Guara glabra Vahl	lt: G; ilt: BM G K NY P	Roraima

780	1509	1842–43	SPT	Mimusops balata (Aubl.) Gaertn.	Manilkara bidentata (A. DC.) A. Chev. subsp. bidentata	lt: P; ilt: BM K NY	Roraima
781	1534		STR	Sterculia ivira Sw.	Sterculia pruriens (Aubl.) K. Schum. var. pruriens	BM	
782			MIM	Inga sertulifera DC.	Inga jenmanii Sandw.	K P	
783	1431		ACA	Mendoncia schomburgkiana Nees	Mendoncia hoffmannseggiana Nees	P W	Roraima
784	1430		EUP	Dalechampia büttnerioides Klotzsch [n.n.]	Dalechampia micrantha Poepp. & Endl.	BM	
785	1426	1842–43	APO	Rauvolfia micrantha Klotzsch [n.n.] ('Rauwolfia')	Condylocarpon intermedium Muell. Arg. var. intermedium	lt: G; ilt: BM	Roraima
786	1387		MLS	Miconia erythropila Steud.	Miconia ceramicarpa (DC.) Cogn. var. ceramicarpa		
787	1407		MLV	Hibiscus verbasciformis Klotzsch	Hibiscus verbasciformis Klotzsch ex Hochr.	BM	
788	1378		APO	Malouetia guianensis Klotzsch [n.n.]	Malouetia tamaquarina (Aubl.) A. DC.		
789	1340		ANA	Tapirira guianensis Aubl.	Tapirira guianensis Aubl.		
790	1350	1841	CLU	Clusia insignis Mart.	Clusia grandiflora Splng.	BM	
791	1432		LCY	Lecythis chartacea O. Berg	Lecythis chartacea O. Berg		
792	1435	1841	CUC	Anguira guianensis Klotzsch [n.n.]	Gurania bignoniacea (Poepp. & Endl.) C. Jeffrey	BM	
793	1406	1841	ANA	Tapirira guianensis Aubl.	Tapirira guianensis Aubl.	BM	
794	1421	1841	MEL	Trichilia guianensis Klotzsch [n.n.]	Trichilia rubra C. DC.	lt: G; ilt: BM G K NY P	
795	1417	1842–43	PGL	Moutabea guianensis Aubl.	Moutabea guianensis Aubl.	BM G K P W	
796		1842–43	BIG	Tabebuia rufinervis DC.	Callichlamys latifolia (Rich.) K. Schum.	BM	Roraima
797	1438		APO	Thyrsanthus adenobasis (Muell. Arg.) Miers	Forsteronia adenobasis Muell. Arg.		
798	1424		LAU	Aydendron aciphyllum Nees	Rhodostemonodaphne kunthiana (Nees) Rohwer		
799	1433		MIM	Piptadenia guianensis Benth.	Stryphnodendron pulcherrimum (Willd.) Hochr.		
800	1367		CLU	Vismia macrophylla Kunth	Vismia macrophylla Kunth	BM	
801	1362	1842–43	HUG	Roucheria schomburgkii Planch.	Roucheria schomburgkii Planch.	BM G	
802	1370		SCR	Vandellia diffusa L.	Lindernia diffusa (L.) Wettst.		
803	1504		LOR	Struthanthus flexistylus Miq.	Phthirusa retroflexa (Ruiz & Pav.) Kuijt		
804	1535	1842–43	CNN	Connarus punctatus Planch.	Connarus punctatus Planch.	G K P W	Roraima
805	1488		EUP	Stilaginella oblonga Tul.	Hyeronima oblonga (Tul.) Muell. Arg.	ht: P; it: F	
806	1512		RUB	Gonzalea spicata DC.	Gonzalagunia spicata (Lamb.) Gómez	BM	
807	1420		CSL	Schnella			
808	1493		MLS	Rhynchanthera	Rhynchanthera dichotoma (Desr.) DC.		
809	1392		LAU	Oreodaphne costulata Nees	Ocotea aciphylla (Nees) Mez	BM K P MEL NY	
810	1427		MIM	Inga sertulifera DC.	Inga jenmanii Sandw.	lt: K; ilt: BM NY P	
811	1408	1843	MLP	Byrsonima altissima DC. var. occidentalis Nied.	Byrsonima aerugo Sagot	BM	Roraima
812	1487		MRT	Calyptranthes obtusa O. Berg	Marliera montana (Aubl.) Amshoff	BM	
813	1507	1842–43	SPT	Chrysophyllum auratum Miq. var. majus Miq.	Chrysophyllum argenteum Jacq. subsp. auratum (Miq.) T.D. Penn.	tc: BM K NY U	Roraima
813*			SAP	Paullinia bipinnata Poir.	Paullinia leiocarpa Griseb.		

Index 3. Robert Schomburgk's second collection series (1841–1844) – continued

No.	Coll.	Year	Code	Name	Determination	Herbaria	Roraima
814	1499		SAP	Monopteris guianensis Klotzsch [n.n.]	Matayba arborescens (Aubl.) Radlk.	BM	Roraima
815	1330	1842–43	CLU	Vismia macrophylla Kunth	Vismia macrophylla Kunth	BM	Roraima
815*	1481	1842–43	MEL	Carapa	Machaerium quinata (Aubl.) Sandw. var. quinata	BM G GH K NY P W	
816	1479		FAB	Machaerium ferrugineum Pers.	Hevea pauciflora (Spruce ex Benth.) Muell. Arg.		
817	1381		EUP	Siphonia schomburgkii Klotzsch [n.n.]	Hippocratea volubilis L.	BM W	Roraima
818	1379	1842–43	HPC	Hippocratea ovata Lam.	Stigmaphyllon puberum (Rich.) A. Juss.	BM W	Roraima
819	1500	1842–43	MLP	Stigmaphyllon puberum (Rich.) A. Juss. var. schomburgkianum Benth.	Zapoteca portoricensis (Jacq.) H. Hern.	ht: K: it: BM G	
820	1515		MIM	Calliandra portoricensis (Willd.) Benth.	Forsteronia guyanensis Muell. Arg.		
821	1466	1842–43	APO	Thyrsanthus guyanensis (Muell. Arg.) Miers	Zygia latifolia (L.) Fawc. & Rendle var. latifolia	tc: BM NY W	Roraima
822	1415		MIM	Pithecellobium cauliflorum (Willd.) Mart. ('Pithecolobium')	Licania affinis Fritsch		
822*	1361		CHB	Licania schomburgkii Klotzsch [n.n.]	Psychotria	ht: W: it: CGE G K P	
823	1439		RUB	Psychotria	Licania heteromorpha Benth. var. perplexans Sandw.	BM	
824	1443		CHB	Licania heteromorpha Benth.	Humiriastrum obovatum (Benth.) Cuatrec.	ht: K: it: BM NY	
825	1359		HUM	Humiria obovata Benth. ('Humirium obovatum')			
826	1395		VIO	Amphirrhox	Sida urens L.		
827	1376		MLV	Sida urens L.	Ocotea cernua (Nees) Mez	BM	
828	1404		LAU	Oreodaphne caudata Nees	Inga leiocalycina Benth.		
829	1391	1842–43	MIM	Inga leiocalycina Benth.	Malouetia flavescens (Willd. ex Roem. & Schult.) Muell. Arg.	lt: P; ilt: BMG, K MEL MO US	
830	1386		APO	Malouetia schomburgkii Muell. Arg.	Inga meissneriana Miq.		
831	1423	1845	MIM	Inga spuria Willd.	Aniba riparia (Nees) Mez	K MEL P	
832	1405		LAU	Aydendron riparium Nees	Pterocarpus rohrii Vahl	BM G K NY	Roraima
833	1329		FAB	Pterocarpus rohrii Vahl	Bacopa aquatica Aubl.	K P	
834	1369		SCR	Bacopa aquatica Aubl.	Rourea frutescens Aubl.	BM F K NY P W	
835	1416		CNN	Rourea frutescens Aubl.	Guatteria schomburgkiana Mart.		
836	1334	1844	ANN	Guattaria vestita Klotzsch var. latifolia Klotzsch [n.n.]	Vismia laxiflora Reidhardt	BM K P	
837	1456	1842–43	CLU	Vismia falcata Rusby	Swartzia grandifolia Bong. ex Benth.	ht: W: it: BM NO	Roraima
838	1441		FAB	Tournatea grandifolia (Bong. ex Benth.) Taub.	Inga nobilis Willd.	BM CGE K W	
839	1419		MIM	Inga corymbifera Benth.	Hirtella racemosa Lam. var. racemosa	K P	
840			CHB	Hirtella strigulosa Steud.	Crudia glaberrima (Steud.) J.F. Macbr.	BM F K NY P W	
841	1453		CSL	Crudia falcata Klotzsch [n.n.] ('Crudya')		BM	

No.	No.	Family	Original name	Accepted name	Herbarium / type	Locality
842	1398	BOR	Cordia melanoneura Klotzsch in M.R. Schomb. [n.n.]	Cordia exaltata Lam. var. melanoneura I.M. Johnst.	K	Roraima
843	1425	GSN	Episcia mimuloides Benth.	Nautilocalyx mimuloides (Benth.) C. Morton		Roraima
844	1496	MLS	Miconia alternans Naudin	Miconia alternans Naudin		
845	1454	FAB	Drepanocarpus schomburgkii Klotzsch [n.n.]	Machaerium inundatum (Mart. ex Benth.) Ducke		
845*	1328	VRB	Avicennia nitida Jacq.	Avicennia germinans (L.) Stearn	BM F FI G NY P R	Roraima
846	1418	CMB	Terminalia tanibouca Rich.	Terminalia dichotoma G. Mey.	BM	Roraima
847	1449 (1842–43)	FLC	Casearia javitensis Kunth	Casearia javitensis Kunth		
848	1347	MRS	Weigeltia guianensis Klotzsch [n.n.]	Cybianthus surinamensis (Spreng.) G. Agostini		
849	1331	FAB	Clitoria arborescens R. Br.	Clitoria arborescens R. Br.	lt: NY; ilt: BM P	Roraima
850	1413 (1842–43)	DSC	Dioscorea schomburgkiana Kunth [n.n.]	Dioscorea pilosiuscula Bert. ex Spreng.	BM P	Roraima
851	1471 (1842–43)	DSC	Dioscorea riparia Kunth & M.R. Schomb. [n.n.]	Dioscorea samydea Mart. ex Griseb.		
852	1474	MIM	Entada myriadena Benth.	Mimosa myriadenia (Benth.) Benth. var. myriadenia	BM	Roraima
853	1473 (1842–43)	RHZ	Cassipourea guianensis Aubl.	Cassipourea guianensis Aubl.	BM K	Roraima
854	1530 (1842–43)	AST	Vernonia scorpioides (Lam.) Pers.	Cyrtocymura scorpioides (Lam.) H. Rob.		
855	1490	LOR		Gonzalagunia	BM	Roraima
856	1529 (1842–43)	RUB	Gonzalea	Corchorus aestuans L.		
857	1486	TIL	Corchorus acutangulus Lam.	Casearia decandra Jacq.	BM G K NY P W	Roraima
858	1528 (1842–43)	FLC	Casearia parvifolia Willd.			
859	1526	SOL	Solanum	Canna glauca L.		
860	1321	CAN	Canna glauca L.	Vismia guianensis (Aubl.) Choisy	BM	Roraima
861	1475 (1842–43)	CLU	Vismia guianensis (Aubl.) Choisy	Clusia myriandra (Benth.) Planch. & Triana	BM	Roraima
862	1354 (1842–43)	CLU	Quapoya ligulata Klotzsch [n.n.]	Casearia decandra Jacq.	BM	Roraima
863	1429 (1842–43)	FLC	Casearia parvifolia Vahl	Casearia guianensis (Aubl.) Urb.	BM FI G K NY W	Roraima
863*	1429* (1842–43)	FLC	Casearia ramiflora Vahl	Chrysophyllum argenteum Jacq. subsp. auratum (Miq.) T.D. Penn.	tc: BM BR F G GH K NY P US W	Roraima
864	1389	SPT	Chrysophyllum auratum Miq.	Melochia melissaefolia Benth.	BM	Roraima
865	1405* (1842–43)	STR	Melochia melissaefolia Benth.	Calycolpus goetheanus (Mart. ex. DC.) O. Berg	BM	Roraima
866	1476 (1842–43)	MRT	Campomanesia glabra Benth.			
867	1384	FAB	Centrosema	Licania leptostachya Benth.	ht: NY; it: K IAN P	
868	1279	CHB	Licania axilliflora Hochr.	Ilex umbellata Klotzsch ex Reissek var. humirioides (Reissek) Loes.	st: BM K NY P W	
869	1358 (1841)	AQF	Ilex umbellata Klotzsch [n.n.]			
870	1379* (1841)	MLP	Byrsonima rugosa Benth.	Byrsonima stipulacea A. Juss.	ht: K; it: BM G K P	Roraima
871	1480	CMB	Combretum glabrum DC.	Combretum glabrum DC.	BM G K P W	
872	1498	CMB	Combretum laxum Jacq.	Combretum laxum Jacq.	BM CGE G K P W	
873	1501	VRB	Aegiphila macrantha Ducke	Aegiphila macrantha Ducke		
874	1505 (1842–43)	MLP	Banisteria lobulata E. Mey.	Banisteriopsis lucida (Rich.) Small	BM K	

Index 3. Robert Schomburgk's second collection series (1841–1844) – continued

875	1510	1841	BOR	Cordia guianensis Klotzsch [n.n.]	Cordia fallax I.M. Johnst.	BM K P	
876	1366	1841	CEC	Coussapoa fagifolia Klotzsch	Coussapoa microcephala Trécul	ht: P; it: BM F Fl G NY P	
877	1393	1841	SAP	Cupania velutina Klotzsch [n.n.]	Cupania hirsuta Radlk.	tc ?: GH	
878	1365		CLU	Tovomita macrophylla Klotzsch [n.n.]	Tovomita obovata Engl.	tc: B BM K	
879	1521		EUP	Stilaginella laxiflora Tul.	Hyeronima alchorneoides Allemão var. alchorneoides	st: K P	
880	1522		PLG	Triplaris vahliana Fisch. & Mey. ex C.A. Mey.	Triplaris weigeltiana (Rchb.) Kuntze		
881	1523		CSL	Cassia bacillaris L.f.	Senna sandwithiana H.S. Irwin & Barneby		Roraima
882	1492		EBN	Diospyros paralea Steud.	Diospyros guianensis (Aubl.) Gürke		
883	1463	1841	FLC	Homalium puberulum Klotzsch [n.n.]	Homalium guianense (Aubl.) Oken	tc: BM G K P W	
884	1352	1841	MRT	Eugenia tapacumensis O. Berg	Eugenia tapacumensis O. Berg	BM	
885	1460	1841	DSC	Dioscorea megalobotrya Kunth ex M.R. Schomb. ex Kunth	Dioscorea amazonum Mart. var. consanguinea (Kunth) Uline	BM P	
886	1462	1841	PON	Heteranthera grandiflora Klotzsch [n.n.]	Eichhornia diversifolia (Vahl) Urb.	BM G NY P	
887	1485	1841	MRS	Weigeltia schomburgkiana Mez	Cybianthus schomburgkianus (Mez) G. Agostini	tc: G	
888	1461		ACA	Aphelandra deppeana Schltdl. & Cham.	Aphelandra scabra (Vahl) Sm.		
889	1467		MLS	Diplochita fothergilla DC.	Miconia mirabilis (Aubl.) L.O. Williams		
890	1450		MLS	Melastoma chrysophyllum Rich.	Miconia chrysophylla (Rich.) Urb.	ht: P	
891	1459	1841	MNS	Anomospermum schomburgkii Miers	Orthomene schomburgkii (Miers) Barneby & Krukoff	BM	
892	1442	1842–43	ANA	Tapirira marchandii Engl.	Tapirira obtusa (Benth.) J.D. Mitch.	lt: K; ilt: NY	
893	1401	1842–43	CLU	Tovomita guianensis Aubl.	Tovomita brevistaminea Engl.	tc: BM	Roraima
894	1422		FAB	Swartzia schomburgkii Benth.	Swartzia schomburgkii Benth. var. schomburgkii	ht: K; it: BM CGE FI G NY W	Roraima
895	1374		LAU	Nectandra leucantha Nees	Nectandra globosa (Aubl.) Mez	st: BM G K NY P	
896	1452		LAU	Nectandra vaga Meisn.	Nectandra globosa (Aubl.) Mez	st: G K L	
897	1368		MLS	Clidemia divaricata Naudin	Leandra divaricata (Naudin) Cogn.	ht: P	
898	1468	1842–43	BRS	Icica decandra Aubl.	Protium decandrum (Aubl.) Marchand	BM	
899	1323		PED	Sesamum			
899*		1841	CLU	Quapoya panapanari Aubl.	Clusia panapanari (Aubl.) Choisy		Roraima
900	1455		CLU	Quapoya microphylla Klotzsch [n.n.]	Clusia panapanari (Aubl.) Choisy	BM	
901	1333	1842–43	RUB	Sabicea velutina Benth.	Sabicea velutina Benth.		
902	1360	1841	VOC	Vochysia schomburgkiana Klotzsch [n.n.]	Vochysia schomburgkii Warm.	BM	
903	1397	1842–43	PIP	Piper arboreum Aubl.	Piper arboreum Aubl.	ht: K; it: BM G P W	Roraima
904	1371		AST	Sparganophorus vaillantii Gaertn.	Struchium sparganophorum (L.) Kuntze	BM	Roraima

905	1382		CHB	Moquilea guianensis Aubl.	Licania guianensis (Aubl.) Griseb.	BM K NY P	
906	1414	1841	MLV	Hibiscus furcellatus Lam.	Hibiscus furcellatus Desr.	BM	
907	1402		MYS	Myristica sebifera Sw.	Virola sebifera Aubl.		
908	1357	1842–43	CHB	Licania heteromorpha Benth.	Licania heteromorpha Benth. var. heteromorpha	BM K NY P	Roraima
909							
910	1470	1841	SPT	Sideroxylon durum Klotzsch [n.n.]	Pouteria cuspidata (A. DC.) Baehni subsp. dura (Eyma) T.D. Penn.	BM	
911	1513	1841	ASC	Asclepias	Matelea delascioi Morillo	K P	
912	1336*	1842–43	RUB	Amaioua surinamensis Steud.	Duroia eriopila L.f.	BM	Roraima
913	1337	1841	CEL	Goupia glabra Aubl.	Goupia glabra Aubl.	BM	
914	1337		CSL	Tachigali paniculata Aubl. ('Tachigalia')	Tachigali paniculata Aubl. var. alba (Ducke) Dwyer		
915	1483		ANA	Tapirira multiflora Mart.	Tapirira guianensis Aubl.		
916	1482		ANA	Tapirira multiflora Mart.	Tapirira guianensis Aubl.		
917			CLU	Vismia sessilifolia (Aubl.) Choisy	Vismia sessilifolia (Aubl.) Choisy	BM CGE Fl G	Roraima
918	1443*		MIM	Inga marginata Willd.	Inga semialata (Vell.) Mart.	K P	Roraima
919	1525		ANA	Spondias lutea L.	Spondias mombin L.		
920	1237	1841	EUP	Discocarpus essequiboensis Klotzsch	Discocarpus essequiboensis Klotzsch	F G K P W	
920*	1503		MLS	Miconia	Miconia prasina (Sw.) DC.		
921	1236		MLS	Miconia aplostachya (Bonpl.) DC.	Miconia aplostachya (Bonpl.) DC.		
922	1278	1841	MRT	Eugenia incanescens Benth.	Eugenia incanescens Benth.	BM	
923	1256		VRB	Vitex umbrosa Sw.	Vitex compressa Turcz.		
924	1268	1841	RUB	Chomelia tenuiflora Benth.	Chomelia schomburgkii Steyerm.	ht: NY; it: BM	
925	1270	1841	ERX	Erythroxylum ectinocalyx Klotzsch ('Erythroxylon')	Erythroxylum divaricatum Peyr.	BM	
926		1841	RUB	Chomelia angustifolia Benth.	Chomelia angustifolia Benth.	BM	
927	1272	1842–43	MRT	Calycampe latifolia O. Berg	Myrcia calycampa Amshoff	BM	Roraima
927*	1271		RUB	Chomelia angustifolia Benth.	Chomelia angustifolia Benth.	ht: P; it: NY	
928	1286		LAU	Oreodaphne fasciculata Nees	Ocotea fasciculata (Nees) Mez		
929	1265		PLG	Coccoloba striata Benth.	Coccoloba striata Benth.	tc: NY	Roraima
930	1289	1841	RUB	Coussarea schomburgkiana (Benth.) Benth. & Hook.f.	Coussarea violacea Aubl.	BM	
931	1300	1842–43	SYM	Symplocos ciponima L'Hér.	Symplocos guianensis (Aubl.) Guerke	BM	Roraima
932	1297	1842–43	RUB	Chomelia	Chomelia	BM	Roraima
933	1290	1842–43	FLC	Casearia spinosa (L.) Willd.	Casearia aculata Jacq.	BM K P W	Roraima
934	1311	1842–43	MLS	Miconia plebeia Naudin	Miconia brevipes Benth.	ht: P; it: NY	Roraima
935	1301	1842–43	CLU	Hypericum cayennense Jacq.	Vismia cayennensis (Jacq.) Pers.	BM	Roraima
936	1306	1842–43	MRT	Myrcia kegeliana O. Berg	Myrcia fallax (Rich.) DC.	BM	Roraima
937	1282	1841	VIO	Alsodeia laxiflora Benth.	Rinorea brevipes (Benth.) S.F. Blake	CGE G K P W	Roraima

Index 3. Robert Schomburgk's second collection series (1841–1844) – continued

938	1283	1842–43	FLC	Casearia densiflora Benth.	Casearia commersoniana Cambess.	BM F G GH NY P W	Roraima
939	1273	1842–43	EUP	Dactylostemon schomburgkii Klotzsch	Actinostemon schomburgkii (Klotzsch) Hochr.	BM K NY	Roraima
940	1253	1842–43	SOL	Solanum microcalyx Klotzsch [n.n.]	Solanum campaniforme Roem. & Schult.	BM	Roraima
941	1252	1842–43	MRT	Psidium parviflorum Benth.	Psidium sartorianum (O. Berg) Nied.	BM	Roraima
942	1302		ANN	Rollinia tinifolia Klotzsch [n.n.]	Rollinia exsucca (DC. ex Dunal) A. DC.		
943	1102		MRT	Eugenia			
944	1292		MRT	Eugenia roraimana O. Berg	Eugenia roraimana O. Berg	BM	
945	1246	1842–43	MRT	Aulomyrcia curatellaefolia (DC.) O. Berg var. parvifolia O. Berg	Myrcia tomentosa (Aubl.) DC.	tc: BM G	
946	1298		RUB	Genipa caruto Kunth	Genipa americana L.	BM	
947	1262		PLG	Coccoloba lucidula Benth.	Coccoloba lucidula Benth.	tc: GH	
948	1279		CHB	Licania leptostachya Benth.	Licania leptostachya Benth.	BM	
949			RUB	Faramea	Faramea	BM	
950	1257	1843	MYS	Myristica fatua Sw.	Virola surinamensis (Rol.) Warb.	G P	
951	1238		AMA	Bucholtzia philoxeroides Mart.	Alternanthera philoxeroides (Mart.) Griseb.		
951*	1251		TRG	Trigonia laevis Aubl.	Trigonia laevis Aubl. var. microcarpa (Sagot ex Warm.) Sagot	BM	
952	1269		MRT	Psidium ciliatum Benth.	Psidium sartorianum (O. Berg) Nied.	tc: BM G	
953	1284	1843	APO	Thyrsanthus gracilis Benth.	Forsteronia gracilis (Benth.) Muell. Arg.	BM	Rupununi R.
953*	1317		FAB	Deguelia scandens Aubl.	Derris pterocarpa (DC.) Killip		
954	1264		MLS	Miconia mucronata (Desr.) Naudin	Miconia holosericea (L.) DC.		
954*	1258		SPT	Mimusops balata (Aubl.) Gaertn.	Manilkara bidentata (A. DC.) A. Chev. subsp. bidentata	BM	
955	1235		AQF	Ilex macoucou Pers.	Ilex guianensis (Aubl.) Kuntze	BM P	
955*	1303		MLS	Miconia mucronata (Desr.) Naudin	Miconia holosericea (L.) DC.		
956	1299		ANA				
956*			MLS	Davya sclerophylla Naudin	Meriania sclerophylla (Naudin) Triana	ht: P	near Mt. Roraima
957	1318		CSL	Tachigali ('Tachigalia')			
958	1316		MRT	Myrciaria verticillata O. Berg	Myrciaria floribunda (West ex Willd.) O. Berg	tc: G	
959	1288	1845	ANN	Duguetia quitarensis Benth.	Duguetia quitarensis Benth.	G P S W	
960	1248		MRT	Myrcia schomburgkiana O. Berg	Myrcia servata McVaugh	BM	
961							
962			RUB	Psychotria	Psychotria	BM	
963		1841	ERI	Ledothamnus guyanensis Meisn.	Ledothamnus guyanensis Meisn.		Roraima
964		1841	ERI	Cavendishia duidae A.C. Sm.	Cavendishia callista J.D. Sm.	BM	

No.	Coll.	Year	Fam.	Name (as collected)	Accepted name	Herbarium	Region
964*	1541	1842–43	MRT	Myrcia bracteata (Rich.) DC.	Myrcia bracteata (Rich.) DC.	BM	Roraima
965			MRS	Cybianthus	Bonyunia minor N.E. Br.	P	Roraima
966			LOG	Bonyunia minor N.E. Br.			
967							
968	1553	1842–43	RUB	Perama dichotoma Poepp. & Endl.	Perama dichotoma Poepp. & Endl.	BM	Roraima
969		1841	OCH	Leitgebia guianensis Eichl.	Sauvagesia guianensis (Eichl.) Sastre	BM	
970			ROS				
971			ORC				
972							
973							
974		1841	VOC	Vochisia costata Warm.	Vochysia costata Warm.	lt: P; ilt: BM G	
975		1842–43	CLU	Caraipa tereticaulis Tul.	Caraipa tereticaulis Tul.	BM	Roraima
976		1842–43	CHB	Licania macrophylla Klotzsch [n.n.]	Licania laxiflora Fritsch	ht: W; it: BM K P	Roraima
977							
978							
979	1605	1841	MRT	Calyptranthes fasiculata O. Berg	Calyptranthes fasiculata O. Berg	BM	
980			CHB	Parinari campestris Aubl. ('Parinarium campestre')	Parinari campestris Aubl. ('Parinarium campestre')	BM P	
981							
982	1581		HUM	Vantanea guianensis Aubl.	Vantanea guianensis Aubl.	BM NY P	Roraima
983	1582		MEL	Moschoxylon propinquum Miq.	Trichilia quadrijuga Kunth subsp. quadrijuga	BM	
984							
985							
986	1599	1842–43	CHB	Licania heteromorpha Benth.	Licania heteromorpha Benth. var. heteromorpha	BM K NY P W	Roraima
987	1604		CNN	Bernardia guianensis Schellenb.	Cnestidium guianense (Schellenb.) Schellenb.	lt: G; ilt: G K P W	
988	1722		CSL	Parivoa	Eperua grandiflora (Aubl.) Benth. subsp. guyanensis R.S. Cowan		
989	1701	1842–43	RUB	Palicourea guianensis Aubl.	Palicourea guianensis Aubl.	BM	Roraima
990	1702		CLU	Garcinia macrophylla Mart.	Rheedia benthamiana Planch. & Triana	st: BM	
991	1713	1843	MYS	Myristica sebifera Sw.	Virola sebifera Aubl.	G NY	
992	1710		SAP	Urvillea schomburgkii Klotzsch [n.n.]	Serjania paucidentata DC.		
993	1716	1844	ANN	Guatteria vestita Klotzsch var. angustifolia Klotzsch [n.n.]	Guatteria schomburgkiana Mart.	B BM K NY P	
994	1704	1842–43	CUC	Anguria polyanthos Klotzsch [n.n.]	Gurania subumbellata (Miq.) Cogn.	BM	Roraima
995	1721		MRS	Conomorpha magnoliifolia Mez	Cybianthus fulvopulverulentus (Mez) G. Agostini subsp. magnoliifolius (Mez) Pipoly		
996	1712	1842–43	CNN	Omphalobium opacum Klotzsch [n.n.]	Connarus coriaceus Schellenb.	lt: W; ilt: BM F G K M P W	Roraima
997	1733	1842–43	SCR	Alectra brasiliensis Benth.	Melasma melampyroides (Rich.) Pennell	BM	

132

Index 3. Robert Schomburgk's second collection series (1841–1844) – continued

998	1718		ACA	Trichanthera gigantea (Humb. & Bonpl.) Nees	Trichanthera gigantea (Humb. & Bonpl.) Nees var. guianensis Gleason	BM	Roraima
999	1732	1842–43	GUT	Renggeria montana Klotzsch [n.n.]	Quapoya bracteolata Sandw.	BM	Roraima
1000	1723		FAB	Clitoria javitensis (Kunth) Benth.	Clitoria javitensis (Kunth) Benth. var. javitensis		
1001	1737*		RUB	Commianthus schomburgkii Benth.	Retiniphyllum schomburgkii (Benth.) Muell. Arg.	BM	Roraima
1002	1729	1842–43	RUB	Malanea angustifolia Bartl. [n.n.]	Malanea obovata Hochr.	ht: NY; it: US	Roraima
1003	1720		LAU	Aiouea demerarensis Kosterm.	Aiouea guianensis Aubl.	tc: BM F K NY US	
1004	1703		LAU	Nectandra rodiei R.H. Schomb.	Chlorocardium rodiei (R.H. Schomb.) Rohwer, Richter & van der Werff	ht: K	
1005	1735	1842–43	MEL	Trichilia roraimana C. DC.	Trichilia micrantha Benth.	ht: G; it: BM G K NY P	Roraima
1006	1719		CHB	Moquilea guianensis Aubl.	Licania guianensis (Aubl.) Griseb.		
1007	1717	1842–43	PGL	Bredemeyera bracteata Klotzsch ex Hassk.	Bredemeyera bracteata Klotzsch ex Hassk.	ht: K; it: BM G K NY P W	Roraima
1008	1727		MLS	Acinodendron plukenetii Kuntze	Miconia plukenetii Naudin	st: P	
1009	1730	1842–43	LAU	Aydendron firmulum Nees	Beilschmiedia curviramea (Meisn.) Kosterm.	st: BM	Roraima
1010	1706		ANA	Tapirira guianensis Aubl.	Tapirira guianensis Aubl.	BM	
1011	1739	1842–43	MLP	Banisteria	Mascagnia guianensis W.R. Anderson	BM	
1012	1738		LAU	Oreodaphne schomburgkiana Nees	Ocotea schomburgkiana (Nees) Mez	BM K	
1013	1737		GNE	Gnetum nodiflorum Brongn.	Gnetum nodiflorum Brongn.		

INDEX 4. ROBERT SCHOMBURGK'S RORAIMA COLLECTION SERIES (1842–1843).

Nr.	Year	Fam.	Original name	Current name	Herbarium	Location
1	Dec 1842	LAU	Acrodiclidium camara R.H. Schomb.			savanna near Roraima
2	Dec 1842	VOC				savanna near Roraima
3	Dec 1842	ANA	Tapirira guianensis Aubl.	Tapirira guianensis Aubl.		savanna near Roraima
4	Dec 1842	CLU	Caraipa			savanna near Roraima
5	Dec 1842	ANA	Anacardium rhinocarpus DC.	Anacardium giganteum Hanc. ex Engl.		savanna near Roraima
6						savanna near Roraima
7						savanna near Roraima
8						savanna of Annawai valley
9	Dec 1842	MIM	Calliandra hookeriana R.H. Schomb.	Calliandra rigida Benth.	K	savanna of Annawai valley
10		HUM	Vantanea emarginata Klotzsch [n.n.]	Vantanea minor Benth.		savanna of Annawai valley
11	Dec 1842	AST	Gongylolepis benthamiana R.H. Schomb.		U	savanna of Annawai valley
12		TNR	Turnera ulmifolia L.	Turnera ulmifolia L. var. intermedia Urb.		savanna of Annawai valley
13		MLS			ht: K	savanna NW of Roraima
14	Dec 1842	EPR	Lightia guianensis R.H. Schomb.	Euphronia guianensis (R.H. Schomb.) Hallier.f.	lt: K; ilt: NY	Roraima
15	Dec 1842	ERI	Ledothamnus guyanensis Meisn.	Ledothamnus guyanensis Meisn.		savanna of Annawai valley
16						savanna of Annawai valley
17		SAP				savanna of Annawai valley
18						savanna of Annawai valley
19		VOC	Qualea muelleriana M.R. Schomb. [n.n.]	Ruizterania rigida (Stafleu) Marc.-Berti		savanna of Annawai valley
20		BML	Anoplophytum guianense Beer	Lindmania guianensis (Beer) Mez var. guianensis		savanna of Annawai valley
21						forest North of Roraima
22						forest North of Roraima
23		ORC				forest North of Roraima
24		CHB	Licania			forest North of Roraima
25		FAB	Dipteryx			forest North of Roraima
26						forest North of Roraima
27		RUB	Psychotria			between Roraima and Cuyuni R.
28	Jan 1843	GEN	Tachia schomburgkiana Benth.	Tachia schomburgkiana Benth.	syntype	near Roraima
29		LYT	Cuphea			near Roraima
30		ORC				near Roraima
31						near Roraima
32						near Roraima
33						near Roraima

Index 4. Robert Schomburgk's Roraima collection series (1842–1843) – continued

No.	Date	Code	Schomburgk name	Current determination	Type/K	Locality
34						
35		ERI	Orthaea hispida A.C. Sm.	Orthaea hispida A.C. Sm.		near Roraima
36		LCY	Lecythis		type	valley
37						near Roraima
38						near Roraima
39						near Roraima
40	Sep 1843					Wanamu R. and Upper Courantyne R.
41	Sep 1843	SAP	Cupania affinis Klotzsch [n.n.]	Matayba macrostylis Radlk.		Wanamu R.
42	Sep 1843	CEL	Goupia glabra Aubl.	Goupia glabra Aubl.		Wanamu R.
43	Sep 1843	PLG	Coccoloba grandis Benth.	Coccoloba latifolia Lam.		Wanamu R.
44	Sep 1843	CHB	Parinari campestris Aubl. ('Parinarium campestre')			Wanamu R.
45	Sep 1843	MNS	Anomospermum schomburgkii Miers	Orthomene schomburgkii (Miers) Barneby & Krukoff		Upper Courantyne R.
46	Sep 1843	LOG	Strychnos			Upper Courantyne R.
47	1843	HUM	Vantanea guianensis Aubl.	Vantanea guianensis Aubl.	K	Cutari R. or Courantyne R.
48	Sep 1843	LOR	Loranthus smythii R.H. Schomb.			Upper Courantyne R.
49	1843	BIG	Bignonia preurei DC.	Mussatia prieurei (DC.) Bureau ex K. Schum.	K	R. Cafuini (Caphiwuin)
50	Sep 1843	MRT				Wanamu R.
51		MIM				Upper Essequibo R.

INDEX 5. RICHARD SCHOMBURGK'S COLLECTION SERIES (1841–1844).

Rich. =Rob. ser. 2	Year	Fam.	Original name	Current name	Herbarium	Location
1		LYT				
2 12		RUB	Coffea	Rudgea hostmanniana Benth.		
3		LAU				
4 24		POA	Panicum pilosum Sw.	Panicum pilosum Sw.		
5 28		MRN				
6		ADI	Doryopteris lomariacea Kunze ex Klotzsch	Doryopteris lomariacea Kunze ex Klotzsch	P US	Roraima
6*		ADI	Doryopteris lomariacea Kunze ex Klotzsch	Doryopteris conformis K.U. Kramer & R. Tryon	BM	
7 39		FAB	Nicolsonia cayennensis DC.	Desmodium barbatum (L.) Benth. & Oerst.		
8 44		RPT	Rapatea			
9		DST	Lindsaea trapeziformis Dryand.	Lindsaea lancea (L.) Bedd. var. lancea	BM	
10 81		TIL	Triumfetta althaeoides Lam.	Triumfetta althaeoides Lam.		Roraima
11						
12 75		CLU	Vismia macrophylla Kunth	Vismia macrophylla Kunth		
13 22		LAU	Oreodaphne costulata Nees	Ocotea aciphylla (Nees) Mez		
14 67		CMM	Tradescantia			
15		CYP	Cyperus odoratus L.	Cyperus odoratus L.		
16		OCH	Leitgebia guianensis Eichl.	Sauvagesia guianensis (Eichl.) Sastre		
17 59		SOL	Solanum callicarpifolium Kunth & Bouché	Solanum callicarpifolium Kunth & Bouché		
18	1841	GLC	Mertensia pectinata Willd.	Dicranopteris pectinata (Willd.) Underw.	B	
19 43		MRN	Maranta divaricata Roscoe			
19* 48		MLS	Miconia barbigera DC.	Miconia ciliata (Rich.) DC.		
19** 98		CNV				
20 54		HMP	Trichomanes polyanthos Sw.	Hymenophyllum decurrens (Jacq.) Sw.	P	
20* 93		AMA	Pupalia densiflora (Humb. & Bonpl.) Mart.	Cyathula achranthoides (Kunth) Moq.		
21 138		APO	Peschiera lorifera Miers	Tabernaemontana lorifera (Miers) Leeuwenb.		
22 29	1841	AST	Hebeclinium macrophyllum (L.) DC.	Hebeclinium macrophyllum (L.) DC.	BM	
23 117		APO	Malouetia	Tabernaemontana rupicola Benth.		
23* 85		ARS	Aristolochia rumicifolia M.R. Schomb. [n.n.]	Aristolochia rugosa Lam.		
24 83		MLS	Clidemia dentata D. Don	Clidemia dentata D. Don		
25		LNT	Utricularia adpressa Salzm. ex A. St.-Hil. & Girard	Utricularia alpina Jacq.		
26						
27 137		LYT	Crenea repens G. Mey.	Crenea maritima Aubl.		

Index 5. Richard Schomburgk's collection series (1841–1844) – continued

28					
29	149	LYG	Lygodium volubile Sw.	Lygodium volubile Sw.	
30	94	SAP	Paullinia podocarpa Klotzsch [n.n.]	Paullinia pinnata L.	
31	99	RUB	Cephaëlis		
32	84	RHZ	Cassipourea guianensis Aubl.	Cassipourea guianensis Aubl.	
33	82	RUB	Bertiera guianensis Aubl.	Bertiera guianensis Aubl.	
34					
35	101	CNV	Ipomoea guyanensis Choisy	Jacquemontia guyanensis (Aubl.) Meisn.	
36	13	ONA	Jussiaea latifolia Benth. ('Jussieua')	Ludwigia latifolia (Benth.) Hara	B
37		moss	Octoblepharum cylindricum Mont.	Octoblepharum cylindricum Mont.	
38	109	ACA	Aphelandra pulcherrima (Jacq.) Kunth	Aphelandra pulcherrima (Jacq.) Kunth	
39	92	CHB	Chrysobalanus icaco L.	Chrysobalanus icaco L.	BR K
40		PGL	Securidaca latifolia Benth.	Securidaca paniculata Rich.	P
41	121	CHB	Licania leptostachya Benth.	Licania leptostachya Benth.	
42					
43		MLP	Hiraea fulgens A. Juss. var. demerarensis A. Juss.	Hiraea faginea (Sw.) Nied.	
44	21	MLS	Spennera aquatica (Aubl.) Mart.	Nepsera aquatica (Aubl.) Naudin	
45		GMM	Polypodium confusum J. Sm.	Grammitis suspensa (L.) Proctor	
46	110	CNV	Maripa scandens Aubl.	Maripa scandens Aubl.	
47	112	PHT	Microtea debilis Sw.	Microtea debilis Sw.	
48		CYP	Scleria hirtella Sw.	Scleria hirtella Sw.	
49	133	SAP	Serjania micrantha Klotzsch [n.n.]	Serjania oblongifolia Radlk.	
50	124	MIM	Inga pezizifera Benth.	Inga pezizifera Benth.	
51		GMM	Polypodium confusum J. Sm.	Grammitis suspensa (L.) Proctor	tc: K US
52	136	MRS	Icacorea guianensis Aubl.	Ardisia guianensis (Aubl.) Mez	
53	105	MIM	Pentaclethra filamentosa Benth.	Pentaclethra macroloba (Willd.) Kuntze	
54		CHB	Licania macrophylla Klotzsch [n.n.]	Licania laxiflora Fritsch	K
55	107	RUT	Monnieria trifolia L.	Ertela trifolia (L.) Kuntze	
56		FAB	Machaerium leiophyllum (DC.) Benth.	Machaerium leiophyllum (DC.) Benth.	
57	126	MLP	Byrsonima spicata (Cav.) DC.	Byrsonima spicata (Cav.) DC.	GH
58	119	CNV	Ipomoea tammifolia L.	Jacquemontia tammifolia (L.) Griseb.	BM K
59					
60		CMM	Dichorisandra schomburgkiana Klotzsch [n.n.]	Dichorisandra hexandra (Aubl.) Standley	
61					
62		SAP	Cupania schomburgkii Klotzsch [n.n.]	Cupania hirsuta Radlk.	

No.	Ref	Year	Fam			Notes
63			IRI	Sisyrinchum brevifolium Klotzsch	Sisyrinchum vaginatum Spreng.	
64	111		APO	Odontadenia speciosa Benth.	Odontadenia macrantha (Roem. & Schult.) Markgr.	
65	123		SOL	Solanum pensile Sendtn.	Solanum pensile Sendtn.	
66						
67						
68	135		CHB	Hirtella paniculata Sw.	Hirtella paniculata Sw.	K
69	88		EUP	Euphorbia thymifolia L.	Euphorbia thymifolia L.	
70	128		MLS	Clidemia dentata D. Don	Clidemia dentata D. Don	
71	41		ULM	Sponia micrantha (L.) Decne.	Trema micrantha (L.) Blume	Demerara R.
72	104		MLS	Clidemia	Clidemia japurensis DC. var. japurensis	
73	141		MLS	Spennera circaeoides Mart.	Aciotis purpurascens (Aubl.) Triana	
74	97		AST	Clibadium surinamense L.	Clibadium sylvestre (Aubl.) Baill.	
75			PAS	Cieca appendiculata Roem.	Passiflora auriculata Kunth	
76						
77	87		MRN	Phrynium pumilum Klotzsch [n.n.]	Calathea micans (Mathieu) Körn.	
78	125		GSN	Episcia mimuloides Benth.	Nautilocalyx mimuloides (Benth.) C. Morton	
79						
80	116		RUB	Psychotria fastigiata Spreng.	Palicourea croceoides Desv.	
81	131		SML	Smilax schomburgkiana Kunth	Smilax schomburgkiana Kunth	tc: B
82	103	1841	MLP	Heteropterys platyptera DC.	Heteropterys multiflora (DC.) Hochr.	BM K
83	96	1841	MLV	Sida acuta Burm.f. or S. rhombifolia L.	Sida acuta Burm.f. or S. rhombifolia L.	Barama R.
84	132		MLS	Spennera circaeoides Mart.	Aciotis purpurascens (Aubl.) Triana	
85						
86			HMP	Hymenophyllum polyanthos (Sw.) Sw.	Hymenophyllum polyanthos (Sw.) Sw.	
87	134		RUB	Posqueria trinitatis DC.	Posqueria trinitatis DC.	
88	155		MLS	Clidemia pustulata DC.	Clidemia capitellata (Bonpl.) D. Don var. dependens (D. Don) J.F. Machr.	
89						
90	120	1841	AST	Wulffia stenoglossa (Cass.) DC.	Tilesia baccata (L.) Pruski	BM
91						
92						
93	127	1841	AST	Mikania racemosa Benth.	Mikania psilostachya DC.	BM
94	106		AST	Schomburgkia caleoides DC.	Calea caleoides (DC.) H. Rob.	
95						
96	139		ERO	Paepalanthus fasciculatus (Rottb.) Körn.	Paepalanthus bifidus (Schrad.) Kunth	tc: F K
97	108		FAB	Machaerium floribundum Benth.	Machaerium floribundum Benth.	P
98	115		MLS	Clidemia benthamiana Miq.	Clidemia hirta (L.) D. Don var. hirta	
99	129		STR	Waltheria americana L.	Waltheria indica L.	

Index 5. Richard Schomburgk's collection series (1841–1844) – continued

No.	Ref	Year	Code	Name	Current name	Note
100	200		OCH	Sauvagesia erecta L.	Sauvagesia erecta L.	
101	122		CMM	Commelina ('Commelyna')		
102						
103			EUP	Acalypha scandens Benth.	Acalypha scandens Benth.	
104	187		LAM	Hyptis atrorubens Poit.	Hyptis atrorubens Poit.	
105			SCR	Achetaria guianensis Pennell	Achetaria guianensis Pennell	
106						
107			MEL	Trichilia brachystachya Klotzsch ex C. DC.	Trichilia pallida Sw.	
108	173		VRB	Petrea schomburgkiana Schauer	Petrea bracteata Steud.	tc: K
109	209		RUB	Coccocypselum tontanea Kunth	Coccocypselum guianense (Aubl.) K. Schum.	
110	157		ACA	Thyrsacanthus schomburgkianus Nees	Odontonema schomburgkianum (Nees) Kuntze	ht: K
111			MLP	Banisteria schomburgkiana Benth.	Banisteriopsis muricata (Cav.) Cuatrec.	
112	219		MLS	Clidemia divaricata Naudin	Leandra divaricata (Naudin) Cogn.	
113	130		MLS	Miconia myriantha Benth.	Miconia myriantha Benth.	
114			CAM	Centropogon surinamensis (L.) C. Presl	Centropogon cornutus (L.) Druce	
115			EUP	Phyllanthus conami Sw.	Phyllanthus brasiliensis (Aubl.) Poir.	
116						
117	178		APO	Tabernaemontana rupicola Benth.	Tabernaemontana rupicola Benth.	
118	167		CEC	Coussapa fagifolia Klotzsch	Coussapa microcephala Trécul	B K
118*		1841				
119	206		VRB	Clerodendron longicolle G. Mey.	Clerodendrum indicum (L.) Kunze	
			VRB	Vitex umbrosa Sw.	Vitex compressa Turcz.	
120						
121						
122	207		LOG	Spigelia schomburgkiana Benth.	Spigelia flemmingiana Cham. & Schltdl.	
123	207*		MLP	Tetrapteris styloptera A. Juss.	Hiraea	
123*			MLP	Hiraea chrysophylla A. Juss.	Hiraea faginea (Sw.) Nied.	K
124						
125	168		CHB	Parinari guyanensis Fritsch	Hirtella guyanensis (Fritsch) Sandw.	
126	184	1841	AST	Mikania hookeriana DC.	Mikania hookeriana DC.	BM
127						
128	218		PLG	Coccoloba excelsa Benth.	Coccoloba excelsa Benth.	
129	182		MLS	Tococa aristata Benth.	Tococa aristata Benth.	
130			LNT	Utricularia concinna N.E. Br.	Utricularia jamesoniana Oliver	
131	175		CLU	Caraipa richardiana Cambess.	Caraipa richardiana Cambess.	K
132	156		FAB	Lonchocarpus		

No.		Year	Fam.	Name (1)	Name (2)	Herb.	Locality
133	210		CSL	Elizabetha princeps M.R. Schomb. ex Benth.	Elizabetha coccinea M.R. Schomb. ex Benth. var. oxyphylla (Harms) R.S. Cowan	K	Upper Essequibo R.
134	190		ACA	Rhytiglossa secunda Nees	Justica secunda Vahl	ht: B	
135	166	1841	HUM	Humiria obovata Benth. ('Humirium obovatum')	Humiriastrum obovatum (Benth.) Cuatrec.		
136	222		FAB	Amphymenium	Pterocarpus amazonum (Benth.) Amshoff		
137			LAU				
138	169		MLS	Clidemia surinamensis Miq.	Clidemia pustulata DC.		
139							
140	140		EUP	Phyllanthus microphyllus Kunth	Phyllanthus microphyllus Kunth		
141	176		ACA	Rhytiglossa pectoralis Nees	Justicia pectoralis Jacq.		
142	177		MLS	Miconia alata (Aubl.) DC.	Miconia alta (Aubl.) DC.		
143			SCZ	Schizaea pennula Sw.	Actinostachys pennula (Sw.) Hook.		
144	195	1841	AST	Acanthospermum xanthioides DC.	Acanthospermum australe (Loefl.) Kuntze	BM	
145	212	1841	AST	Vernonia tricholepis DC.	Lepidaploa remotiflora (Rich.) H. Rob.	BM	
146	183	1841	VRB	Aegiphila cuspidata Mart.	Aegiphila racemosa Vell.		
147	172		SPT	Sideroxylon cuspidatum A. DC. var. ellipticum Miq.	Pouteria cuspidata (A. DC.) Baehni subsp. cuspidata	tc: BR K P	
148			BMN				
149	201		LNT	Polypompholyx schomburgkii Klotzsch [n.n.]	Utricularia longeciliata A. DC.		
150	202		BMN	Burmannia bicolor Mart.	Burmannia bicolor Mart.		
151	186		RUB	Psychotria			
152	180		EUP	Amanoa guianensis Aubl.	Amanoa guianensis Aubl.		
153	38		MLS	Diplochita parviflora Benth.	Miconia pubipetala Miq.		
154			PAS	Decaloba hemicycla Roem.	Passiflora vespertilio L.		
155							
156	49	1841	CSL	Cassia multijuga Rich.	Senna multijuga (Rich.) H.S. Irwin & Barneby var. multijuga	BM	
157	191		SCR	Capraria biflora L.	Capraria biflora L.		
158	194		ERO	Tonina fluviatalis Aubl.	Tonina fluviatalis Aubl.		
159	220		SML	Smilax schomburgkiana Kunth	Smilax latipes Gleason		
160			LYT				
161	193		AQF	Ilex martiniana D. Don	Ilex martiniana D. Don	BM G P W	
162			MLP	Hiraea chrysophylla A. Juss. var. argentata Griseb.	Hiraea faginea (Sw.) Nied.	type	Essequibo R.
163	192		MRT	Eugenia incanescens Benth.			
163*	196		SAP	Cupania retusa Klotzsch [n.n.]	Cupania scrobiculata Rich.		
164	223		LOG				
165							
166							

Index 5. Richard Schomburgk's collection series (1841–1844) – continued

No.	No.	Year	Fam.	Name	Accepted name	Herb.	Locality
167	199		MLS	Henriettea multiflora Naudin	Henriettea multiflora Naudin	ht: P	
167*			MLP	Byrsonima crassifolia (L.) Kunth	Byrsonima crassifolia (L.) Kunth	W	
168	197		CSL	Cynometra spruceana Benth.	Cynometra spruceana Benth.		
169	205		APO	Malouetia tamaquarina (Aubl.) A. DC.	Malouetia tamaquarina (Aubl.) A. DC.		
170	204	1841	FLC	Homalium racoubea Sw.	Homalium guianense (Aubl.) Oken	BM P U W	
171							
172		1841	MIM	Inga tenuifolia Salzm.	Inga thibaudiana DC.	K P	
173		1841	VRB	Petrea macrostachya Benth.	Petrea macrostachya Benth.	F G W	
174		1841	VRB	Petrea schomburgkiana Schauer	Petrea bracteata Steud.	BM	
175	102		FAB	Swartzia triphylla Willd.	Swartzia arborescens (Aubl.) Pittier		
176			LOG	Spigelia schomburgkiana Benth.	Spigelia flemmingiana Cham. & Schltdl.		
177							
178			PLG				
179			AST				
180							
181			MLS	Clidemia sericea D. Don	Clidemia sericea D. Don		
182							
183			APO	Echites subcarnosa Benth.	Mandevilla subcarnosa (Benth.) Woodson		
184			BIG	Bignonia aequinoctialis L.	Cydista aequinoctialis (L.) Miers		Essequibo R.
185	122		MNY	Limnanthemum humboldtianum Griseb.	Nymphoides humboldtianum (Kunth) Kuntze		
186	165		MLS	Chaetogastra glomerata (Rottb.) Benth.	Pterolepis glomerata (Rottb.) Miq.		
187							
188	189		EUP	Euphorbia thymifolia L.	Euphorbia thymifolia L.		
189	211	1842	CMM	Tradescantia schomburgkiana Kunth	Tripogandra serrulata (Vahl) Handlos	ht: B: it: B BM	
189*	211*		POA	Panicum trichoides Sw.	Panicum trichoides Sw.		
190			GSN	Gesneria guianensis Benth.	Sinningia schomburgkiana (Kunth & Bouché) Chautems	B †	
191			LAU	Nectandra leucantha Nees	Nectandra globosa (Aubl.) Mez	type	
192							
193	217		SCR	Vandellia diffusa L.	Lindernia diffusa (L.) Wettst.		
194			MLP	Spachea elegans (G. Mey.) A. Juss.	Spachea elegans (G. Mey.) A. Juss.	syntype	Essequibo R.
195			GEN	Coutoubea ramosa Aubl.	Coutoubea ramosa Aubl.		
196			HLC	Heliconia schomburgkiana Klotzsch [n.n.]	Heliconia psittacorum L.f.	type	
197			EUP	Jatropha elegans (Pohl) Klotzsch	Jatropha gossypiifolia L.		
198	164		MLS	Mouriria guianensis Aubl.	Mouriri guianensis Aubl.		

No.	Coll.	Fam.	Name as collected	Type	Accepted name	Locality
199		MIM	Pithecellobium stipulare Benth.		Zygia latifolia (L.) Fawc. & Rendle var. communis Barneby & J.W. Grimes	
200	216	APO	Echites nitida Vahl		Odontadenia nitida (Vahl) Muell. Arg.	
201						
202						
203						
204	160	MLV	Pavonia cancellata (L.) Cav.		Pavonia cancellata (L.) Cav.	
205	163	MLV	Urena americana L.		Urena lobata L.	
206		BOR	Heliophytum indicum (L.) DC.	G	Heliotropium indicum L.	
207	386	MRT	Eugenia subobliqua Benth.		Myrcia subobliqua (Benth.) Nied.	
208		PGL	Badiera diversifolia (L.) DC.	P	Securidaca diversifolia (L.) S.F. Blake	
209		MLS	Spennera disophylla Benth.		Aciotis annua (DC.) Triana	
210		MLP	Brachypteris borealis A. Juss.	K	Stigmaphyllon bannisterioides (L.) C. Anderson	
211						
212						
213		CNN	Connarus ruber (Poepp. & Endl.) Planch.	K	Connarus ruber (Poepp. & Endl.) Planch. var. sprucei (Baker) Forero	
214						
215						
216	118	PLG	Coccoloba marginata Benth.		Coccoloba marginata Benth.	
217		PLP	Polypodium incanum Sw.	K	Polypodium polypodioides (L.) Watt	
218		ADI	Pteris collina Raddi	K	Doryopteris collina (Raddi) J. Sm.	
219		MLV	Malachra fasciata Jacq.		Malachra fasciata Jacq.	
220						
221		CYP	Cyperus aurantiacus Kunth		Cyperus amabilis Vahl	
222		VRB	Volkameria aculeata L.		Cleodendrum aculeatum (L.) Schltdl.	
223		BRS				
224		OXL	Oxalis schomburgkiana Prog.		Oxalis frutescens L.	
225		PON	Eichhornia azurea Kunth	G	Eichhornia azurea (Sw.) Kunth	
226		PON	Heteranthera grandiflora Klotzsch [n.n.]		Eichhornia diversifolia (Vahl) Urb.	
227		LYT	Lawsonia alba Lam.		Lawsonia inermis L.	
227*		PIP				
228						
229		RUB	Malanea glabrescens Bartl. [n.n.]	ht: K; it: F NY	Malanea schomburgkii Steyerm.	Barima R.
230						
231						
232						
233		ADI	Gymnogramma calomelanos (L.) Kaulf.		Pityrogramma calomelanos (L.) Link	

Index 5. Richard Schomburgk's collection series (1841–1844) – continued

No.		Year	Code	Original name	Current name	Herbarium	Locality
234	170		MLS	Diplochita fothergilla DC.	Miconia mirabilis (Aubl.) L.O. Williams		
235				Eugenia			
236			MRT	Calycolpus schomburgkianus O. Berg	Calycolpus goetheanus (Mart. ex. DC.) O. Berg	tc: B †	Essequibo R.
237			MRT				
238	215						
239			ORC	Oncidium iridifolium Kunth	Psygmorchis pusilla (L.) Dodson & Dressler		
240	171	1841	LAU	Goeppertia multiflora Miq.	Endlicheria multiflora (Miq.) Mez		
241			VTT	Antrophyum cajenense (Desv.) Spreng.	Antrophyum cajenense (Desv.) Spreng	UC	
242			LOM	Acrostichum herminieri Bory & Fée ex Fée	Elaphoglossum herminieri (Bory & Fée) T. Moore	B	
243			PLP	Taenitis desvauxii Klotzsch	Dicranoglossum desvauxii (Klotzsch) Proctor	B	
244			OLN	Nephrolepis exaltata (Sw.) Schott	Nephrolepis rivularis (Vahl) Mett. ex Krug	B BM UC	
245		1842	CTH	Alsophila ferox C. Presl	Cyathea microdonta (Desv.) Domin	B P	
246		1842	CTH	Hemitelia hostmannii Hook.	Cyathea cyatheoides (Desv.) K.U. Kramer	B	
247			PLP	Polypodium lycopodioides L.	Microgramma lycopodioides (L.) Copeland	B	
248		1841	CTH	Hemitelia hostmannii Hook.	Cyathea cyatheoides (Desv.) K.U. Kramer	B	
248*		1841	PLP	Polypodium persicariifolium Schrad. ('persicariaefolium')	Microgramma persicariifolium (Schrad.) C. Presl	B	
249		1842	LYG	Lygodium volubile Sw.	Lygodium volubile Sw.	B	Essequibo R.
250			DST	Lindsaea falcata Dryand.	Lindsaea lancea (L.) Bedd. var. falcata (Dryand.) Rosenstock	P	Amatu
251			DST	Lindsaea trapeziformis Dryand.	Lindsaea lancea (L.) Bedd. var. lancea	B L	
252		1842	ADI	Adiantum triangulatum Kaulf.	Adiantum latifolium Lam.	B	Essequibo R.
253		1842	DST	Lindsaea pendula Klotzsch	Lindsaea pendula Klotzsch	ht: B; ii: BM	
253*			DST	Lindsaea raddiana Klotzsch	Lindsaea stricta (Sw.) Dryand. var. stricta	ht: B	
254			SCZ	Schizaea trilateralis Schkuhr	Actinostachys pennula (Sw.) Hook.	B	
255		1842	HMP	Trichomanes schomburgkianum J.W. Sturm	Trichomanes pinnatum Hedw.	st: B BR	Essequibo R.
256			DST	Lindsaea dubia Spreng.	Lindsaea dubia Spreng.	B BM K	Essequibo R.
257			SCZ	Schizaea flabellum Mart.	Schizaea elegans (Vahl) Sw.	B	
258		1842	GMM	Polypodium confusum J. Sm.	Grammitis suspensa (L.) Proctor	B	Essequibo R.
259			fern				
260		1842	HMP	Hymenophyllum schomburgkii C. Presl ex J.W. Sturm	Hymenophyllum decurrens (Jacq.) Sw.	B	
261			GMM	Xiphopteris serrulata (Sw.) Kaulf.	Cochlidium serrulatum (Sw.) L.E. Bishop	B	
262*			LOM	Acrostichum alatum Fée	Elaphoglossum pteropus C. Chr.	B	
262**			LOM	Acrostichum glabellum J. Sm.	Elaphoglossum glabellum J. Sm.	B	

No.	Year	Code	Schomburgk name	Current name	Distribution	Locality
263	1841	GLC	Mertensia pectinata Willd.	Dicranopteris pectinata (Willd.) Underw.	B	Essequibo R.
264		SCZ	Schizaea elegans (Vahl) Sw.	Schizaea elegans (Vahl) Sw.	B	Essequibo R.
265		ASL	Asplenium schomburgkianum Klotzsch	Asplenium serratum L.	st: B	Essequibo R.
266	1842	ADI	Adiantum serratodentatum Willd.	Adiantum serratodentatum Willd.	ht: B	Essequibo R.
266*		ADI	Adiantum serratodentatum Willd.	Adiantum latifolium Lam.	B	
267		DST	Lindsaea reniformis Dryand.	Lindsaea reniformis Dryand.		Essequibo R.
268		DST	Lindsaea divaricata Klotzsch	Lindsaea divaricata Klotzsch	B	
269		HMP	Trichomanes heterophyllum Willd.	Trichomanes humboldtii (Bosch) Lellinger	B K	
270		OLN	Aspidium nodosum Willd.	Oleandra articulata (Sw.) C. Presl	lt: PRC, ilt: B	Essequibo R.
271		HMP	Trichomanes pellucens Kunze	Trichomanes accedens C. Presl	st	
272		DST	Lindsaea moritziana Klotzsch	Lindsaea stricta var. moritziana (Klotzsch) K.U. Kramer		
273		DST	Lindsaea moritziana Klotzsch	Lindsaea stricta var. moritziana (Klotzsch) K.U. Kramer	st	
273*	1842	DST	Lindsaea gracilis Klotzsch	Lindsaea stricta Dryand. var. parvula (Fée) K.U. Kramer	ht: B; it: BM	
274		DST	Lindsaea crenata Klotzsch	Lindsaea portoricensis Desv.	ht: B	
274*		DST	Lindsaea guianensis (Aubl.) Dryand.	Lindsaea guianensis (Aubl.) Dryand. subsp. guianensis	B	
275	1842	DRY	Aspidium fraxinifolium Schrad.	Tectaria incisa Cav.	B BM	Essequibo R.
276		DST	Lindsaea stricta Dryand.	Lindsaea portoricensis Desv.	B	
277	1842	HMP	Trichomanes plumula C. Presl	Trichomanes martiusii C. Presl	ht: B	
278		DST	Lindsaea schomburgkii Klotzsch	Lindsaea schomburgkii Klotzsch	B G K	
279	1841	MTX	Amphidesmium blechnoides (Sw.) Klotzsch	Metaxya rostrata (Kunth) C. Presl	B	
280	1841	CTH	Hemitelia parkeri Hook.	Cyathea surinamensis (Miq.) Domin	B BR G	
280*		CTH	Hemitelia hostmannii Hook.	Cyathea cyatheoides (Desv.) K.U. Kramer	st: B	
280**		CTH	Alsophila multiflora (Sw.) J. Sm.	Cyathea cyatheoides (Desv.) K.U. Kramer	B	
281		PLP	Polypodium ciliatum Willd.	Microgramma reptans (Cav.) A.R. Sm.	B	
282		ADI	Adiantum schomburgkianum Klotzsch	Adiantum terminatum Kunze ex Miq.	B	
283	1843	ADI	Gymnogramma calomelanos (L.) Kaulf.	Pityrogramma calomelanos (L.) Link	B G K UC	
284		PLP	Polypodium aureum L.	Phlebodium aureum (L.) J. Sm.	B † K	
285		OLN	Nephrolepis exaltata (Sw.) Schott	Nephrolepis rivularis (Vahl) Mett. ex Krug		
286		CYP	Diplasia karataefolia Rich.	Diplasia karataefolia Rich.		
287		MLS	Tococa guianensis Aubl.	Tococa guianensis Aubl.		
288	1842	MRT	Campomanesia glabra Benth.	Calycolpus goetheanus (Mart. ex. DC.) O. Berg	st	Essequibo R.
289 302		LNT	Utricularia calycifida Benj.	Utricularia calycifida Benj.	ht: B †; it: K M	
290 304		ACA	Hygrophila guianensis Nees ex Benth.	Hygrophila costata Nees		
291 331						

Index 5. Richard Schomburgk's collection series (1841–1844) – continued

292	225		SCR	Bacopa aquatica Aubl.	Bacopa aquatica Aubl.		
293	305		XYR	Xyris surinamensis Miq.	Xyris jupicai Rich.		
294	159		RUB	Malanea angustifolia Bartl. [n.n.]	Malanea obovata Hochr.		
295	303		SCR	Conobea aquatica Aubl.	Conobea aquatica Aubl.		
296	306		MRT	Aulomyrcia pirarensis O. Berg	Myrcia inaequiloba (DC.) D. Legrand		
297							
298	231		RUB	Cephaëlis bracteocardia DC.	Psychotria bracteocardia (DC.) Muell. Arg.		
299			GEN				
300	317		RUB	Psychotria inundata Benth.	Psychotria capitata Ruiz & Pav. subsp. inundata (Benth.) Steyerm.		
301	323		LOG	Spigelia humilis Benth.	Spigelia humilis Benth.		
302	227		OCH	Gagemia essequiboensis Klotzsch & M.R. Schomb. [n.n.]	Elvasia essequibensis Engl.	type	Essequibo R.
303	229		MLS	Tococa subnuda Benth.	Tococa subciliata (DC.) Triana		
304		Dec 1841	AST	Vernonia gracilis Kunth var. villosa Less.	Vernonia gracilis Kunth		Essequibo R.
305	310		OXL	Oxalis schomburgkiana Prog.	Oxalis frutescens L.		
306			TNR	Turnera aurantiaca Benth.	Turnera aurantiaca Benth.		
307	392		MLS	Comolia veronicaefolia Benth.	Comolia villosa (Aubl.) Triana var. C		
308	308		EUP	Phyllanthus guianensis Klotzsch	Phyllanthus caroliniensis Walter subsp. guianensis (Klotzsch) Webster	ht: K	
309	321		CHB	Moquilea multiflora Benth.	Couepia multiflora Benth.		
310			RUB	Faramea amplexicaulis Bartl. [n.n.]	Faramea sessilifolia (Kunth) A. DC.	BR K US	Essequibo R.
311	318		CHB	Licania incana Aubl.	Licania incana Aubl.		
312			PGL	Securidaca marginata Benth.	Securidaca marginata Benth.	P	
313	235		CSL	Martia excelsa Benth.	Martiodendron excelsum (Benth.) Gleason		
314			MNM	Citriosma guianensis (Aubl.) Tul.	Siparuna guianensis Aubl.		
315	224		TRG	Trigonia hypoleuca Griseb.	Trigonia hypoleuca Griseb.		
316	226	Dec 1841	MRT	Eugenia subobliqua Benth.	Myrcia subobliqua (Benth.) Nied.	ht: GOET; it: K	Essequibo R.
317							
318			CPP	Cleome latifolia Vahl ex DC.	Cleome latifolia Vahl ex DC.		
319			PON	Eichhornia azurea Kunth	Eichhornia azurea (Sw.) Kunth		
320							
321	228	Jan 1842	BOR	Heliotropium helophilum Mart.	Heliotropium filiforme Lehm.	tc: B †	
322	332		CSL	Tachigali pubiflora Benth. ('Tachigalia')	Tachigali pubiflora Benth. ('Tachigalia')	P	Essequibo R.
323	329	Jan 1842	AQF	Ilex laurina Klotzsch [n.n.]	Ilex daphnogenea Reissek	tc: B † BR F K NY P	

No.	Ref	Date	Fam	Name as labelled	Current name	Notes	Locality
324			PLP	Polypodium phyllitidis L.	Campyloneurum phyllitidis (L.) C. Presl	B U	
325	307		APO	Malouetia gracilis (Benth.) A. DC	Malouetia gracilis (Benth.) A. DC		
326	248		RUB	Palicourea riparia Benth.	Palicourea croceoides Desv.		
327	234		FAB	Andira retusa Kunth	Andira surinamensis (Bondt) Splitg. ex Amshoff		
327*	309		RUB	Psychotria rubra (Willd.) Muell. Arg.	Psychotria hoffmannseggiana (Willd. ex Roem. & Schult.) Muell. Arg.		
328			MLP	Heteropterys cristata Benth.	Heteropterys cristata Benth.		
329			EUP				
330			RUB	Posoqueria longifolia Aubl.	Posoqueria longifolia Aubl.		
331	238		OLC	Heisteria cauliflora J.E. Sm.	Heisteria cauliflora J.E. Sm.		
332	251		CSL	Elizabetha princeps M.R. Schomb. ex Benth.	Elizabetha coccinea M.R. Schomb. ex Benth. var. oxyphylla (Harms) R.S. Cowan		Essequibo R.
333			BIG	Tecoma floccosa Klotzsch [n.n.]	Xylophragma seemannianum (Kuntze) Sandw.	tc: B † K	Essequibo R.
334			LAM	Hyptis parkeri Benth.	Hyptis parkeri Benth.		
335	330		FAB	Etaballia guianensis Benth.	Etaballia dubia (Kunth) Rudd		
336	319		VIO	Corynostylis hybanthus (Aubl.) Mart. & Zucc.	Corynostylis arborea (L.) S.F. Blake		
337			CYP	Calyptrocarya angustifolia Nees	Calyptrocarya glomerulata (Brongn.) Urb.	BR U US	
338			PAS	Passiflora pedata L.	Passiflora pedata L.		
339							
340							
341			CHB	Licania incana Aubl.	Licania leptostachya Benth.	BR	
342	320		MRT	Calycampe angustifolia O. Berg	Myrcia calycampa Amshoff		
343			TRG	Trigonia macrostachya Klotzsch [n.n.]	Trigonia villosa Aubl. var. macrocarpa (Benth.) Lleras		Essequibo R.
344			RUB				
345			STR				
346							
347			EUP	Mabea taquari Aubl.	Mabea taquari Aubl.	ht: B †; lt: P	
348	411	Jan 1842	SAP	Lamprospermum schomburgkii Klotzsch [n.n.]	Matayba camptoneura Radlk.	K	Essequibo R.
349	230		CHB	Licania aperta Benth.	Licania apetala (E. Mey.) Fritsch var. aperta (Benth.) Prance		Essequibo R.
350	236		DLL	Doliocarpus calinea J.F. Gmel.	Doliocarpus spraguei Cheesm.		
351			OCH	Ouratea longifolia (DC.) Engl. var. microcalyx Engl.	Ouratea longifolia (DC.) Engl.	type	Essequibo R.
352			AMA	Serturnera schomburgkii Klotzsch [n.n.]	Pfaffia glomerata (Spreng.) Pedersen		
353	387		MIM	Pithecellobium adiantifolium Benth. var. multipinnum Benth.	Macrosamanea pubiramea (Steud.) Barneby & J.W. Grimes var. pubiramea		
354	241	1842	CSL	Cynometra bauhiniifolia Benth. ('bauhiniaefolia')	Cynometra bauhiniifolia Benth. var. bauhiniifolia	BM K	Essequibo R.

Index 5. Richard Schomburgk's collection series (1841–1844) – continued

No.	Coll.	Date	Fam.	Field name	Accepted name	Herb.	Locality
355	264*		SAP	Cupania			
356	244		CHB	Moquilea comosa Benth.	Coupeia comosa Benth.	K	
357			CNN				
358		Jan 1842	BOR	Cordia rufa Klotzsch [n.n.]	Cordia grandiflora (Desv.) Kunth	tc: B	Essequibo R.
359							
360			MLP	Heteropterys candolleana A. Juss.	Heteropterys macradena (DC.) W.R. Anderson		
361	260		FAB	Dioclea guianensis Benth.	Dioclea guianensis Benth.		Essequibo R.
362			VRB	Aegiphila arborenscens (Aubl.) Vahl	Aegiphila integrifolia (Jacq.) B.D. Jacks.		
363	245		FAB	Aeschynomene sensitiva Sw.	Aeschynomene sensitiva Sw.		
364	246		AST	Eupatorium ivaefolium L.	Chromolaena ivaefolium (L.) R.M. King & H. Rob.		
365							
366	250		AST	Gnaphalium americanum Mill.	Gamochaeta americana (Mill.) Wedd.		
367	247	Jan 1842	AST	Trichospira menthoides Kunth	Trichospira verticillata (L.) S.F. Blake	tc: G	Rupununi R.
367*			DST	Lindsaea reniformis Dryand.	Lindsaea reniformis Dryand.	B BR	
368			DST	Lindsaea divaricata Klotzsch	Lindsaea divaricata Klotzsch	ht: B; it: BM	
369			FAB	Ecastophyllum ferrugineum Klotzsch [n.n.]	Dalbergia riedelii (Radlk.) Sandw.		
370			TNR	Piriqueta stenophylla Klotzsch [n.n.]	Piriqueta cistioides (L.) Griseb. subsp. cistioides		
371	237		VIO	Ionidium oppositifolium (L.) Roem. & Schult.	Hybanthus oppositifolius (L.) Taub.		
372	440*		CLU	Vismia cayennensis (Jacq.) Pers.	Vismia cayennensis (Jacq.) Pers.		
373	249		TRG	Trigonia subcymosa Benth.	Trigonia subcymosa Benth.		
374	242		OCH	Gomphia dura Klotzsch [n.n.]	Ouratea rigida Engl.	GOET K	
375			RUB	Coffea tenuiflora Benth.	Morinda tenuiflora (Benth.) Steyerm. var. tenuiflora		
376			LOR	Psittacanthus guianensis Klotzsch [n.n.]	Psittacanthus cordatus (Hoffmanns.) Blume	L	
377	403		PAS	Tacsonia spinescens Klotzsch [n.n.]	Passiflora securiclata Mast.		
378			CCR	Caryocar glabrum (Aubl.) Pers.	Caryocar glabrum (Aubl.) Pers.		
379	239		FAB	Leptolobium nitens Vogel	Acosmium nitens (Vogel) Yakovlev		
380	243	Jan 1842	OCH	Gomphia rupununiensis Klotzsch [n.n.]	Ouratea rupununiensis Engl.	st: B	
381			SAP	Matayba inelegans (Spruce) Radlk.	Barhamia macrostachya Klotzsch		
382	316	1841–42	EUP	Croton essequeboensis Klotzsch			
383							
384	233		AST	Vernonia tricholepis DC.	Lepidaploa remotiflora (Rich.) H. Rob.		Rupununi R.
385	254		MIM	Machaerium schomburgkii Benth.	Mimosa annularis Benth. var. odora Barneby	F frag K	Prara
386	417		MLV	Sida angustifolia Lam.	Sida serrata Willd. ex Spreng.		
387	325		PHT	Microtea maypurensis (Kunth) G. Don	Microtea maypurensis (Kunth) G. Don		
387*			MLP	Byrsonima coccolobifolia Kunth	Byrsonima coccolobifolia Kunth	K	

No.	Coll. no. / Date	Fam.	Name (as written)	Accepted name	Status	Locality
388	327	RUB	Palicourea rigida Kunth	Palicourea rigida Kunth		
389	266	MLP	Byrsonima crassifolia (L.) Kunth	Byrsonima crassifolia (L.) Kunth	K	
390						
391	1841–42	CSL	Cassia	Chamaecrista nictitans (L.) Moench subsp. patellaria (Collad.) H.S. Irwin & Barneby	NY	Pirara
392				Myrcia guianensis (Aubl.) DC. var. guianensis	tc: B †	
393		MRT	Aulomyrcia conduplicata O. Berg	Diodia hyssopifolia (Willd. ex. Roem. & Schult.) Cham. & Schltdl. var. articulata (Pohl ex DC.) Steyerm.		
394	333	RUB	Diodia articulata DC.			
395		STR		Merremia aturensis (Kunth) Hallier f.		
396	324	CNV	Ipomoea juncea Choisy	Coutoubea ramosa Aubl.		
397	300	GEN	Coutoubea ramosa Aubl.	Buchnera palustris (Aubl.) Spreng.		
398	1841–42	SCR	Buchnera palustris (Aubl.) Spreng.	Polygala trichosperma L.	L	
399	262	PGL	Polygala variabilis Kunth	Distictis granulosa Bureau & K. Schum.	ht: K	
400		BIG	Pithecoctenium granulosum Klotzsch [n.n.]	Turnera guianensis Aubl.		
401		TNR	Turnera guianensis Aubl.	Licania apetala (E. Mey.) Fritsch var. aperta (Benth.) Prance	K	
402	252	CHB	Licania aperta Benth.			
403				Lippia origanoides Kunth	tc: K	
404	326	VRB	Lippia schomburgkiana Schauer			
405		PIP	Peperomia schomburgkii C. DC.	Peperomia schomburgkii C. DC.	tc: B	
406						
407		MIM	Pentaclethra filamentosa Benth.	Pentaclethra macroloba (Willd.) Kuntze	US	
408		LYT				
409		CSL	Cassia lotoides Kunth	Chamaecrista hispidula (Vahl) H.S. Irwin & Barneby		Rupununi R.
410	261	RUB	Calycophyllum sp. nov.	Calycophyllum stanleyanum R.H. Schomb.	type	
411	356	LOG	Spigelia polystachya Klotzsch [n.n.]	Spigelia polystachya Klotzsch ex Prog.		
412						
413						
414	364	MRT	Eugenia pyrroclada O. Berg	Eugenia punicifolia (Kunth) DC.		
415		ORC				
416						
417	412	FAB	Neurocarpum angustifolium Kunth	Clitoria guianensis (Aubl.), Benth.		
418						
419		LYT				
420		STR		Vitex schomburgkiana Schauer		
421	1843 288	VRB	Vitex schomburgkiana Schauer	Vitex schomburgkiana Schauer	ht: K NY frag	

No.	Year	Family	Name (as published)	Current determination	Type / Herbaria	Coll. no.	Locality
453		RUB	Diodia setigera DC.	Diodia apiculata (Willd. ex Roem. & Schult.) K. Schum.		257	Warrewarrema
454		FLC	Casearia carpinifolia Benth.	Casearia silvestris Sw. var. lingua (Cambess.) Eichl.		256	
455		CNN	Omphalobium opacum Klotsch [n.n.]	Connarus coriaceus Schellenb.	ht: B †	264	
456		MLS	Clidemia pustulata DC.	Clidemia pustulata DC.	K	368	
457		AST	Elephantopus angustifolius Sw.	Orthopappus angustifolius (Sw.) Gleason		268	
458		FAB	Tephrosia brevipes Benth.	Tephrosia brevipes Benth.		286	
459							
460							
461		STR	Helicteres guazumaefolia Kunth	Helicteres guazumaefolia Kunth		269	
462		HYD	Hydrolea spinosa L.	Hydrolea spinosa L.		265	
463		MRT	Eugenia dipoda DC. var. brachypoda DC.	Eugenia punicifolia (Kunth) DC.			
464							
465							
466		AST	Wulffia platyglossa DC.	Tilesia baccata (L.) Pruski		271	
467							
468		VRB	Amasonia erecta L.f.	Amasonia campestris (Aubl.) Moldenke	BR G L P W	272	Pirara
469	1841–42	PGL	Polygala adenophora DC.	Polygala adenophora DC.			
470		MIM	Pithecellobium glomeratum Benth.	Zygia cataractae (Kunth) L. Rico			
471		TNR	Turnera opifera Mart.	Turnera caerulea Moç. & Sessé ex DC. var. surinamensis (Urb.) Arbo		273	
472							
473		MLS	Clidemia rubra (Aubl.) Mart.	Clidemia rubra (Aubl.) Mart.			
474				Acisanthera uniflora (Vahl) Gleason			
475		MLS	Microlicia recurva (Rich.) DC.	Ruellia geminiflora Kunth var. angustifolia (Nees) Griseb.		274	
476		ACA	Dipteracanthus canescens Nees		syntype	291	
477							
478		PGL	Polygala timoutou Aubl.	Polygala timoutou Aubl.	G K L P	395	Pirara
479		VRB	Stachytarpheta elatior Schrad. ex Roem. & Schult. ('Stachytarpha')	Stachytarpheta angustifolia (Mill.) Vahl			
480		EUP	Tragia			369	
481						793	
482							
483		FAB	Tephrosia toxicaria Pers.	Tephrosia sinapou (Bucholz) A. Chev.		267	
484							
485							

Index 5. Richard Schomburgk's collection series (1841–1844) – continued

No.	No. 2	Date	Fam.	Name	Current name	Herbarium	Locality
486							
487	255		AMA	Serturnera guianensis Klotzsch [n.n.]	Pfaffia glomerata (Spreng.) Pedersen		
488	353*		AST	Ageratum scorpioideum Baker	Ageratum scorpioideum Baker		
489	354	Feb 1842	ONA	Jussiaea repens L. ('Jussieua')	Ludwigia inclinata (L.f.) Gómez		
490	394		CMM	Commelina schomburgkiana Klotzsch var. latifolia Klotzsch [n.n.] ('Commelyna')	Commelina schomburgkiana Klotzsch ('Commelyna')	B tc: B	Rupununi R.
491							
492			GEN	Schultesia brachyptera Cham.	Schultesia brachyptera Cham.		
493	396		MIM	Piptadenia peregrina (L.) Benth.	Adenanthera peregrina (L.) Speg.	lt: B; ilt: BM L F MO US	
494	348		HPC	Raddia pachyphylla Miers	Salacia pachyphylla Peyr.		
494*	348*		HPC	Tontelea	Tontelea sandwithii A.C. Sm.		
495	353	Feb 1842	PON	Eichhornia azurea Kunth	Eichhornia azurea (Sw.) Kunth	B	Rupununi R.
496	347		HPC	Salacia guianensis Klotzsch ex Peyr.	Salacia elliptica (Mart. ex Roem. & Schult.) G. Don	MO	
497	313		MIM	Mimosa schomburgkii Benth.	Mimosa schomburgkii Benth.		
498	365		MRT	Psidium ciliatum Benth.	Psidium salutare (Kunth) O. Berg		
499			BIG	Spathodea ovata Klotzsch [n.n.]	Memora schomburgkii (DC.) Miers		
500	295		MLP	Heteropterys candolleana A. Juss.	Heteropterys macradena (DC.) W.R. Anderson	K	Takutu R.
501	409		BIG	Tecoma nigricans Klotzsch [n.n.]	Tabebuia serratifolia (Vahl) G. Nichols.		Upper Rupununi R.
502							
503			SCR	Buchnera pusilla Kunth	Buchnera pusilla Kunth		
503*			SCR	Buchnera palustris (Aubl.) Spreng.	Buchnera palustris (Aubl.) Spreng.		
504			BRS	Icica heptaphylla Aubl.	Protium heptaphyllum (Aubl.) Marchand subsp. heptaphyllum		
505	336		SAP	Schmidelia edulis A. St.-Hil.	Allophylus edulis (A. St.-Hil.) Radlk.		
506			MLP	Spachea tenuifolia Griseb.	Spachea elegans (G. Mey.) A. Juss.		
507	399		SML	Smilax pirarensis Kunth & M.R. Schomb. ex Kunth	Smilax cumanensis Willd.	ht: GOET; it: K	Takutu R.
507*	281		EUP	Adenogyne guyanensis Klotzsch [n.n.]	Gymnanthes guyanensis Klotzsch ex Muell. Arg.		
508	344		FAB	Swartzia microstylis Benth.	Swartzia dipetala Willd. ex Vogel		
509	339		AST	Eupatorium subobtusum DC.	Ayapana amygdalina (Lam.) R.M. King & H. Rob.		
510							
511			VOC	Vochysia tetraphylla (G. Mey.) DC.	Vochysia tetraphylla (G. Mey.) DC.		
512			BIG	Bignonia chamissonis Klotzsch [n.n.]	Macfadyena unguis-cati (L.) A. Gentry		
513	335		ERX	Erythroxylum rufum Cav. ('Erythroxylon')	Erythroxylum rufum Cav.		Takutu R.

No.	Date	Fam.	Schomburgk name	Current name	Notes	Locality
514	337	RUB	Randia spinosa (Jacq.) K. Schum. var. nitida K. Schum.	Randia armata (Sw.) DC.		Takutu R.
515						
516						
517						
518	416	1841–42	PGL	Polygala hygrophila Kunth	Polygala hygrophila Kunth	G L
519			MRN	Calathea villosa Lindl.	Calathea villosa Lindl.		Takutu R.
520		Mar 1842	MNS	Cissampelos pareira L.	Cissampelos pareira L.	
521	276		FAB	Tephrosia penicillata Benth.	Tephrosia adunca Benth.	
522			FAB			
523	280	Mar 1842	SAP	Schmidelia mollis Klotzsch [n.n.]	Allophylus racemosus Sw.	st: B	Takutu R.
524	350		BRS	Protium	Protium	
525	362		FAB	Zornia		
526	338		CNV	Evolvulus sericeus Sw.	Evolvulus sericeus Sw.	
527						
528	311		MRT	Campomanesia coaetanea O. Berg	Campomanesia aromatica (Aubl.) Griseb.	LE	Takutu R.
529						
530	341	Mar 1842	AST	Vernonia odoratissima Kunth	Vernonia scabra Pers.	
531						
532	363		MLS	Miconia revoluta Benth.	Miconia prasina (Sw.) DC.	
533	366		MLS	Miconia hypargyrea Miq.	Miconia stenostachya DC.	
534						
535	358	Apr 1842	EUP	Mabea schomburgkii Benth.	Mabea taquari Aubl.	st
536	343		SMR	Simaba guianensis Aubl.	Simaba guianensis Aubl.	
537			BIG			
538	340		MRT	Eugenia incanescens Benth.	Eugenia incanescens Benth.	
539			MRT	Psidium aquaticum Benth. var. triflorum O. Berg	Psidium aquaticum Benth.	tc: B †
540			FLC			
541	345		PLG	Ruprechtia brachystachya Benth.	Ruprechtia brachystachya Benth.	
542	334		LYT	Cuphea antisyphilitica Kunth	Cuphea antisyphilitica Kunth	
543	278	Apr 1842	ERX	Erythroxylum campestre A. St.-Hil. ('Erythroxylon')	Erythroxylum suberosum A. St.-Hil.	
543*	290		PGL	Polygala mollis Kunth	Polygala hebeclada DC.	BM K P	Pirara
544	275		MLP	Bunchosia schomburgkiana Nied.	Bunchosia mollis Benth.	ht:?; it: K NY	Takutu R.
545						
546			CNV	Ipomoea potentilloides Meisn.	Merremia potentilloides (Meisn.) Hallier f.	
547	277		APO	Thevetia humboldtii M.R. Schomb. [n.n.]	Aspidosperma macrophyllum Muell. Arg.		Takutu R.

Index 5. Richard Schomburgk's collection series (1841–1844) – continued

No.	Date	Family	Name	Accepted name	Type/herbarium	Locality
548		ERX	Erythroxylum campestre A. St.-Hil. ('Erythroxylon')	Erythroxylum suberosum A. St.-Hil.		Takutu R.
549		MLS	Rhynchanthera grandiflora (Aubl.) DC.	Rhynchanthera grandiflora (Aubl.) DC.		
550		MLV	Pavonia bracteosa Benth.	Peltaea trinervis (C. Presl) Krapov. & Cristóbal		
551						
552		RUB	Perama hirsuta Aubl.	Perama hirsuta Aubl.		
553		RUB	Mitracarpus scabrellus Benth.	Mitracarpus scabrellus Benth.		
554		LOR	Loranthus smithii R.H. Schomb.	Phthirusa cucullaris (Lam.) Blume		
555		MLS	Comolia hirtella Naudin	Comolia villosa (Aubl.) Triana var. villosa		Rupununi R.
556						
557		CNV	Evolvulus glomeratus Nees & Mart.	Evolvulus glomeratus Nees & Mart.		
558		PAS	Dysosmia foetida (L.) Roem.	Passiflora foetida L. var. foetida		
559		BIG	Lundia schomburgkii Klotzsch [n.n.]	Callichlamys latifolia (Rich.) K. Schum.		
560	1842	HUM	Humiria laurina Klotzsch ex Urb. ('Humirium laurinum')	Humiria balsamifera (Aubl.) J. St.-Hil. var. laurina (Urb.) Cuatrec.	ht: M; it: NY US	
561		FAB	Bowdichia			Cuyuni R.
562		ANN	Guatteria guyanensis Klotzsch [n.n.]	Guatteria schomburgkiana Mart.	B	
563						
564						
565		MRN	Calathea macrostachya Klotzsch [n.n.]	Ischnosiphon foliiosus Gleason		
566		STR				
567		MRT	Eugenia flavescens DC.	Eugenia flavescens DC.		
568	298	MLS	Chaetogastra lasiophylla Benth.	Macairea lasiophylla (Benth.) Wurdack		
569		RUT	Esenbeckia pilocarpoides Kunth var. guianensis Engl.	Esenbeckia pilocarpoides Kunth		
570		KRM	Krameria spartioides Klotzsch ex O. Berg	Krameria spartioides Klotzsch ex O. Berg	type	
571	359	ASC	Ditassa pauciflora Decne.	Ditassa pauciflora Decne.	G	
572	284	CPP	Physostemon intermedium Moric.	Cleome guianensis Aubl.		
573	May 1842	BOR	Heliophytum passerinoides Klotzsch [n.n.]	Heliotropium ternatum Vahl		
574	312	RUB	Brignolia pubigera Benth.	Isertia parviflora Vahl		
574*		BRS	Protium	Protium guianense (Aubl.) Marchand	tc: B †	Rupununi R.
575	1842	AST	Piptocarpha polycephala Baker	Piptocarpha polycephala Baker		
575*	286	CNV			tc: G	
576	282	RUB	Psychotria cordifolia Kunth	Psychotria cordifolia Kunth subsp. perpusilla Steyerm.		

Coll. no.	No. 2	Family	Name as published	Current name	Type data	Date	Locality
577		TNR	Turnera chamaedrys Klotzsch [n.n.]	Turnera ulmifolia L.			
578	361	STR	Helicteres brevispica A. St.-Hil. or H. urbani K. Schum.	Helicteres brevispica A. St.-Hil. or H. urbani K. Schum.	UPS		
579		BIG	Bignonia aequinoctialis L.	Cydista aequinoctialis (L.) Miers			
580		VOC	Vochisia curvata Klotzsch [n.n.]	Vochysia crassifolia Warm.			
581							
582		RUB	Coccocypselum canescens Willd.	Coccocypselum aureum (Spreng.) Cham. & Schltdl. var. capitatum (Benth.) Steyerm.			
583	292						
584		HUM	Humiria obovata Benth. ('Humirium obovatum')	Humiriastrum obovatum (Benth.) Cuatrec.			
585	293	FAB	Swartzia alterna Benth.	Bocoa alterna (Benth.) R.S. Cowan			
586	299	MIM	Acacia paniculata Willd.	Acacia tenuifolia (L.) Willd.			
587		TNR	Turnera corchorifolia Willd. ex Schult.	Turnera odorata Vahl ex DC.	ht: B; it: B K	1843	
588	342	NYC	Pisonia schomburgkiana Heimerl [n.n.]	Guapira cuspidata (Heimerl) Lundell			
589	389	FAB	Desmodium axillare (Sw.) DC.	Desmodium axillare (Sw.) DC.			
590	314	FAB	Centrolobium robustum Mart. ex Benth.	Centrolobium paraense Tul.	ht: B †; lt: K;		
591	294	MRT	Eugenia perforata O. Berg	Eugenia egensis DC.			
592	401	VRB	Aegiphila laevis Willd.	Aegiphila laxiflora Benth.	ht: GOET; it: K		Pirara
593	402	STR	Helicteres althaeifolia Benth.	Helicteres baruensis Jacq.	B		Pirara
594		MLP	Banistera cristata Griseb.	Banisteriopsis cristata (Griseb.) Cuatrec.			
595		NYC	Pisonia guianensis Klotzsch [n.n.]	Guapira eggersiana (Heimerl) Lundell			
596		DRO					
597		SML	Smilax pirarensis Kunth & M.R. Schomb. ex Kunth	Smilax pirarensis Kunth & M.R. Schomb. ex Kunth	tc : B	Jun 1842	Pirara
598		MNM	Siparuna guianensis Aubl.	Siparuna guianensis Aubl.			
599	349	MOR	Brosimum aubletii Poepp. & Endl.	Brosimum guianense (Aubl.) Huber	B		Pirara
600		NYC	Pisonia guianensis Klotzsch [n.n.]	Guapira eggersiana (Heimerl) Lundell	B	1842	Pirara
601		BOR	Cordia sericicalyx A. DC.	Cordia sericicalyx A. DC.	ht: W; it: B † L	Jan 1842	
602	357	ERX	Erythroxylum amazonicum Peyr.	Erythroxylum amazonicum Peyr.			
603	393	FAB	Rhynchosia schomburgkii Benth.	Rhynchosia schomburgkii Benth.			
604	388	MRT	Psidium ciliatum Benth.	Psidium sartorianum (O. Berg) Nied.			Rupununi R.
605		VRB	Lantana canescens Kunth	Lantana canescens Kunth			
606	415						
607	398	CSL	Cassia disadena Steud.	Chamaecrista nictitans (L.) Moench var. disadena (Steud.) H.S. Irwin & Barneby			
608		AST	Apeiba tibourbou Aubl.	Apeiba schomburgkii Szyszyl.			Pirara
609	400	TIL		Trichilia pallida Sw.			
610	315	MEL	Trichilia brachystachya Klotzsch ex C. DC.				

No.	Coll. no. / Date	Fam.	Name (as recorded)	Name (accepted)	Herb.	Loc.
641*	371	CMM	Tradescantia			
642		PAS				
643	385	EUP	Euphorbia hypericifolia L.	Euphorbia hyssopifolia L. subsp. hyssopifolia		
644		MIM	Schrankia leptocarpa DC.	Mimosa quadrivalvis L. var. leptocarpa (DC.) Barneby		
645	380	FAB	Trifolium guianense Aubl.	Stylosanthes guianensis (Aubl.) Sw.		
646	381	ASC	Oxypetalum capitatum Mart.	Oxypetalum capitatum Mart.		
647	408	MLV	Sida ciliaris var. guianensis K. Schum.	Sida ciliaris L.		
648	405	APO	Prestonia latifolia Benth.	Prestonia tomentosa R. Br.		
649	374	FLC	Casearia densiflora Benth.	Casearia commersoniana Cambess.	B	
650	424	EUP	Traganthus sidioides Klotzsch	Bernardia sidoides (Klotzsch) Muell. Arg.		
651	407	EUP	Brachystachys hirta Klotzsch	Croton glandulosus L.		
652	406	RUB	Richardsonia divergens (Pohl) DC.	Richardia		
653	435	CNV	Jacquemontia hirsuta Choisy	Jacquemontia sphaerostigma (Cav.) Rusby		
654	370	SAP	Cardiospermum halicacabum L.	Cardiospermum halicacabum L.		
655	382	AST	Synedrella nodiflora (L.) Gaertn.	Synedrella nodiflora (L.) Gaertn.		
656	376	CLU	Clusia nemorosa G. Mey.	Clusia nemorosa G. Mey.	K	
657		EUP	Euphorbia pilulifera L.	Euphorbia hirta L.		
658	377 / 1843	TEO	Clavija ornata D. Don	Clavija macrophylla (Link ex Roem. & Schult.) Radlk.	B GH	
659	419	FAB	Zornia			Pirara
660	375	IRI	Cipura palludosa Aubl.	Cipura palludosa Aubl.		Pirara
661	384	RUB	Mitracarpum puberulum Benth.	Borreria ocymoides (Burm.f.) DC.		
662		RUB	Declieuxia chiococcoides Kunth	Declieuxia fruticosa (Willd. ex Roem. & Schult.) Kuntze		
663	373	VIO	Ionidium itoubou Kunth	Hybanthus calceolaria (L.) G.K. Schulze		Pirara
664	372*	PGL	Polygala longicaulis Kunth	Polygala longicaulis Kunth	K L P W	
665		PGL	Polygala trichosperma L.	Polygala trichosperma L.	BR G L	
666	1841–42					
667	432 / 1841–42					
667*						
668	427 / Feb 1842	BOR	Tournefortia spigeliiflora A. DC. ('spigeliaeflora')	Tournefortia paniculata Cham. var. spigeliiflora (A. DC.) I.M. Johnst.	B †	Pirara
669						
670		SOL	Solanum julocrotonoides Klotzsch [n.n.]		B †	Pirara
671	436	BIG	Amphilophium paniculatum (L.) Kunth	Amphilophium paniculatum (L.) Kunth		
672	428	SOL	Solanum radula Vahl	Solanum asperum Rich.		
673		EUP	Dalechampia tiliifolia Lam.	Dalechampia tiliifolia Lam.		

Index 5. Richard Schomburgk's collection series (1841–1844) – continued

No.	Coll.	Date	Fam.	Name	Accepted name	Herbarium	Locality
674	434	July 1842	TIL	Corchorus argutus Kunth	Corchorus orinocensis Kunth		
675	433		LAU	Oreodaphne glomerata Nees	Ocotea glomerata (Nees) Mez		
676	437		BIG	Arrabidaea cordifolia Klotzsch [n.n.]	Arrabidaea pubescens (L.) A. Gentry	ht: B; it: B	Pirara
677							Pirara
678	429		RUB	Cephaëlis			
679							
680	420		SPT	Chrysophyllum sparsiflorum Klotzsch ex Miq.	Chrysophyllum sparsiflorum Klotzsch ex Miq.	tc: F K frag	
681	421		APO	Secondatia densiflora A. DC.	Secondatia densiflora A. DC.		Pirara
682	422		RUB	Amaioua corymbosa Kunth	Amaioua corymbosa Kunth		
683	431		MLS	Miconia rufescens (Aubl.) DC.	Miconia rufescens (Aubl.) DC.		
684						K	Pirara
685			SAP	Cardiospermum halicacabum L.	Cardiospermum halicacabum L.		
686			CNN	Connarus incomptus Planch.	Connarus incomptus Planch.		
687			CNV	Ipomoea evolvuloides Moric.	Jacquemontia evolvuloides (Moric.) Meisn.		
688	430		LIL	Bomarea fuscata Klotzsch [n.n.]	Bomarea edulis (Tussac) Herbert		
689	441		CSL	Cassia patellaria DC.	Chamaecrista nictitans (L.) Moench subsp. patellaria (Collad.) H.S. Irwin & Barneby	B	
690	425	Aug 1842	FAB	Aeschynomene hystrix Poir.	Aeschynomene hystrix Poir.		Rupununi R. and
691	423		DSC	Dioscorea syringifolia Kunth & M.R. Schomb. [n.n.] ('syringaefolia')	Helmia syringifolia Kunth & M.R. Schomb. ('syringaefolia')	ht: B	Pirara
692			ORC	Galeandra juncea Lindl.	Galeandra stylomisantha (Vell.) Hoehne		
693	444		RUB	Chiococca			
694	426		EUP	Manihot guianensis Klotzsch [n.n.]	Manihot esculenta Crantz	B F NY P	Pirara
695	445		OXL	Oxalis barrelieri L.	Oxalis barrelieri L.		
696	440	Aug 1842	MNS	Trichoa guianensis Klotzsch [n.n.]	Abuta grandifolia (Mart.) Sandw.		
697						ht: B	Pirara
698	442		AST	Porophyllum latifolium Benth.	Porophyllum ruderale (Jacq.) Cass. var. ruderale		Upper Rupununi R.; dry savanna
699			MLV	Gaya subtriloba Kunth	Herissantia crispa (L.) Brizicky		
700			ORC	Habenaria seticanda Lindl. ex Benth.	Habenaria seticanda Lindl. ex Benth.		
701		Aug 1842	POA	Panicum bergii Arechav.	Panicum bergii Arechav.		
702							
703			POA	Olyra ciliatifolia Raddi	Olyra ciliatifolia Raddi	B K	Pirara
704			CYP	Cyperus elegans L.	Cyperus laxus Lam.	K US	Pirara
705			ARE	Geonoma maxima (Poit.) Kunth	Geonoma maxima (Poit.) Kunth		

No.	Coll. no.	Fam.	Collected name	Accepted name	Herb.	Loc.
706						
707						
708		CYP	Rhynchospora	Rhynchospora nervosa (Vahl) Boeck. subsp. nervosa		
709		PAS	Passiflora glandulosa Cav.	Passiflora glandulosa Cav.	P	
710						
711		CYP	Psilocarya rufa Nees	Rhynospora velutina (Kunth) Boeck.		
712						
713						
714		CYP	Cyperus haspan L.	Cyperus haspan L. var. riparius (Nees) Kük.		
715		CYP	Cyperus corymbosus Rottb.	Cyperus corymbosus Rottb.		
716		CYP	Mariscus elatus Vahl	Cyperus aggregatus (Willd.) Endl.		
717						
718						
719						
720		CYP	Rhynchospora polycephala Wydler	Rhynchospora holoschoenoides (Rich.) Herter		Roraima
721		POA	Axonopus scoparius (Flüggé) Hitchc.	Axonopus pubivaginatus Henr.		
722	453	AST	Pectis elongata Kunth	Pectis elongata Kunth var. elongata		
723	450	MLV	Sida althaeifolia Sw. var. aristosa DC.	Sida cordifolia L.		
724	451	MLV	Abutilon lucianum Sweet	Wissadula contracta (Link) R.E. Fr.		
725	454	CSL	Cassia hispida Collad.	Chamaecrista hispidula (Vahl) H.S. Irwin & Barneby	US	
726	449	MLS	Tibouchina aspera Aubl.	Tibouchina aspera Aubl. var. aspera		
727	1842–43	PGL	Catocoma lucida Benth.	Bredemeyera altissima A.W. Benn.		
728	455	MLV	Fugosia guianensis Klotzsch [n.n.]	Cienfuegosia phlomidifolia (A. St.-Hil.) Garcke		
729	458	LCY	Lecythis schomburgkii O. Berg	Lecythis schomburgkii O. Berg	MEL	
730	457	FAB	Leptolobium nitens Vogel	Acosmium nitens (Vogel) Yakovlev		
731		lichen	Strigula elegans (Fée) Muell. Arg.	Strigula elegans (Fée) Muell. Arg. var. subceliata Muell. Arg.	tc: UPS	
732						
733	459	FAB	Amphymenium rohrii (Vahl) Kunth	Pterocarpus rohrii Vahl		
734	461	CSL	Outea multijuga DC.	Macrolobium multijugum (DC.) Benth. var. multijugum		
735	462	RUB	Cordiera latifolia Benth.	Alibertia latifolia Benth. var. latifolia	tc: K	Roraima
736	460	CSL	Outea multijuga DC.	Macrolobium multijugum (DC.) Benth. var. multijugum		
737	456, 1842–43	CSL	Outea acaciifolia Benth. ('acaciaefolia')	Macrolobium acaciifolium (Benth.) Benth.		
738	463	EUP	Amanoa guianensis Aubl.	Amanoa guianensis Aubl.	K	Roraima

Index 5. Richard Schomburgk's collection series (1841–1844) – continued

			Collected name	Revised name	Notes
739	477	RUB	Randia mussaendae (Thunb.) DC.	Randia formosa (Jacq.) K. Schum. var. densiflora Bartl. ex K. Schum.	It: B †; ilt: F
740	466	CPP	Cleome stenophylla Klotzsch ex Urb.	Cleome stenophylla Klotzsch ex Urb.	
741	483	CNV	Evolvulus linifolius L.	Evolvulus filipes Mart.	
742	487	GEN	Schultesia benthamiana Klotzsch [n.n.]	Schultesia benthamiana Klotzsch ex Griseb.	syntype
743					
744	464	RUB	Sipanea sp. nov.	Limnosipanea schomburgkii Hook.f.	
745					
746	486	LYT	Maja hypericoides Klotzsch [n.n.]	Cuphea anagalloidea A. St.-Hil.	near Mt. Roraima
747					
748					
749	484	FAB	Desmodium pachyrhiza Vogel	Desmodium pachyrhiza Vogel	
750	467	SPT	Lucuma glomerata Miq.	Pouteria glomerata (Miq.) Radlk. subsp. glomerata	
751		FLC	Casearia celastroides Klotzsch [n.n.]	Casearia zizyphoides Kunth	
752	722	GEN	Schultesia brachyptera Cham.	Schultesia brachyptera Cham.	B †
		1843			
753	501	STR	Waltheria involucrata Benth.	Waltheria involucrata Benth.	
754	489	CSL	Cassia obtusifolia L.	Senna obtusifolia (L.) H.S. Irwin & Barneby	
755		STR	Byttneria divaricata Benth. ('Büttneria')	Byttneria divaricata Benth.	
755*		ANA	Tapirira guianensis Aubl.	Tapirira guianensis Aubl.	
756	490	MIM	Pithecellobium multiflorum (Kunth) Benth. ('Pithecolobium')	Albizia subdimidiata (Splitg.) Barneby & J.W. Grimes var. subdimidiata	
757		VRB	Volkameria aculeata L.	Cleodendrum aculeatum (L.) Schltdl.	
758	469	MIM	Pithecellobium glomeratum Benth. ('Pithecolobium')	Zygia cataractae (Kunth) L. Rico	MEL
759	474	TIL	Carpodiptera schomburgkii Baill.	Christiana africana DC.	BM K P
760	476	STR	Byttneria divaricata Benth. ('Büttneria')	Byttneria divaricata Benth.	
761	497	STR	Waltheria americana L.	Waltheria indica L.	
761*		MLP	Tetrapterys discolor (G. Mey.) DC.	Tetrapterys discolor (G. Mey.) DC.	
762					
763	493	MIM	Acacia polyphylla DC.	Acacia polyphylla DC.	
764	468	MLV	Pavonia sp. nov.	Pavonia geminiflora Moric.	
765	470	MIM	Mimosa schrankioides Benth.	Mimosa schrankioides Benth. var. schrankioides	
766	472	CSL	Peltogyne pubescens Benth.	Peltogyne paniculata Benth. subsp. pubescens (Benth.) M.F. Silva	R. Cotinga

159

No.	Date	Coll.	Fam.	Name (as labelled)	Determination	Type/Herb.	Locality
767		494	MIM	Entada polystachya (L.) DC.	Entada polystachya (L.) DC.		
768		495	MIM	Acacia westiniana DC.	Acacia riparia Kunth		
769		499	VRB	Stachytarpheta mutabilis Vahl ('Stachytarpha')	Stachytarpheta sprucei Moldenke		
770		480	SOL	Schwenckia hirta Klotzsch var. angustifolia Benth. ('Schwenckia')	Schwenckia americana L. var. hirta (Klotzsch) Carvalho		
771		498	CSL	Cassia filipes Benth.	Chamaecrista rotundifolia (Pers.) Greene var. grandiflora (Benth.) H.S. Irwin & Barneby		
772		478	RUB	Tocoyena neglecta N.E. Br.	Tocoyena neglecta N.E. Br.	K	R. Surumu
773		465	MLS	Microlicia	Acisanthera limnobios (DC.) Triana	tc: GOET K NY	Roraima
773*		488	MLP	Banisteria corymbosa Griseb.	Banisteriopsis cinerascens (Benth.) B. Gates		R. Surumu or R. Cotinga
774		500	STR	Waltheria paniculata Benth.	Waltheria paniculata Benth.		
775		473	CSL	Cassia prostrata L.	Chamaecrista serpens (L.) Greene var. serpens	U	
776		471	CSL	Cassia arowanna R.H. Schomb. [n.n.]	Senna velutina (Vogel) H.S. Irwin & Barneby	st: K NY	R. Surumu or R. Cotinga
777		507	MLP	Byrsonima schomburgkiana Benth.	Byrsonima schomburgkiana Benth.		
778		509	CMB	Combretum laxum Jacq.	Combretum laxum Jacq.		
779							
780		512	FAB	Andira retusa Kunth	Andira surinamensis (Bondt) Splitg. ex Amshoff		R. Surumu
781		511	SAP	Cupania quercifolia Klotzsch [n.n.]	Cupania rubiginosa (Poir.) Radlk.		
782			CNV	Quamoclit coccinea Choisy	Ipomoea angulata Mart. ex Choisy		
783							
784		502	FAB	Crotalaria stipularis Desv.	Crotalaria stipularis Desv.		
785		479	EUP	Euphorbia amoena Klotzsch [n.n.]	Euphorbia dioeca Kunth	K	
785*		503	SCR	Buchnera palustris (Aubl.) Spreng.	Buchnera palustris (Aubl.) Spreng.		
786			ACA	Mendoncia hoffmannseggiana Nees	Mendoncia hoffmannseggiana Nees		
787		513	SCR	Stemodia foliosa Benth.	Stemodia pratensis (Aubl.) C.P. Cowan		
788		508	MIM	Mimosa pudica L.	Mimosa pudica L.		
789	Sep 1842		SOL	Schwenckia chenopodiacea Klotzsch [n.n.] ('Schwenkia')	Schwenckia mollissima Nees & Mart.		R. Cotinga
790		496	KRM	Krameria spartioides Klotzsch ex O. Berg	Krameria spartioides Klotzsch ex O. Berg		
791		492	MLS	Microlicia brevifolia (Rich.) DC.	Acisanthera bivalvis (Aubl.) Cogn.		
792		482	LOG	Strychnos schomburgkiana Klotzsch [n.n.]	Strychnos bredemeyeri (Schult.) Sprague & Sandw.	F	R. Cotinga
793		481	GEN	Schultesia subcrenata Klotzsch ex Griseb.	Schultesia subcrenata Klotzsch ex Griseb.		
794		491	SCR	Ilysanthes gratioloides (L.) Benth.	Lindernia dubia (L.) Pennell		
795		510	FAB	Copaifera	Andira surinamensis (Bondt) Splitg. ex Amshoff		
796	Sep 1842	517	ERX	Erythroxylum orinocense Klotzsch ('Erythroxylon')	Erythroxylum schomburgkii Peyr.	ht: B †	

Index 5. Richard Schomburgk's collection series (1841–1844) – continued

No.	Field no.	Date	Family	Determination	Accepted name	Type	Locality
797	505		CNV	Ipomoea evolvuloides Moric.	Jacquemontia evolvuloides (Moric.) Meisn.		Roraima Mts.
798			ADI	Doryopteris euchlora Kunze ex Klotzsch	Doryopteris collina (Raddi) J. Sm.	tc: B G K	Cotinga R.; savanna
799		1842–43	SCZ	Anemia villosa Humb. & Bonpl. ex Willd.	Anemia villosa Humb. & Bonpl. ex Willd.	BM	
800							
801	475		LAU	Goeppertia reflectens Nees	Endlicheria reflectens (Nees) Mez	ht: B †; it: US	R. Cotinga
802	506	1842–43	FAB	Crotalaria anagyroides Kunth	Crotalaria micans Link		
803	504	1842–43	PGL	Catocoma cuneata Klotzsch	Bredemeyera cuneata Klotzsch ex Hassk.	ht: L; it: K L P	R. Cotinga
804	542		MLV	Pavonia speciosa Kunth	Peltaea speciosa (Kunth) Standley		
805	518		CLU	Mahurea exstipulata Benth.	Mahurea exstipulata Benth.		
806	571		MIM	Pithecellobium aff. polycephalum Benth. ('Pithecolobium')	Pithecollobium aff. polycephalum Benth. ('Pithecolobium')		
807			SCZ	Lygodium polymorphum Kunth	Lygodium venustum Sw.		R. Cotinga
808							
809	537		CYP	Rhynchospora exaltata Kunth	Rhynchospora exaltata Kunth var. microcephala Kük.	syntype	
810			MRT	Eugenia ochra O. Berg	Eugenia tapacumensis O. Berg		
811	514		CSL	Cassia polystachya Benth.	Chamaecrista polystachya (Benth.) H.S. Irwin & Barneby		
812	519		CHB	Licania flavicans Klotzsch [n.n.]	Licania compacta Fritsch		Roraima
813	526		RUB	Coutarea			
814	524		MIM	Mimosa microcephala Humb. & Bonpl. ex Willd.	Mimosa microcephala Humb. & Bonpl. ex Wildl.		
815	724		RUB	Patima laxiflora Benth.	Retiniphyllum laxiflorum (Benth.) N.E. Br. var. laxiflorum		
815*	536	Oct 1842	ONA	Jussieua nervosa Poir. ('Jussieua')	Ludwigia rigida (Miq.) Sandw.	B	
816	523		FAB	Stylosanthes angustifolia Vogel	Stylosanthes angustifolia Vogel		
817	540		MLS	Rhynchanthera serrulata (Rich.) DC.	Rhynchanthera serrulata (Rich.) DC.		
818			LNT	Polypompholyx bicolor Klotzsch [n.n.]	Utricularia simulans Pilg.		
819	538	Oct 1842	MLV	Sida pitifera Klotzsch [n.n.]	Sida aggregata C. Presl	tc: B	
820	539		STR	Waltheria viscosissima A. St.-Hil.	Waltheria viscosissima A. St.-Hil.		
821	527		TIL	Luehea rufescens A. St.-Hil.	Luehea speciosa Willd.		
822	520		LOG	Antonia pilosa Hook.	Antonia ovata Pohl		
823	521		MLS	Macairea pachyphylla Benth.	Macairea pachyphylla Benth.		
824	525		MIM	Pithecellobium polycephalum Benth. ('Pithecolobium')	Albizia glabripetala (H.S. Irwin) G.P. Lewis & P.E. Owen		Mt. Roraima
825	522		FAB	Dioclea guianensis Benth.	Dioclea guianensis Benth.		

				Name (as written)	Current name	Codes	Locality
826	515		MLS	Dicrananthera hedyotidea C. Presl	Acisanthera hedyotidea (C. Presl) Triana		Roraima
827	541		FAB	Nicolsonia cayennensis DC.	Desmodium barbatum (L.) Benth. & Oerst.		
828	516	1842–43	EUP	Caperonia angustissima Klotzsch [n.n.]	Caperonia stenophylla Muell. Arg.		
829			TNR	Priqueta lanceolata Benth.	Piriqueta cistioides (L.) Griseb. subsp. cistioides		
830			NYC	Boerhavia surinamensis Miq.	Boerhavia diffusa L.		
831	535		CNV	Lysiostyles scandens Benth.	Lysiostyles scandens Benth.	BM G	Roraima
831*				Swartzia latifolia Benth.	Swartzia latifolia Benth. var. sylvestris R.S. Cowan	tc: LE W	
832	534		FAB	Aeschynomene trisperma Klotzsch [n.n.]	Aeschynomene brasiliana (Poir.) DC.		
833	530		FAB	Tephrosia leptostachya DC.	Tephrosia purpurea (L.) Pers.		
834	529		FAB	Ooclinium villosum DC.	Praxelis pauciflora (Kunth) R.M. King & H. Rob.		
835	532		AST	Cassia prostrata L.	Chamaecrista serpens (L.) Greene var. serpens		
836	531		CSL	Pectis elongata Kunth	Pectis elongata Kunth var. elongata		
837	533		AST	Batatas cissoides Choisy	Merremia cissoides (Lam.) Hallier		
837*			CNV	Caperonia angustissima Klotzsch [n.n.]	Caperonia stenophylla Muell. Arg.		
838			EUP	Cassia pulchra Kunth	Cassia tetraphylla Desv. var. tetraphylla	BM	
839	528	1842–43	CSL	Cassia roraimae Benth.	Chamaecrista roraimae (Benth.) Gleason	K LE	
840	582		CSL	Vochisia lucida Klotzsch [n.n.]	Vochysia glaberrima Warm.		
841	642		VOC	Sacoglottis guianensis Benth.	Sacoglottis guianensis Benth. var. guianensis	M US	
842	574	Oct 1842	HUM	Cyrilla antillana Michx.	Cyrilla racemiflora L.		
843	596		CYR	Ternstroemia roraimae Klotzsch [n.n.]	Ternstroemia laevigata Wawra		
844	573		TEA	Humiria elliptica Klotzsch ex Urb. ('Humirium ellipticum')	Humiria balsamifera (Aubl.) J. St.-Hil. var. savannarum (Gleason) Cuatrec.	pt: B P K	Roraima
845	576		HUM	Roupala complicata Kunth	Roupala montana Aubl. var. montana	B	
846	572		PRT	Myrciaria uliginosa O. Berg	Myrciaria uliginosa O. Berg		
847	562		MRT	Parinari campestris Aubl. ('Parinarium campestre')	Parinari campestris Aubl. ('Parinarium campestris')	BR	
848	564		CHB	Eupatorium martiusii DC.	Chromolaena squalida (DC.) R.M. King & H. Rob.		
849	565		AST	Clibadium surinamense L.	Clibadium surinamense L.		
850	563		AST	Podostachys guianensis Klotzsch [n.n.]	Croton sclerocalyx Muell. Arg var. pubescens Muell. Arg.	tc: B	
851			EUP	Swartzia oblonga Benth.	Swartzia oblonga Benth.	ht: K	Mt. Roraima
852	548		FAB	Hyptis arborea Benth.	Hyptidendron arboreum (Benth.) Harley	st: B †	
853	546	Oct 1842	LAM	Forsteronia diospyrifolia Muell. Arg.	Forsteronia diospyrifolia Muell. Arg.	ht: B †; nt: G	Roraima
854	725		APO	Roupala suaveolens Klotzsch	Roupala suaveolens Klotzsch	tc: B	
855	544	Oct 1842	PRT	Chrysophyllum guyanense Klotzsch ex Miq. [n.n.]	Chrysophyllum argenteum Jacq. subsp. auratum (Miq.) T.D. Penn.	lt: K; ilt: BR P	
856	550		SPT	Melastoma dodecandra Desv.	Miconia dodecandra (Desv.) Cogn.		
857	557		MLS	Nectandra salicifolia (Kunth) Nees	Nectandra sanguinea Rol. ex Rotth.	st: B	Rupununi R.
858	549	1842	LAU				

Index 5. Richard Schomburgk's collection series (1841–1844) – continued

No.	Date	Fam.	Name (as labelled)	Current name	Types	Locality
859		SAP	Talisia megaphylla Sagot ex Radlk.	Talisia megaphylla Sagot ex Radlk.		
860		FLC				
861		VIO				
862						
863	545	MLV	sp. nov.			
864						
865	570	CHB	Hirtella scabra Benth.	Hirtella scabra Benth.	F K	
866	636	BNN	Bonnetia sessilis Benth.	Bonnetia sessilis Benth.		
867	566	ERI	Thibaudia nutans Klotzsch [n.n.]	Thibaudia nutans Klotzsch ex Mansf.		
868	559 Oct 1842	ERI	Vaccinium puberulum Klotzsch [n.n.]	Vaccinium puberulum Klotzsch ex Meisn. var. puberulum	ht: B †; it: NY	
869						
870	578	RUB	Pagamea capitata Benth.	Pagamea capitata Benth. subsp. capitata		
871	577 Oct 1842	ERI	Befaria schomburgkiana Klotzsch [n.n.]	Bejaria sprucei Meisn.	lt: K	
872	569	OCH	Poecilandra retusa Tul.	Poecilandra retusa Tul.		Roraima
873	567	ERI	Thibaudia nutans Klotzsch [n.n.]	Thibaudia nutans Klotzsch ex Mansf.		
874						
875		RUB	Synisoon schomburgkiana Baill.	Retiniphyllum laxiflorum (Benth.) N.E. Br. var. laxiflorum	ht: B; it: NY	
876	556	BNN	Archytaea multiflora Benth.	Archytaea triflora Mart.	ht: ?; it: GH	
877	558	MRT	Myrcia subcordata DC.	Marlierea lituatinervia (O. Berg) McVaugh	K	
878	597	ASC	Ditassa taxifolia Decne.	Ditassa taxifolia Decne.		
879	551 1842–43	PGL	Polygala paniculata L.	Polygala paniculata L.		
880	552	CLU	Vismia guianensis (Aubl.) Choisy	Vismia guianensis (Aubl.) Choisy		
881		CNV	Maripa densiflora Benth.	Maripa densiflora Benth.		
882	555	AQF	Ilex laureola Triana & Planch.	Ilex laureola Triana & Planch.		
883	579	RUB	Cephaëlis tomentosa Willd.	Psychotria poeppigiana Muell. Arg.	B G K W	Roraima
884	551*	CNV	Quamoclit coccinea Choisy	Ipomoea angulata Mart. ex Choisy		
885	568	MLS	Graffenrieda ovalifolia Naudin	Graffenrieda weddelii Naudin		
886		CNV	Ipomoea glabra (Aubl.) Choisy	Merremia macrocalyx (Ruiz & Pav.) O'Donell		
887	554	MRS	Badula schomburgkiana A. DC.	Stylogyne schomburgkiana (DC.) Mez		
887*	560	MLS	Rhexia taxifolia A. St.-Hil.	Marcetia taxifolia (A. St.-Hil.) DC.		
888	547	RUB	Cascarilla schomburgkii Klotzsch	Ladenbergia lambertiana (A. Br. ex Mart.) Klotzsch	P	near Mt. Roraima
889	543	AST	Stiffia condensata Baker	Stomatochaeta condensata (Baker) Maguire & Wurdack		

No.	Coll.	Date	Fam.	Name (as labelled)	Accepted name	Type	Locality
890	553		CSL	Cassia uniflora Spreng. var. ramosa (Vogel) Benth.	Chamaecrista ramosa (Vogel) H.S. Irwin & Barneby var. ramosa		Roraima
891			PGL	Bredemeyera moritziana Klotzsch ex Hassk.	Bredemeyera floribunda Willd.	K F US	
892	581		RUB	Psychotria inundata Benth.	Psychotria capitata Ruiz & Pav. var. roraimensis (Wernham) Steyerm.		
893	584	Oct 1842	VOC	Qualea schomburgkiana Warm.	Qualea schomburgkiana Warm.	ht: B; it: GH	Roraima
894	583		MEL	Guarea aubletti A. Juss.	Guarea guidonia (L.) Sleumer		
895							
896			RUB	Chiococca nitida Benth.	Chiococca nitida Benth. var. amazonica Muell. Arg		
897	917	Oct 1842	XYR	Xyris fontanesiana Kunth	Xyris seubertii A. Nilsson	ht: B; it: US	Roraima
898							
899	676	Sep 1842	ERO	Paepalanthus dichotomus Klotzsch [n.n.]	Paepalanthus dichotomus Klotzsch ex Körn.	tc: B	Roraima
900	575	Nov 1842	AST	Vernonia opaca Benth.	Piptocarpha opaca (Benth.) Baker		Roraima
901	594		EUP	Peridium schomburgkianum Benth.	Pera bicolor (Klotzsch) Muell. Arg.		Roraima
902							
903	595		BRS	Trattinickia guianensis Klotzsch [n.n.]	Trattinickia burserifolia Mart.		
904	693		CLU	Clusia insignis Mart.	Clusia grandiflora Splitg.		
905	580		EUP	Peridium schomburgkianum Benth.	Pera bicolor (Klotzsch) Muell. Arg.	st: B †	Roraima
906	588	Nov 1842	MYS	Myristica sebifera Sw.	Virola sebifera Aubl.	GH	
907	613		MRT	Myrcia ferruginea (Poir.) DC.	Marlierea ferruginea (Poir.) McVaugh		
908			CLU	Clusia schomburgkii Vesque	Clusia schomburgkii Vesque		
909	591		TEA	Lettsomia guianensis Klotzsch [n.n.]	Freziera roraimensis Tul.		Roraima
910	592		LAU	Oreodaphne guianensis (Aubl.) Nees	Ocotea guianensis Aubl.		
911	589	Nov 1842	STY	Styrax subleprosum Klotzsch [n.n.]	Styrax roraimae Perkins	ht: B	Mt. Roraima
912	587		MLP	Byrsonima concinna Benth.	Byrsonima concinna Benth.	ht: K; it: CGE K NY	Mt. Roraima
913	590		LAU	Aiouea guianensis Aubl.	Aiouea guianensis Aubl.		
914	657		CLU	Clusia nemorosa G. Mey.	Clusia nemorosa G. Mey.		Roraima
915	627		ASC	Ditassa angustifolia Decne.	Ditassa angustifolia Decne.		
916	680		SOL				
917							
918	653		CLU	Clusia macropoda Engl.	Clusia macropoda Engl.	type	Roraima
919	610	Nov 1842	AST	Vernonia tricholepis DC.	Lepidaploa remotiflora (Rich.) H. Rob.		Roraima
920			SYM	Symplocos schomburgkii Klotzsch [n.n.]	Symplocos guianensis (Aubl.) Guerke		
921	626	Nov 1842	AST	Oliganthes schomburgkii Sch. Bip.	Piptocoma schomburgkii (Sch. Bip.) Pruski	tc: B †	Roraima
922	624	Nov 1842	TNR	Turnera schomburgkiana Urb.	Turnera schomburgkiana Urb.		
923	609		SYM	Symplocos schomburgkii Klotzsch [n.n.]	Symplocos guianensis (Aubl.) Guerke		
924	625		CYP	Hypolytrum pungens (Vahl) Kunth	Hypolytrum pulchrum (Rudge) H. Pfeiff.	tc: B	
925	682		CLU	Caraipa tereticaulis Tul.	Caraipa tereticaulis Tul.		

Index 5. Richard Schomburgk's collection series (1841–1844) – continued

No.	Coll. no.	Date	Fam.	Name as written	Current name	Type	Locality
926	623		MLS	Miconia schomburgkii Benth.	Miconia lepidota DC.		
927	658		CYP	Hypolytrum pungens (Vahl) Kunth	Hypolytrum pulchrum (Rudge) H. Pfeiff.		
928	615		BIG	Tecoma dura Bureau & K. Schum.	Tabebuia insignis (Miq.) Sandw. var. insignis	K	Roraima
929	611		LAU	Oreodaphne costulata Nees	Ocotea aciphylla (Nees) Mez	st: US	
930	598		FAB	Aeschynomene hystrix Poir.	Aeschynomene hystrix Poir.		
931	616		CSL	Amorphocalyx roraimae Klotzsch [n.n.]	Sclerolobium guianense Benth.		
932	717		FLC	Patrisia bicolor A. DC.	Ryania speciosa Vahl var. tomentella Sleumer		
933	603		CLU	Clusia schomburgkii Vesque	Clusia schomburgkii Vesque		
934	601	Nov 1842	LAU	Oreodaphne crassifolia Nees	Ocotea crassifolia (Nees) Mez	ht: B; it: B	
935	599		CHB	Licania rufescens Klotzsch [n.n.]	Licania rufescens Klotzsch ex Fritsch	tc: K	Roraima
936	600	1842	LAU	Oreodaphne guianensis (Aubl.) Nees	Ocotea guianensis Aubl.		
937	614		TEA	Ternstroemia punctata (Aubl.) Sw.	Ternstroemia punctata (Aubl.) Sw.		
938	617		CLU	Clusia nemorosa G. Mey.	Clusia nemorosa G. Mey.	G	Roraima
939	655		LOG	Bonyunia superba M.R. Schomb. ex Progel	Bonyunia superba M.R. Schomb. ex Progel	tc: U	Mt. Roraima
940	602		ANN	Guatteria ouregou (Aubl.) Dunal	Guatteria	B P	
941	605		MLS	Fothergilla mirabilis Aubl.	Miconia mirabilis (Aubl.) L.O. Williams		
942	619		TEA	Ternstroemia suborbicularis Klotzsch [n.n.]	Ternstroemia crassifolia Benth.		
943	604		AQF	Ilex schomburgkii Klotzsch [n.n.]	Ilex thyrsifolia Klotzsch ex Reissek var. schomburgkii (Klotzsch ex Reissek) Loes.	st: B G K W	Roraima
944	607		TEA	Ternstroemia punctata (Aubl.) Sw.	Ternstroemia punctata (Aubl.) Sw.		Roraima
945	612		AQF	Ilex thyrsifolia Klotzsch [n.n.]	Ilex thyrsifolia Klotzsch ex Reissek var. thyrsifolia	st: BR K	Roraima
946	593		AQF	Ilex retusa Klotzsch [n.n.]	Ilex retusa Klotzsch ex Reissek	st: B BR G U W	Roraima
947	652		CLU	Calophyllum lucidum Benth.	Calophyllum lucidum Benth.		
948	634	1842–43	EUP	Tragia grandifolia Klotzsch	Adenophaedra grandifolia (Klotzsch) Muell. Arg.		Roraima
949	606	Nov 1842	AST	Vernonia schomburgkiana Sch. Bip. var. lanceolata Sch. Bip.	Vernonia ehretifolia Benth.	ht: B †; it: GH NY	Roraima
950	618		PRT	Andripetalum sessilifolium (Rich.) Klotzsch	Panopsis sessilifolia (Rich.) Sandw.	syntype	
951	622		FAB	Dipteryx reticulata Benth.	Dipteryx reticulata Benth.		
952	621	Nov 1842	ERX	Erythroxylum roraimae Klotzsch ex O.E. Schulz	Erythroxylum roraimae Klotzsch ex O.E. Schulz	st: B U	Mt. Roraima
953			SAP	Cupania			
953*			TRG	Trigonia laevis Aubl.	Trigonia laevis Aubl. var. microcarpa (Sagot ex Warm.) Sagot		
954			BRS	Protium plagiocarpum Benoist	Protium plagiocarpum Benoist	G	
955		1842	RUB	Aspidanthera klotzschiana M.R. Schomb. [n.n.]	Ferdinandusa goudotiana K. Schum.		Roraima
956	630	Nov 1842	CLE	Clethra guianensis Klotzsch ex Meisn.	Clethra guianensis Klotzsch ex Meisn.	tc: B	

No.	Ref	Fam	Name	Date	Accepted name	Herbarium	Locality
957	608	MRT	Myrciaria cordata O. Berg		Myrciaria cordata O. Berg	tc: K L W	Roraima
958	632	MLP	Heteropterys carinata Benth.		Heteropterys cristata Benth.	ht: K; it: BM	Roraima
959	586	MIM	Inga setifera DC.		Inga pilosula (Rich.) J.F. Macbr.		
960		PAS	Passiflora auriculata Kunth		Passiflora auriculata Kunth		
961	633	CLU	Quapoya robusta Klotzsch [n.n.]	Nov 1842	Clusia schomburgkiana (Planch. & Triana) Benth. ex Engl.	tc: B	Roraima
962	635	CSL	Dimorphandra macrostachya Benth.	1842–43	Dimorphandra macrostachya Benth. subsp. macrostachya	BM F G K U	
963		PRT	Roupala montana Aubl.		Roupala montana Aubl. var. montana	B	Roraima
964	585	VOC	Vochisia curvata Klotzsch [n.n.]	Nov 1842	Vochysia crassifolia Warm.	K	Roraima
965	631	OCH	Ouratea roraimae Engl.		Ouratea roraimae Engl.		
966	629	TEA	Ternstroemia schomburgkiana Benth.	1842	Ternstroemia schomburgkiana Benth.	tc: B G	near Roraima: dry savanna
967	628	HUM	Humiria floribunda Mart. (Humirium floribundum')		Humiria balsamifera (Aubl.) J. St.-Hil. var. floribunda (Mart.) Cuatrec.	B BR G N US	Upper Mazaruni R. or Potaro R.
968		BIG	Bignonia tubulosa Klotzsch [n.n.]		Pyrostegia dichotoma Miers ex K. Schum.	type	Kanuku Mts.
969		AQF	Ilex umbellata Klotzsch [n.n.]		Ilex umbellata Klotzsch ex Reissek var. humirioides (Reissek) Loes.		
970		CUN	Weinmannia ovalis Ruiz & Pav.		Weinmannia balbisiana Kunth var. roraimensis (Pampan.) Bernardi		
971	662	LOR	Gaiadendron tagua (Kunth) G. Don		Gaiadendron punctatum (Ruiz & Pav.) G. Don		
972	669	AST	Vernonia schomburgkiana Sch. Bip. var. elliptica Sch. Bip.		Vernonia ehretifolia Benth.		
973	656	ERI	Thibaudia guianensis Klotzsch [n.n.]		Psammisia guianensis Klotzsch		Roraima
974	670	ERI	Sphyrospermum roraimae Klotzsch		Sphyrospermum buxifolium Poepp. & Endl.	B K	Roraima
975	720	SPT	Lucuma rigida Mart. & Eichl. ex Miq.	1842	Pouteria rigida (Mart. & Eichl.) Radlk. subsp. rigida	ht: B † NY	Roraima
976	637	CPR	Viburnum			ht: ?: it: K	
977	644	MRT	Aulomyrcia triflora O. Berg		Myrcia citrifolia (Aubl.) Urb.		
978	639	ERI	Gaultheria roraimae Klotzsch ex Meisn.	Nov 1842	Gaultheria erecta Vent.	st: BR NY W	Mt. Roraima
979	641	BIG	Tabebuia triphylla (L.) DC.		Tabebuia insignis (Miq.) Sandw. var. insignis		Mt. Roraima
980	640	PLG	Coccoloba schomburgkii Meisn.		Coccoloba schomburgkii Meisn.		
981	643	AST	Baccharis roraimae M.R. Schomb. [n.n.]		Baccharis brachylaenoides DC. var. ligustrina (DC.) Maguire & Wurdack		
982	645	SAR	Heliamphora nutans Benth.		Heliamphora nutans Benth.		
983	646	RUB	Utricularia humboldtii R.H. Schomb.				
984	649	LNT	Utricularia humboldtii R.H. Schomb.		Utricularia humboldtii R.H. Schomb.		
985		PAS	Astrophea emarginata Roem.	Nov 1842	Passiflora sclerophylla Harms		Roraima
986							

Index 5. Richard Schomburgk's collection series (1841–1844) – continued

No.	No.	Date	Fam.	Name as written	Accepted name	Specimens	Locality
987	672	Nov 1842	RPT	Stegolepis guianensis Klotzsch ex Körn.	Stegolepis guianensis Klotzsch ex Körn.	ht: B; it: G K NY	Roraima
988							
989	664		MLS	Arthrostema bonplandii DC.	Monochaetum bonplandii (Kunth) Naudin	K	Roraima
990	660		AST	Achyrocline			
991	647		MLS	Ossaea coriacea Naudin	Clidemia tepuiensis Wurdack	K P	
992			MRS	Grammadenia lineata Benth.	Cybianthus lineatus (Benth.) Pipoly		
993	654	Nov 1842	AST	Verbesina schomburgkii Sch. Bip. [n.n.]	Verbesina schomburgkii Sch. Bip. ex Klatt		
994	663		MIM	Pithecellobium ferrugineum Benth. ('Pithecolobium')	Abarema ferruginea (Benth.) Pittier	tc: B †	
995			RUB	Cosmibuena trifolia (Benth.) Klotzsch	Cosmibuena grandiflora (Ruiz & Pav.) Rusby		
996	648		MLS	Clidemia lutescens Naudin var. lindeniana Naudin	Leandra lindenliana (Naudin) Cogn.	K	near Mt. Roraima
997							
998	650		MLP	Hiraea oleifolia Benth.	Tetrapterys oleifolia (Benth.) Nied.	ht: K; it: K NY	Roraima
999	651		MLP	Banisteria leptocarpa Benth.	Banisteria martiniana (A. Juss.) Cuatrec. var. martiniana	st: GOET K NY W	Roraima
1000	666		AST	Mikania	Mikania pannosa Baker		
1001							
1002		1842–43	RUB	Malanea roraimensis Wernham [n.n.]	Malanea obovata Hochr.	ht: K; it: NY US	Roraima
1003	675		MLS	Meisneria microlicioides Naudin	Siphanthera cordifolia (Benth.) Gleason		
1004	671		MLS	Davya crassiramis Naudin	Meriania crassiramis (Naudin) Wurdack		
1005	661	Nov 1842	CUN	Weinmannia guyanensis Klotzsch ex Engl.	Weinmannia guyanensis Klotzsch ex Engl.	B	Roraima
1006			EUP	Phyllanthus pycnophyllus Muell. Arg.	Phyllanthus pycnophyllus Muell. Arg.	ht: ?; it: K	Mt. Roraima
1007	659		EUP	Glochidion vacciniifolius Muell. Arg.	Phyllanthus vacciniifolius (Muell. Arg.) Muell. Arg. subsp. vacciniifolius	ht: B †	Roraima
1008			DSC	Dioscorea polygonoides Humb. & Bonpl. ex Willd.	Dioscorea polygonoides Humb. & Bonpl. ex Willd.		
1009	667	Nov 1842	LIL	Nietneria corymbosa Klotzsch & M.R. Schomb. [n.n.]	Nietneria corymbosa Klotzsch & M.R. Schomb. ex Jackson	BM	Roraima
1010							
1011							
1012							
1013	663		MLS	Chaetolepis anisandra Naud.	Chaetolepis anisandra Naudin	ht: K	Roraima
1014			MLS	Graffenrieda obliqua Triana	Graffenrieda obliqua Triana		
1015		Nov 1842	AQF	Ilex vacciniifolia Klotzsch [n.n.]	Ilex vacciniifolia Klotzsch ex Reissek	tc: B BM K W	near Mt. Roraima
1016			RUB	Cinchona roraimae Benth.	Remijia roraimae (Benth.) K. Schum.	tc: B	Roraima
1017			MLS	Graffenrieda sessilifolia Triana	Graffenrieda sessilifolia Triana	ht: K	near Mt. Roraima

No.	Coll.	Date	Fam.	Original determination	Current name	Herb./type	Locality
1018			RUB	Psychotria oblita Wernham	Psychotria oblita Wernham		Roraima
1019							
1020							
1021	687				Connellia augustae (M.R. Schomb.) N.E. Br.		Roraima
1022		Nov 1842	BML	Encholirium augustae M.R. Schomb.			
1023			CLR	Hedyosmum racemosum (Ruiz & Pav.) G. Don	Hedyosmum tepuiense Todzia		Roraima
1024			GEN	Leiothamnus elisabethae M.R. Schomb.	Symbolanthus elisabethae (M.R. Schomb.) Gilg		
1025			CYP	Hypolytrum pungens (Vahl) Kunth	Hypolytrum pulchrum (Rudge) H. Pfeiff.		
1025*			BML	Tillandsia usneoides (L.) L.	Aechmea bromeliifolia (Rudge) Baker	K	
1026	707		AST	Baccharis schomburgkii Baker	Baccharis schomburgkii Baker		
1026*	713	Nov 1842	ERO	Paepalanthus schomburgkii Klotzsch [n.n.]	Paepalanthus schomburgkii Klotzsch ex Körn.	ht: B	Roraima
1027	681		MRS	Cybianthus crotonoides M.R. Schomb. [n.n.]	Cybianthus crotonoides (M.R. Schomb. ex Mez) G. Agostini		
1028		1842-43	EUP	Croton subincanus Muell. Arg.	Croton subincatus Mull. Arg.	st: B †	Roraima
1029	665		MRT	Myrcia phaeoclada O. Berg var. guyanensis O. Berg	Myrcia magnolifolia DC.	type	
1030							
1031			NYC	Pisonia eggersiana Heimerl	Guapira eggersiana (Heimerl) Lundell	B	
1032	678*	Nov 1842	BOR	Cordia dichotoma Klotzsch [n.n.]	Cordia bicolor A. DC.	tc: B	
1033			AST	Pterocaulon alopecuroides (Lam.) DC.	Pterocaulon alopecuroides (Lam.) DC.	K	Mt. Roraima
1034			DRO	Drosera roraimae Klotzsch [n.n.]	Drosera roraimae (Klotzsch ex Diels) Maguire & Laundon	tc: B	
1035			CLU	Clusia crassifolia Planch. & Triana	Clusia crassifolia Planch. & Triana	tc: B	Roraima
1036	709		CLU	Clusia sessilis Klotzsch ex Engl.	Clusia sessilis Klotzsch ex Engl.	tc: NY frag	
1037		Nov 1842	ROS	Rubus guyanensis Focke	Rubus guyanensis Focke		
1038	688	Nov 1842	ERI	Thibaudia formosa Klotzsch [n.n.]	Psammisia formosa Klotzsch		Roraima
1039	668		ERI	Befaria guyanensis Klotzsch [n.n.]	Bejaria sprucei Meisn.		
1040			CYP	Rhynchospora cephalotes (L.) Vahl	Rhynchospora cephalotes (L.) Vahl	lt: B †; ilt: CGB K	
1041	673		MLP	Coleostachys hypoleuca Benth.	Blephandra hypoleuca (Benth.) Griseb.		
1042	692		RUB	Isertia hypoleuca Benth.	Isertia hypoleuca Benth.	ht: K; it: K NY	Roraima
1043	677		PRT	Roupala schomburgkii Klotzsch	Roupala schomburgkii Klotzsch	ht: B	
1044	674		IXO	Ochthocosmus parvifolius Hallier f.	Ochthocosmus roraimae Benth. var. parvifolius (Hallier f.) Steyerm. & Luteyn	tc: B † NY	Roraima
1045							
1046							
1047	694		RUB	Cordiera	Alibertia myrciifolia K. Schum.	K L P	
1048		1842-43	PGL	Polygala galioides Poir.	Polygala galioides Poir.		
1049			OCH	Ouratea nitida (Sw.) Engl.	Ouratea nitida (Sw.) Engl.		

Index 5. Richard Schomburgk's collection series (1841–1844) – continued

No.	Coll. no.	Year	Family	Name as given	Accepted name	Herbarium	Locality
1050							
1051					Hirtella glandulosa Spreng.	ht: B †	
1052	721		CHB	Hirtella glandulosa Spreng.	Tapirira guianensis Aubl.		
1053	683		ANA	Tapirira guianensis Aubl.	Senna multijuga (Rich.) H.S. Irwin & Barneby var. multijuga		
			CSL	Cassia calliantha G. Mey.			
1054	698		MRT	Myrcia hostmanniana Kiaersk.	Myrcia amazonica DC.		
1055	678		MRT	Myrcia			
1056	689		MLS	Miconia sp. nov.	Miconia macrothyrsa Benth.		
1057							
1058							
1059	638	1842–43	EUP	Maprounea guianensis Aubl.	Maprounea guianensis Aubl.		
1060			GEN	Coutoubea spicata Aubl.	Coutoubea reflexa Benth.		
1061	679		CNN	Rourea subtriplinervis Radlk.	Pseudoconnarus subtriplinervis (Radlk.) Schellenb.	GOET	R. Cukenam
1062			BML	Pitcairnia guyanensis Baker	Puya floccosa (Linden) E. Morren ex Mez var. floccosa	tc: US	Roraima
1063							
1064							
1065	718		ORC	Cypripedium klotzscheanum Rchb.f.	Phragmipedium klotzscheanum (Rchb.f.) Rolfe		
1066	690		MLS	Miconia fallax Benth.	Miconia fallax DC.		Mt. Roraima; swampy savanna
1067	708		TEA	Ternstroemia suborbicularis Klotzsch [n.n.]	Ternstroemia crassifolia Benth.		
1068			MLS				
1069	691		MLS	Miconia heterochroa Miq.			
1070	686	1842–43	EUP	Peridium bicolor Klotzsch var. nitidum Benth.	Miconia albicans (Sw.) Triana	K US	
1071	685	1842–43	EUP	Peridium bicolor Klotzsch var. nitidum Benth.	Pera decipiens (Muell. Arg.) Muell. Arg.	lt: K; ilt: B † G	
1072	684		ANN	Xylopia sericea A. St.-Hil.	Pera decipiens (Muell. Arg.) Muell. Arg.	st: B † K	Roraima
1073	706		MLS	Macairea multinerva Benth.	Xylopia sericea A. St.-Hil.		Roraima
					Macairea multinerva Benth.		
1074			RUB				
1075							
1076	716		MLS	Clidemia capitata Benth.	Clidemia capitata Benth.		
1077	715		AST	Gnaphalium simplicicaule Willd. ex Spreng.	Gamochaeta simplicicaulis (Willd. ex Spreng.) Cabrera		Roraima; moist savanna
1078	713		MEL	Moschoxylon propinquum Miq.	Trichilia quadrijuga Kunth subsp. quadrijuga		

No.	Coll.	Date	Fam.	Name as written	Current name	Type / Herbaria	Locality
1079	697		FAB	Machaerium ferrugineum Pers.	Machaerium quinata (Aubl.) Sandw. var. parviflorum (Benth.) Rudd	BM P	
1080	695		MIM	Inga bracteosa Benth.	Inga bracteosa Benth.		
1081	711		RHM	Gouania virgata Reissek	Gouania virgata Reissek		
1082							
1083	710		AST	Elephantopus angustifolius Sw.	Orthopappus angustifolius (Sw.) Gleason		
1084							
1085	704		MRT	Myrcia guianensis (Aubl.) DC.	Myrcia guianensis (Aubl.) DC. var. guianensis		
1086	701		MRT	Myrcia schomburgkiana O. Berg	Myrcia servata McVaugh	st: B	
1087	712	Dec 1842	LAU	Oreodaphne uruphylla Nees	Ocotea cernua (Nees) Mez		
1088	702	Dec 1842	AST	Eupatorium ixodes Benth.	Ayapana amygdalina (Lam.) R.M. King & H. Rob.	ht: K; it: B † BM BR G K NY P US USD W	Mt. Roraima
1089	699		MLS	Macairea parvifolia Benth.	Macairea parvifolia Benth.		
1090	705		RUB	Gonzalea spicata DC.	Gonzalagunia spicata (Lamb.) Gómez		
1091	703		ANN	Xylopia grandiflora A. St.-Hil.	Xylopia aromatica (Lam.) Mart.		
1092	700		MRT	Myrciaria ehrenbergiana O. Berg	Myrcia ehrenbergiana (O. Berg) McVaugh		Roraima
1093							
1094							
1095	723		LOG	Strychnos rhexioides Klotzsch [n.n.]	Strychnos tomentosa Benth.	syntype	Roraima
1096			ORC	Cypripedium lindleyanum R.H. Schomb.	Phragmipedium lindleyanum (R.H. Schomb.) Rolfe		
1097	727		EUP	Astraea lobata (L.) Klotzsch var. pilosa Klotzsch	Croton lobatus L.		
1098							
1099	726		MLP	Heteropterys candolleana A. Juss.	Heteropterys macradena (DC.) W.R. Anderson	K	
1100							
1101			MRT	Eugenia			
1102	943	Dec 1842	TEA	Ternstroemia longipes Klotzsch [n.n.]	Ternstroemia longipes Klotzsch ex Wawra	tc: FM G	
1103							
1104							
1105	730	1842–43	SAP	Cupania subsinuata Klotzsch [n.n.]	Cupania rubiginosa (Poir.) Radlk.	G L P	Roraima
1106	729		PGL	Catocoma lucida Benth.	Bredemeyera lucida (Benth.) Klotzsch ex Hassk.		
1107	728		MNM	Siparuna guianensis Aubl.	Siparuna guianensis Aubl.		
1108			DLL	Davilla rugosa Poir.	Davilla rugosa Poir. var. rugosa	BR	Roraima
1109	731	1843	EUP	Mabea piriri Aubl. var. laevigata Muell. Arg.	Mabea biglandulosa Muell. Arg.	ht: ?; it: G K	
1110	732	Jan 1843	BOR	Tournefortia volubilis L.	Tournefortia volubilis L.	B †	
1111	733		FAB	Lonchocarpus nitidulus Benth.	Lonchocarpus floribundus Benth.		
1112			FLC	Casearia celastroides Klotzsch [n.n.]	Casearia zizyphoides Kunth	ht: B †	
1113			MLP	Byrsonima crassifolia (L.) Kunth	Byrsonima crassifolia (L.) Kunth	K	

Index 5. Richard Schomburgk's collection series (1841–1844) – continued

No.	No. 2	Year	Family	Name (Schomburgk)	Current name	Types	Locality
1114			MLS				
1115			ANA	Anacardia occidentale L.	Anacardia occidentale L.		
1116	735		CNN	Omphalobium opacum Klotsch [n.n.]	Connarus coriaceus Schellenb.	tc: US	
1117	734		MRT	Aulomyrcia roraimensis O. Berg	Aulomyrcia roraimensis O. Berg		
1118			SAP	Urvillea pubescens Klotzsch [n.n.]	Urvillea ulmacea Kunth		
1119	737	1842–43	MLP	Hiraea gracilis Benth.	Tetrapterys maranhamensis A. Juss.	ht: K; it: K	
1120							
1121			CTH	Cyathea vestita Mart.	Cyathea		
1122			fern				
1123			fern				
1124		1842–43	CTH	Alsophila gibbosa Klotzsch	Cyathea gibbosa (Klotzsch) Domin	ht: B; it: BR G GH NY P	
1124*		1842–43	CTH	Alsophila gibbosa Klotzsch	Cyathea delgaddi Sternb.	B F NY P	
1125			CTH	Alsophila oblonga Klotzsch	Cyathea procera (Willd.) Domin	st: B BR G HBG K NY P US	Humirida Mts.
1126			GLC	Mertensia longipinnata (Hook.) Klotzsch	Gleichenia longipinnata Hook.	B	Roraima Mts.
1127			HMP	Hymenophyllum trichomanoides Bosch	Hymenophyllum trichomanoides Bosch		
1128			GLC	Mertensia longipinnata (Hook.) Klotzsch	Gleichenia longipinnata Hook.	B	Roraima Mts.
1128*			TEC	Dryopteris refulgens Klotzsch ex Mett.	Ctenitis refulgens (Klotzsch ex Mett.) C. Chr. ex Vareschi	st: B	Roraima Mts.
1129		1843	CTH	Alsophila marginalis Klotzsch	Cyathea marginalis (Klotzsch) Domin	ht: B; it: B BM BR G HBG K NY	Roraima Mts.
1130		1843	PLP	Polypodium crassifolium L.	Niphidium crassifolium (L.) Lellinger	B	
1131			ADI	Adiantum triangulatum Kaulf.	Adiantum villosum Willd.	B	
1132			ADI	Adiantum radiatum L.	Adiantopsis radiata (L.) Fée	B	Kanuku Mts.
1133			BLE	Blechnum asplenioides Sw.	Blechnum asplenioides Sw.	B	Kanuku Mts.
1134			DST	Lindsaea dubia Spreng.	Lindsaea dubia Spreng.		
1135			HMP	Trichomanes schomburgkianum J.W. Sturm	Trichomanes pinnatum Hedw.	B	Roraima Mts.
1135*			PLP	Polypodium areolatum Humb. & Bonpl. ex Willd.	Polypodium aureum L. var. areolatum (Humb. & Bonpl. ex Willd.) Baker	st: BM BR K	Kanuku Mts.
1136			PLP	Polypodium paradisea Langsd. & Fisch.	Plecuma plumula (Humb. & Bonpl. ex Willd.) Price	B	
1137			PTR	Pteris litobrochioides Klotzsch	Pteris pungens Willd.	tc: B K	
1138		1843	DST	Lindsaea pumila Klotzsch	Lindsaea pumila Klotzsch	ht: B	
1138*			BLE	Blechnum asplenioides Sw.	Blechnum asplenioides Sw.	K	
1139		1843	HMP	Hymenostachys diversifrons Bory	Trichomanes diversifrons (Bory) Mett. ex Sadeb.	B	Roraima Mts.
1140			BLE	Blechnum volubilis Kaulf.	Salpichlaena volubilis (Kaulf.) J. Sm.	B K	Roraima Mts.

No.		Code	Name (original)	Year	Current name	Specimens	Locality
1141		LOM	Acrostichum plumosum Fée		Elaphoglossum plumosum (Fée) Moore	K	Roraima Mts.
1142		BLE	Blechnum asplenioides Sw.		Blechnum asplenioides Sw.	B	
1143		DST	Lindsaea guianensis (Aubl.) Dryand.		Lindsaea guianensis (Aubl.) Dryand. subsp. guianensis	B BM	
1143*		DST	Lindsaea falcata Dryand.		Lindsaea lancea (L.) Bedd. var. falcata (Dryand.) Rosenstock		Kanuku Mts.
1144		ADI	Adiantum hirtum Klotzsch		Adiantum terminatum Kunze ex Miq.	tc: B K	
1145		PLP	Polypodium leucorhizon Kunze ex Klotzsch	1843	Polypodium leucorhizon Kunze ex Klotzsch	tc: B	
1146		GMM	Polypodium pendulum Sw.	1843	Grammitis kaieteura (Jenm.) C. Morton	st: B	
1147		CTH	Alsophila oblonga Klotzsch		Cyathea procera (Willd.) Domin	B	
1147*		LOM	Acrostichum peltatum Sw.		Elaphoglossum peltatum (Sw.) Urb.	B	
1148		GLC	Mertensia pubescens Willd.		Gleichenia lanuginosa Moric.	B	
1149		DST	Pteris elegans Sw.	1843	Histiopteris incisa (Thunb.) J. Sm.	ht: B NY frag	Mt. Roraima
1150		ASL	Asplenium alloeopteron Kunze ex Klotzsch		Asplenium radicans L. var. radicans	st: K P US	
1151		TEC	Dryopteris amplissima (C. Presl) Kunze var. subeffusa C. Chr.		Lastreopsis amplissima (C. Presl) Tindale		
1151*		DAV	Aspidium coriaceum Sw.	1843	Polypodium adiantiforme G. Forst.	B	Mt. Roraima
1152		DRY	Aspidium denticulatum Sw.	1843	Arachniodes denticulata (Sw.) Ching	B	
1153		LOM	Acrostichum peltatum Sw.		Elaphoglossum peltatum (Sw.) Urb.	B	
1154		ADI	Gymnogramma guianensis Klotzsch		Pityrogramma chrysoconia (Desv.) Maxon ex Domin	ht: NY; it: B G K NY	
1154*		ADI	Gymnogramma ornithopteris Klotzsch		Pityrogramma ornithopteris (Klotzsch) Maxon ex Knuth var. guianensis Domin	ht: K	
1155	142	HMP	Trichomanes schomburgkianum J. W. Sturm	1843	Trichomanes pinnatum Hedw.	st: B	Kanuku Mts.
1156		ADI	Adiantum glaucescens Klotzsch	1843	Adiantum glaucescens Klotzsch	ht: B; it: B GH US	
1157		DRY	Aspidium guianense Klotzsch		Cyclodium guianense (Klotzsch) van der Werff ex L.D. Gómez	ht: B; it: K P UC	
1158		ASL	Asplenium serra Langsd. & Fisher	1843	Asplenium serra Langsd. & Fisher	B UC	Mt. Roraima
1159		LYC	Lycopodium carolinianum L.		Lycopodiella alopecuroides (L.) Cranfill	B	
1160		GMM	Polypodium confusum J. Sm.		Grammitis suspensa (L.) Proctor	B	
1160*		PLP	Polypodium confusum J. Sm.	1843	Polypodium subsessile Baker	B	
1161		PLP	Polypodium paradisea Langsd. & Fisch.		Pecluma plumula (Humb. & Bonpl. ex Willd.) Price	ht: B; it: B BR NY	
1162		BLE	Lomaria schomburgkii Klotzsch		Blechnum schomburgkii (Klotzsch) C. Chr.	B	
1163		PLP	Polypodium lepidopteris (Langsd.) Kunze	1843	Polypodium lepidopteris (Langsd.) Kunze	B	
1164		PTR	Pteris deflexa Link	1843	Pteris deflexa Link	B	
1165		THL	Polypodium nervosum Klotzsch		Thelypteris rudis (Kunze) Proctor	ht: B	
1166		DST	Hypolepis guianensis Klotzsch		Hypolepis guianensis Klotzsch	ht: B	
1167		TEC	Aspidium schomburgki Klotzsch		Ctenitis fasciculata (Raddi) Ching	st: B	Roraima Mts.
1168		ASL	Asplenium macilentum Kunze ex Klotzsch		Asplenium auritum Sw.	B	

Index 5. Richard Schomburgk's collection series (1841–1844) – continued

No.	Year	Family	Collected as	Determined as	Herbarium	Locality
1169		BLE	Asplenium auritum Sw.	Blechnum asplenioides Sw.	B UC	
1170	1843	GMM	Polypodium firmum Klotzsch	Grammitis firma (J. Sm.) C. Morton	lt: B; it: CAY K US	
1171*	1843	GMM	Polypodium trichomanoides Sw.	Grammitis trichomanoides (Sw.) Ching	B	
1172	1843	GMM	Polypodium trichomanoides Sw.	Grammitis taenifolia (Jenm.) Proctor	B K	
1173	1843	GMM	Polypodium xanthotrichum Klotzsch	Grammitis xanthotricha (Klotzsch) Duek & Lellinger	st: B	
1173*		HMP	Trichomanes prieurii Kunze	Trichomenes elegans Rich.	B	Roraima
1174		HMP	Trichomanes rigidum Sw.	Trichomanes rigidum Sw.	B	Kanuku Mts.
1175	1843	BLE	Blechnum asplenioides Sw.	Blechnum asplenioides Sw.	B K	Kanuku Mts.
1176		ADI	Adiantum obliquum Willd.	Adiantum obliquum Willd.	B	Roraima Mts.
1177	1843	ASL	Asplenium serra Langsd. & Fisher	Asplenium serra Langsd. & Fisher	B UC	Kanuku Mts.
1178	1843	BLE	Blechnum gracile Kaulf.	Blechnum gracile Kaulf.	B K	
1178*	1843	DST	Lindsaea rigescens Willd.	Lindsaea stricta (Sw.) Dryand. var. stricta	B	
1179	1843	TEC	Aspidium cicutarium Sw.	Triplophyllum angustifolium Holttum	B	
1180	1843	ADI	Adiantum triangulatum Kaulf.	Adiantum latifolium Lam.	B	Kanuku Mts.
1180*	1843	HMP	Hymenophyllum poeppigianum C. Presl	Hymenophyllum decurrens (Jacq.) Sw.	B BR	
1181		HMP	Trichomanes polyanthos Sw.	Hymenophyllum polyanthos (Sw.) Sw.	B	Kanuku Mts.
1182		LYC	Lycopodium subulatum Desv.	Huperzia subulata (Desv. ex Poir.) Holub		
1183		BLE	Blechnum asplenioides Sw.	Blechnus asplenioides Sw.	B BM	Roraima Mts.
1184		TEC	Dryopteris refulgens Klotzsch ex Mett.	Ctenitis refulgens (Klotzsch ex Mett.) C. Chr. ex Vareschi	st: B	
1184*		ADI	Adiantum schomburgkianum Klotzsch	Adiantum cajennense Willd. ex Klotzsch	B	Kanuku Mts.
1185		ADI	Adiantum serratodentatum Willd.	Adiantum serratodentatum Willd.	B	Kanuku Mts.
1186	1843	DST	Lindsaea tenuis Klotzsch	Lindsaea tenuis Klotzsch	ht: B; it: B BR K	Mt. Roraima
1186*		HMP	Trichomanes cellulosum Klotzsch	Trichomanes cellulosum Klotzsch	ht: B; it: B BM BR K	Kanuku Mts.
1187		HMP	Trichomanes polyanthos Sw.	Hymenophyllum polyanthos (Sw.) Sw.	B BR K	
1188	1843	GMM	Grammitis furcata Hook. & Grev.	Cochlidium furcatum (Hook. & Grev.) C. Chr.	st: B	
1189		PLP	Polypodium lepidotum Willd.	Pleopeltis astrolepis (Liebm.) Fourn.		
1190		SCZ	Schizaea dichotoma Sw.	Schizaea poepigiana J.W. Sturm	st: B BR M	Kanuku Mts.
1191	1843	BLE	Lomaria plumieri Desv.	Blechnum	st: B	
		fern				
1192		LYC	Lycopodium intermedium Spring	Huperzia intermedia Trevis.		
1193	1843	LYC	Lycopodium reflexum Lam.	Huperzia reflexa (Lam.) Trevis.	B	
1194	1843	PLP	Polypodium areolatum Humb. & Bonpl. ex Willd.	Polypodium aureum L. var. areolatum (Humb. & Bonpl. ex Willd.) Baker	B	Mt. Roraima

No.	Year	Code	Name (original)	Name (current)	Herbarium	Locality
1195		OLN	Nephrolepis pendula (Raddi) J. Sm.	Nephrolepis rivularis (Vahl) Mett. ex Krug	B UPS	
1196		ADI	Gymnogramma schomburgkiana Kunze ex Klotzsch	Eriosorus hispidulus (Kunze) Vareschi var. hispidulus	ht: B; it: B BM G GH K	
1197		ADI	Doryopteris lomariacea Kunze ex Klotzsch	Doryopteris lomariacea Kunze ex Klotzsch	ht: B; it: BM K P	
1197*		ADI	Doryopteris lomariacea Kunze ex Klotzsch	Doryopteris conformis K.U. Kramer & R. Tyron	B BM G K	
1198		HMP	Trichomanes rigidum Sw.	Trichomanes rigidum Sw.	B	
1199	1843	CTH	Alsophila villosa (Humb. & Bonpl. ex Willd.) Desv.	Cyathea villosa Humb. & Bonpl. ex Willd.	B	Kanuku Mts.
1200		ADI	Adiantum ternatum Humb. & Bonpl. ex Willd.	Adiantum tetraphyllum Humb. & Bonpl. ex Willd.		Kanuku Mts.
1201	1843	ADI	Adiantum cajenense Willd. ex Klotzsch	Adiantum cajenense Willd. ex Klotzsch	ht: B; it: B BM	Kanuku Mts.
1202	1843	ADI	Adiantum polyphyllum Willd.	Adiantum tomentosum Klotzsch	ht: B; it: GH K U US	
1203		TEC	Aspidium funestum Kunze	Triplophyllum funestum (Kunze) Holttum	B	
1204		DST	Lindsaea falcata Dryand.	Lindsaea lancea (L.) Bedd. var. falcata (Dryand.) Rosenstock	B BR LE	
1205		DST	Lindsaea guianensis (Aubl.) Dryand.	Lindsaea guianensis (Aubl.) Dryand. subsp. guianensis	B	
1205*		DST	Lindsaea pallida Klotzsch	Lindsaea pallida Klotzsch	ht: B; it: B BM	
1206	1843	ASL	Asplenium alloeopteron Kunze ex Klotzsch	Asplenium radicans L. var. radicans	st: B G K US	
1207		GMM	Xiphopteris serrulata (Sw.) Kaulf.	Cochlidium serrulatum (Sw.) L.E. Bishop	B K	Kanuku Mts.
1208		HMP	Hymenophyllum poeppigianum C. Presl	Hymenophyllum decurrens (Jacq.) Sw.	B	Roraima Mts.
1209		LYC	Lycopodium robustum Klotzsch	Huperzia robusta (Klotzsch) Holub	ht: B	
1210		PLP	Polypodium subulatum Klotzsch	Polypodium triseriale Sw.	tc: B	
1211		GMM	Mecosorus marginellus Klotzsch var. major Klotzsch	Mecosorus marginellus Klotzsch var. major Klotzsch	type	Roraima Mts.
1212		ASL	Asplenium auriculatum Klotzsch	Asplenium harpeodes Kunze	B G	Roraima
1213	1843	GMM	Polypodium apiculatum Kunze ex Klotzsch	Grammitis apiculata (Kunze ex Klotzsch) Seymour	st: B	
1214		HMP	Hymenophyllum poeppigianum C. Presl	Hymenophyllum decurrens (Jacq.) Sw.	B	Kanuku Mts.
1214*	1843	PTR	Polypodium confusum J. Sm.	Ctenopteris melanosticta (Kunze) Copeland	B	
1214**		GMM	Xiphopteris serrulata (Sw.) Kaulf.	Cochlidium serrulatum (Sw.) L.E. Bishop	B	
1215		HMP	Trichomanes guianense J.W. Sturm	Trichomanes ankersii Parker ex Hook. & Grev.	st: B BR K	Kanuku Mts.
1215*		HMP	Trichomanes commutatum J.W. Sturm	Trichomanes pedicellatum Desv.	st: B BR K	Kanuku Mts.
1216		LOM	Acrostichum cuspidatum Willd.	Elaphoglossum laminarioides (Bory ex Fée) T. Moore	B	
1217		GMM	Polypodium trifurcatum L.	Enterosora trifurcata (L.) L.E. Bishop		Roraima Mts.
1218		BLE	Blechnum asplenioides Sw.	Blechnum asplenioides Sw.		
1219		SCZ	Anemia schomburgkiana C. Presl	Anemia oblongifolia (Cav.) Sw. var. microphylla	B K	
1220		ARE	Hyospathe elegans Mart.	Hyospathe elegans Mart.	ht: B	Kanuku Mts.
1221						

Index 5. Richard Schomburgk's collection series (1841–1844) – continued

No.	Coll.	Date	Fam.	Original name	Current name	Herb.	Locality
1222		Apr 1843	ERO	Paepalanthus capillaceus Klotzsch [n.n.]	Rondonanthus capillaceus (Klotzsch ex Körn.) Hensold. & Giul.	tc: B	Mt. Roraima & Humirida Mts.
1223			POA	Panicum pilosum Sw.	Panicum pilosum Sw.		
1224							
1225							
1226			CYP	Diplasia karataefolia Rich.	Diplasia karataefolia Rich.		
1227			POA	Bambusa surinamensis Rupr.	Bambusa vulgaris Schrad. ex Wendl.	BM K P US	Roraima
1227*			CYP	Cryptangium stellatum Boeck.	Didymiandrum stellatum (Boeck.) Gilly	tc: B	
1228			CYP	Rhynchospora globosa (Kunth) Roem. & Schult.	Rhynchospora globosa (Kunth) Roem. & Schult.		
1229							
1230							
1231							
1232			ORC	Epidendrum durum Lindl.	Epidendrum dendrobioides Thunb.		
1233			ORC	Habenaria demerarensis Rchb.f. [n.n.]	Habenaria parviflora Lindl.		
1234			EUP	Phyllanthus klotzschianus Muell. Arg.			
1235	955		AQF	Ilex macoucou Pers.	Ilex guianensis (Aubl.) Kuntze		
1236	921		MLS	Miconia aplostachya (Bonpl.) DC.	Miconia aplostachya (Bonpl.) DC.		
1237	920	1842–43	EUP	Discocarpus essequiboensis Klotzsch	Discocarpus essequiboensis Klotzsch		
1238	951		AMA	Bucholtzia philoxeroides Mart.	Alternanthera philoxeroides (Mart.) Griseb.	F	Manakobi
1239							
1240							
1241			MRT	Myrcia schomburgkiana O. Berg	Myrcia servata McVaugh	BM	
1242			HLC	Heliconia bicolor Klotzsch [n.n.]	Heliconia hirsuta L.f.	type	vicinity of Pirara
1243			ACA	Pipteracanthus microcalyx Nees	Ruellia microcalyx (Nees) Lindau	ht: B †; it: K	savanna
1244			FLC	Hisingera ciliatifolia Clos	Xylosma ciliatifolium (Clos) Eichl.	B	
1245			MRN	Calathea latifolia (Willd. ex Link) Klotzsch			
1246	945		MRT	Aulomyrcia curatellaefolia (DC.) O. Berg var. parvifolia O. Berg	Myrcia tomentosa (Aubl.) DC.	F	
1247			BIG	Bignonia stricta Klotzsch [n.n.]	Lundia densiflora DC.	type	savanna
1248	960		MRT	Myrcia schomburgkiana O. Berg	Myrcia servata McVaugh	K	Kanuku Mts.
1249		1843	VIO	Alsodeia laxiflora Benth.	Rinorea brevipes (Benth.) S.F. Blake	M	
1250			EUP	Alchornea latifolia Sw.	Aparisthmium cordatum (A. Juss.) Baill.		
1251	951	1843	TRG	Trigonia laevis Aubl.	Trigonia laevis Aubl. var. microcarpa (Sagot ex Warm.) Sagot		
1252	941		MRT	Psidium parviflorum Benth.	Psidium sartorianum (O. Berg) Nied.		

				Original name	Current name	Herbarium	Locality
1253	940	Apr 1843	SOL	Solanum microcalyx Klotzsch [n.n.]	Solanum campaniforme Roem. & Schult.	tc: B †	Kanuku Mts.
1254		Apr 1843	CUC	Anguria triphylla Klotzsch [n.n.]	Psiguria triphylla (Miq.) C. Jeffrey	tc: B †	
1255			MLS	Miconia plebeia Naudin	Miconia brevipes Benth.		
1256	923	Apr 1843	VRB	Vitex umbrosa Sw.	Vitex compressa Turcz.		Kanuku Mts.
1257	950		MYS	Myristica fatua Sw.	Virola surinamensis (Rol.) Warb.	G	Kanuku Mts.
1258	954		SPT	Mimusops balata (Aubl.) Gaertn.	Manilkara bidentata (A. DC.) A. Chev. subsp. bidentata		Kanuku Mts.
1259			EUP	Caperonia cubensis Klotzsch [n.n.]	Caperonia castaneifolia (L.) A. St.-Hil.		
1260			CLU	Clusia nemorosa G. Mey.	Clusia nemorosa G. Mey.		
1261							
1262	947		PLG	Coccoloba lucidula Benth.	Coccoloba lucidula Benth.		
1263							
1264	954	Apr 1843	MLS	Miconia mucronata (Desr.) Naudin	Miconia holosericea (L.) DC.	tc: NY	Roraima
1265	929		PLG	Coccoloba striata Benth.	Coccoloba striata Benth.		
1266			MRT	Psidium araca Raddi	Psidium guineense Sw.		
1267			TNR	Turnera aurantiaca Benth.	Turnera aurantiaca Benth.		
1268	924		RUB	Chomelia tenuiflora Benth.	Chomelia schomburgkii Steyerm.		
1269	952		MRT	Psidium	Psidium sartorianum (O. Berg) Nied.		
1270	925		ERX	Erythroxylum ectinocalyx Klotzsch ('Erythroxylon')	Erythroxylum divaricatum Peyr.		
1271	927*		RUB	Chomelia angustifolia Benth.	Chomelia angustifolia Benth.		
1272	927		MRT	Calycampe latifolia O. Berg	Myrcia calycampa Amshoff		Rupununi R.
1273	939	1842–43	EUP	Dactylostemon schomburgkii Klotzsch	Actinostemon schomburgkii (Klotzsch) Hochr.	ht: B	Rupununi R.
1274		Apr 1843	LAU	Oreodaphne schomburgkiana Nees var. sparsiflora	Ocotea schomburgkiana (Nees) Mez		
1275			MRT				
1276			OCH				
1277							
1278	922		MRT	Eugenia incanescens Benth.	Eugenia incanescens Benth.	BR	
1279	948		CHB	Licania leptostachya Benth.	Licania leptostachya Benth.		
1280			EUP	Adenogyne guyanensis Klotzsch [n.n.]	Gymnantes guyanensis Klotzsch ex Muell. Arg.	ht: B †	
1281			HLC	Heliconia bicolor Klotzsch [n.n.]	Heliconia hirsuta L.f.		
1282	937		VIO	Alsodeia laxiflora Benth.	Rinorea brevipes (Benth.) S.F. Blake	K L M	
1283	938		FLC	Casearia densiflora Benth.	Casearia commersoniana Cambess.	M	
1284	953		APO	Thyrsanthus gracilis Benth.	Forsteronia gracilis (Benth.) Muell. Arg.		
1285							
1286	928	May 1843	LAU	Oreodaphne fasciculata Nees	Ocotea fasciculata (Nees) Mez	ht: B; it: B	Rupununi R.
1287			MRT	Campomanesia poiteaui O. Berg	Campomanesia grandiflora (Aubl.) Sagot		Rupununi R.

Index 5. Richard Schomburgk's collection series (1841–1844) – continued

No.	Series	Date	Fam.	Name	Current name	Status	Locality
1288	959		ANN	Duguetia quitarensis Benth.	Duguetia quitarensis Benth.	B	
1289	930		RUB	Coussarea schomburgkiana (Benth.) Benth. & Hook.f.	Coussarea violacea Aubl.		
1290	933		FLC	Casearia spinosa (L.) Willd.	Casearia aculata Jacq.	K	
1291			ALJ	Sagittaria lancifolia L.	Sagittaria lancifolia L.		
1291*			SAP	Schmidelia conduplicata Klotzsch [n.n.]	Paullinia conduplicata Radlk.	type	
1292	944	May 1843	MRT	Eugenia roraimana O. Berg	Eugenia roraimana O. Berg		Rupununi R.
1293			LOR	Struthanthus amplexicaulis (Kunth) G. Don	Orychanthus alveolatus (Kunth) Kuijt		
1294			MRT				
1295			MRT	Eugenia eurycheila O. Berg	Eugenia eurycheila O. Berg	tc: B †	Rupununi R.
1296			BIG	Bignonia sordida Klotzsch [n.n.]	Roentgenia sordida (Bureau & K. Schum.) Sprague & Sandw.	type	Rupununi R.
1296*			SAP	Paullinia bipinnata Poir.	Paullinia leiocarpa Griseb.		
1297	932		RUB	Chomelia	Chomelia		
1298	946		RUB	Genipa caruto Kunth	Genipa americana L.		
1299	956		ANA				
1299*			RUT	Zanthoxylum perrottetii DC.	Zanthoxylum rhoifolium Lam.		
1300	931		SYM	Symplocos ciponima L'Hér.	Symplocos guianensis (Aubl.) Guerke		
1301	935		CLU	Hypericum cayennense Jacq.	Vismia cayennensis (Jacq.) Pers.		
1302	942		ANN	Rollinia timifolia Klotzsch [n.n.]	Rollinia exsucca (DC. ex Dunal) A. DC.	ht: B †	Rupununi R.
1303	955		MLS	Miconia mucronata (Desr.) Naudin	Miconia holosericea (L.) DC.	K US	Rupununi R.
1304			BOR	Cordia polystachya Kunth	Cordia schomburgkii DC.	B	Rupununi R.
1305		May 1843	MRN		Myrosma cannifolia L.f.		
1306	936		MRT	Thalianthus macropus Klotzsch [n.n.]	Mycia fallax (Rich.) DC.		
1307							
1308							
1309							
1310		May 1843	MRS	Conomorpha robusta Klotzsch [n.n.]	Cybianthus robustus (Mez) G. Agostini	type	Rupununi R.
1311	934		MLS	Miconia plebeia Naudin	Miconia brevipes Benth.		
1312			HAE	Xiphidium floribundum Sw.	Xiphidium caeruleum Aubl.		
1313			MLP	Camarea affinis A. St.-Hil.	Camarea affinis A. St.-Hil.		
1314							
1315							
1316	958		MRT	Myrciaria verticillata O. Berg	Myrciaria floribunda (West ex Willd.) O. Berg		
1317	953*		FAB	Deguelia scandens Aubl.	Derris pterocarpa (DC.) Killip	U	

No.	No.	Date	Fam.	Name	Current name	Notes	Locality
1318	957		CSL	Tachigali ('Tachigalia')	Dichapetalum rugosum (Vahl) Prance	ht: B †; it: F MO	
1319			DCH	Dichapetalum flavicans Engl.			
1320							
1321	860		CAN	Canna glauca L.	Canna glauca L.		
1322	742		PED	Sesamum			
1323	899		PED	Sesamum			
1324	748		MEL	Melia sempervirens Sw.	Melia azedarach L.		
1325							
1326			MRN	Thalia altissima Klotzsch [n.n.]	Thalia geniculata L.	K	Pakaraima Mts.
1327	845		VRB	Avicennia nitida Jacq.	Avicennia germinans (L.) Stearn		
1328	833		FAB	Pterocarpus rohrii Vahl	Pterocarpus rohrii Vahl		
1329	815		CLU	Vismia macrophylla Kunth	Vismia macrophylla Kunth		
1330	849		FAB	Clitoria arborescens R. Br.	Clitoria arborescens R. Br.	tc: K	Lake Tapakuma
1331			BIG	Arrabidaea schomburgkii Klotzsch [n.n.]	Arrabidaea candicans (Rich.) DC.		
1332	741		RUB	Sabicea velutina Benth.	Sabicea velutina Benth.		
1333	901						
1334	836	Aug 1843	ANN	Guatteria vestita Klotzsch var. latifolia Klotzsch [n.n.]	Guatteria schomburgkiana Mart.	ht: B	
1335	743		MLP	Bysonima propinqua Benth.	Byrsonima spicata (Cav.) DC.	ht: K	
1336	749		LAU	Oreodaphne schomburgkiana Nees	Ocotea schomburgkiana (Nees) Mez	st: B	
1336*	912	Aug 1843	RUB	Amaioua surinamensis Steud.	Duroia eriopila L.f.		
1337	914		CSL	Tachigali paniculata Aubl. ('Tachigalia')	Tachigali paniculata Aubl. var. alba (Ducke) Dwyer		
1338	739		CLU	Clusia alba L.	Clusia palmicida Rich.		
1339	746		BRS	Trattinickia schomburgkii Klotzsch [n.n.]	Trattinickia burserifolia Mart.		
1340	789		ANA	Tapirira guianensis Aubl.	Tapirira guianensis Aubl.		
1341	778		ORC	Cypripedium palmifolium Lindl.	Selenipedium palmifolium (Lindl.) Rchb.f.		
1342	750		ANA	Mangifera indica L.	Mangifera indica L.		
1343	744		XYR	Xyris			
1344							
1345	759	Aug 1843	OLC	Olax schomburgkii Klotzsch [n.n.]	Dulacia guianensis (Engl.) Kuntze	ht: B; it: B F K	Lake Tapakuma
1346	752		MEL	Trichilia schomburgkii C. DC.	Trichilia schomburgkii C. DC. subsp. schomburgkii		
1347	848		MRS	Weigeltia guianensis Klotzsch [n.n.]	Cybianthus surinamensis (Spreng.) G. Agostini	B	
1348	753	Aug 1843	CLU	Tovomita schomburgkiana Klotzsch [n.n.]	Tovomita schomburgkii Planch. & Triana		
1349	740		LAU	Persea gratissima Gaertn.	Persea americana P. Mill.		
1350	790		CLU	Clusia insignis Mart.	Clusia grandiflora Splitg.		
1351	738		SAP	Lasianthemum unijugum Klotzsch [n.n.]	Talisia squarrosa Radlk.	lt: K	
1352	884		MRT	Eugenia tapacumensis O. Berg	Eugenia tapacumensis O. Berg		
1353	747	Aug 1843	AQF	Ilex umbellata Klotzsch [n.n.]	Ilex umbellata Klotzsch ex Reissek var. umbellata	tc: B BR F G	Lake Tapakuma

Index 5. Richard Schomburgk's collection series (1841–1844) – continued

No.	No.	Date	Fam.	Original name	Current determination	Notes	Locality
1354	862		CLU	Quapoya ligulata Klotzsch [n.n.]	Clusia myriandra (Benth.) Planch. & Triana		
1355							
1356			ORC	Epidendrum odoratissimum Lindl.	Encycla odoratissima (Lindl.) Schlechter		
1357	908		CHB	Licania heteromorpha Benth.	Licania heteromorpha Benth. var. heteromorpha		
1358	869		AQF	Ilex umbellata Klotzsch [n.n.]	Ilex umbellata Klotzsch ex Reissek var. humirioides (Reissek) Loes.	K P; st: B G K W	
1359	825		HUM	Humiria obovata Benth. ('Humirium obovatum')	Humiriastrum obovatum (Benth.) Cuatrec.		
1360	902		VOC	Vochisia schomburgkiana Klotzsch [n.n.]	Vochisia schomburgkii Warm.		
1361	822		CHB	Licania schomburgkii Klotzsch [n.n.]	Licania affinis Fritsch	BR K	Upper Demerara R.
1362	801		HUG	Roucheria schomburgkii Planch.	Roucheria schomburgkii Planch.		
1363			AQF	Ilex martiniana D. Don	Ilex martiniana D. Don		
1364			ARA	Spathiphyllum saggitifolium (Rodsch.) Schott ('saggitaefolium')	Spathiphyllum saggitifolium (Rodsch.) Schott ('saggitaefolium')	st: B	
1365	878	Aug 1843	CLU	Tovomita macrophylla Klotzsch [n.n.]	Tovomita obovata Engl.		
1366	876	Aug 1843	CEC	Coussapoa fagifolia Klotzsch	Coussapoa microcephala Trécul	B	
1367	800		CLU	Vismia macrophylla Kunth	Vismia macrophylla Kunth	ht: B; it: K	Pomeroon R.
1368	897		MLS	Clidemia divaricata Naudin	Leandra divaricata (Naudin) Cogn.		
1369	834		SCR	Bacopa aquatica Aubl.	Lindernia diffusa (L.) Wettst.		
1370	802		SCR	Vandellia diffusa L.			
1371	904		AST	Sparganophorus vaillantii Gaertn.	Struchium sparganophorum (L.) Kuntze		Pomeroon R.
1372							
1373							
1374	895		LAU	Nectandra leucantha Nees	Nectandra globosa (Aubl.) Mez		
1375	754		APO	Bonafousia undulata (Vahl) A. DC.	Tabernaemontana undulata Vahl	st: B	
1376	827		MLV	Sida urens L.	Sida urens L.		
1377							
1378	788		APO	Malouetia guianensis Klotzsch [n.n.]	Malouetia tamaquarina (Aubl.) A. DC.		
1379	818		HPC	Hippocratea ovata Lam.	Hippocratea volubilis L.		
1379*	870		MLP	Byrsonima rugosa Benth.	Byrsonima stipulacea A. Juss.		
1380	745		LAU	Acrodiclidium guianense Nees	Licaria polyphylla (Nees) Kosterm.	ht: K	
1381	817		EUP	Siphonia schomburgkii Klotzsch [n.n.]	Hevea pauciflora (Spruce ex Benth.) Muell. Arg.		
1382	905		CHB	Moquilea guianensis Aubl.	Licania guianensis (Aubl.) Griseb.	BR	
1383							
1384	867		FAB	Centrosema			
1385							

No.		Date	Fam	Name (as labelled)	Accepted name	Type	Locality
1386	830		APO	Malouetia schomburgkii Muell. Arg.	Malouetia flavescens (Willd. ex Roem. & Schult.) Muell. Arg.	type	
1387	786		MLS	Miconia erythropila Steud.	Miconia ceramicarpa (DC.) Cogn. var. ceramicarpa		
1388	768		ELC	Sloanea dentata L.	Sloanea grandiflora Sm.		
1389	864		SPT	Chrysophyllum auratum Miq.	Chrysophyllum argenteum Jacq. subsp. auratum (Miq.) T.D. Penn.		
1390			AQF	Ilex martiniana D. Don	Ilex martiniana D. Don	MEL	Pomeroon R.
1391	829	Aug 1843	MIM	Inga leiocalycina Benth.	Inga leiocalycina Benth.		Pomeroon R.
1392	809		LAU	Oreodaphne costulata Nees	Ocotea aciphylla (Nees) Mez		
1393	877	1843	SAP	Cupania velutina Klotzsch [n.n.]	Cupania hirsuta Radlk.		
1394	765		VIO	Amphirrhox			Pomeroon R.
1395	826		MIM	Inga graciliflora Benth.	Inga graciliflora Benth.	MEL	Pomeroon R.
1396	756		PIP	Piper arboreum Aubl.	Piper arboreum Aubl.		
1397	903		BOR	Cordia melanoneura Klotzsch [n.n.]	Cordia exaltata Lam. var. melanoneura I.M. Johnst.	ht: B	Pomeroon R.
1398	842	Aug 1843	MIM	Inga myriantha Poepp.	Inga umbellifera (Vahl) Steud. ex DC.	K P	Pomeroon R.
1399			CLU	Tovomita brevistaminea Engl.	Tovomita brevistaminea Engl.	B	Pomeroon R.
1400	751	Aug 1843	MYS	Myristica sebifera Sw.	Virola sebifera Aubl.		Pomeroon R.
1401	893		LAU	Oreodaphne caudata Nees	Ocotea cernua (Nees) Mez	st: GZU	
1402	907	Aug 1843	LAU	Aydendron riparium Nees	Aniba riparia (Nees) Mez	tc: F K	Pomeroon R.
1403			STR	Melochia melissaefolia Benth.	Melochia melissaefolia Benth.		
1404	828		ANA	Tapirira guianensis Aubl.	Tapirira guianensis Aubl.		Pomeroon R.
1405	832		MLV	Hibiscus verbasciformis Klotzsch [n.n.]	Hibiscus verbasciformis Klotzsch ex Hochr.		
1405*	865		MLP	Byrsonima altissima DC. var. occidentalis Nied.	Byrsonima aerugo Sagot	st: K	Pomeroon R.
1406	793		MIM	Pithecellobium trapezifolium (Vahl) Benth. ('Pithecolobium')	Abarema jupunba var. trapezifolia (Vahl) Barneby & J.W. Grimes		
1407	787		CLU	Tovomita schomburgkii Planch. & Triana	Tovomita schomburgkii Planch. & Triana	ht: B †	
1408	811		EUP	Dactylostemon riparium Nees [n.n.]	Actinostemon guyanensis Pax & K. Hoffm.		
1409							
1410							
1411							
1412							
1413	850		DSC	Dioscorea schomburgkiana Kunth [n.n.]	Dioscorea pilosiuscula Bert. ex Spreng.		
1414	906		MLV	Hibiscus furcellatus Desr.	Hibiscus furcellatus Desr.		Pomeroon R.
1415	822		MIM	Pithecellobium cauliflorum (Willd.) Mart. ('Pithecolobium')	Zygia latifolia (L.) Fawc. & Rendle var. latifolia		
1416	835		CNN	Rourea frutescens Aubl.	Rourea frutescens Aubl.		
1417	795	1842–43	PGL	Moutabea guianensis Aubl.	Moutabea guianensis Aubl.	B	
1418	846		CMB	Terminalia tanibouca Rich.	Terminalia dichotoma G. Mey.		

Index 5. Richard Schomburgk's collection series (1841–1844) – continued

No.	No.	Date	Fam.	Name as written	Accepted name	Type	Locality
1448	769		LOG	Strychnos toxifera R.H. Schomb. ex Benth. var. latifolia Klotzsch ex Prog.	Strychnos toxifera R.H. Schomb. ex Benth.	type	Pomeroon R.
1449	847		FLC	Casearia javitensis Kunth	Casearia javitensis Kunth		
1450	890		MLS	Melastoma chrysophyllum Rich.	Miconia chrysophylla (Rich.) Urb.	tc: B †	
1451	760	Sep 1843	LCS	Lacistema macrophylla Klotzsch [n.n.]	Lacistema aggregatum (Bergius) Rusby	syntype	
1452	896		LAU	Nectandra vaga Meisn.	Nectandra globosa (Aubl.) Mez		
1453	841		CSL	Crudia falcata Klotzsch [n.n.] ('Crudya')	Crudia glaberrima (Steud.) J.F. Macbr.	tc: F frag LE P	
1454	845		FAB	Drepanocarpus schomburgkii Klotzsch	Machaerium inundatum (Mart. ex Benth.) Ducke		
1455	900	Sep 1843	CLU	Quapoya microphylla Klotzsch [n.n.]	Clusia panapanari (Aubl.) Choisy		Sururu R.
1456	837		CLU	Vismia falcata Rusby	Vismia laxiflora Reidhardt		
1456*	917		CLU	Vismia sessilifolia (Aubl.) Choisy	Vismia sessilifolia (Aubl.) Choisy		
1457							
1458	755		CNV	Operculina pterodes Choisy	Operculina hamiltonii (G. Don) Austin & Staples		Sururu R.
1459	891	Sep 1843	MNS	Anomospermum schomburgkii Miers	Orthomene schomburgkii (Miers) Barneby & Krukoff		
1460	885		DSC	Dioscorea megalobotrya Kunth ex M.R. Schomb. ex Kunth	Dioscorea amazonum Mart. var. consanguinea (Kunth) Uline	B	
1461	888		ACA	Aphelandra deppeana Schltdl. & Cham.	Aphelandra scabra (Vahl) Sm.		
1462	886		PON	Heteranthera grandiflora Klotzsch [n.n.]	Eichhornia diversifolia (Vahl) Urb.	tc: B †	
1463	883	Sep 1843	FLC	Homalium puberulum Klotzsch [n.n.]	Homalium guianense (Aubl.) Oken		
1464		Sep 1843	SMR	Picramnia macrostachys Klotzsch ex Engl.	Picramnia macrostachys Klotzsch ex Engl.	ht: B †; it: A F MICH MO NY	
1465	770	Sep 1843	LOG	Strychnos toxifera R.H. Schomb. ex Benth. var. obliqua Klotzsch ex Prog.	Strychnos toxifera R.H. Schomb. ex Benth.	type	
1466	821		APO	Thyrsanthus guyanensis (Muell. Arg.) Miers	Forsteronia guyanensis Muell. Arg.		Sururu R.
1467	889		MLS	Diplochita fothergilla DC.	Miconia mirabilis (Aubl.) L.O. Williams		
1468	898	Sep 1843	BRS	Icica decandra Aubl.	Protium decandrum (Aubl.) Marchand	B † K MO US	Pomeroon R.
1469	910		CNV	Maripa scandens Aubl.	Maripa scandens Aubl.	ht: B †; it: K P	
1470			SPT	Sideroxylon durum Klotzsch [n.n.]	Pouteria cuspidata (A. DC.) Baehni subsp. dura (Eyma) T.D. Penn.	lt: B	Pomeroon R.
1471	851	Sep 1843	DSC	Dioscorea riparia Kunth & M.R. Schomb. [n.n.]	Dioscorea samydea Mart. ex Griseb.		
1472	758		CMB	Cacoucia coccinea Aubl.	Combretum cacoucia (Baill.) Exell ex Sandw.		
1473	853		RHZ	Cassipourea guianensis Aubl.	Cassipourea guianensis Aubl.		
1474	852		MIM	Endata myriadena Benth.	Mimosa myriadenia (Benth.) Benth. var. myriadenia	MEL	
1475	861		CLU	Vismia guianensis (Aubl.) Choisy	Vismia guianensis (Aubl.) Choisy		
1476	866		MRT	Campomanesia glabra Benth.	Calycolpus goetheanus (Mart. ex. DC.) O. Berg		
1477		Sep 1843	MIM	Inga tenuifolia Salzm.	Inga thibaudiana DC.		
1478	757		GUT	Mammea americana L.		MEL	
1479	816		FAB	Machaerium ferrugineum Pers.	Machaerium quinata (Aubl.) Sandw. var. quinata		Pomeroon R.

Index 5. Richard Schomburgk's collection series (1841–1844) – continued

No.	Coll.	Date	Fam.	Determination	Accepted name	Type	Locality
1480	871		CMB	Combretum laxum Jacq.	Combretum laxum Jacq.		
1481	815		MEL	Carapa		P	Pomeroon R.
1482	916		ANA	Tapirira multiflora Mart.	Tapirira guianensis Aubl.		
1483	915		ANA	Tapirira multiflora Mart.	Tapirira guianensis Aubl.		
1484							
1485	887		MRS	Weigeltia schomburgkiana Mez	Cybianthus schomburgkianus (Mez) G. Agostini		
1486	857		TIL	Corchorus aestuans L.	Corchorus acutangulus Lam.		Moruca R.
1487	812		MRT	Calyptranthes obtusa O. Berg	Marlierea montana (Aubl.) Amshoff		Moruca R.
1488	805		EUP	Stilaginella oblonga Tul.	Hyeronima oblonga (Tul.) Muell. Arg.		
1489			LOR				
1490	855		MRT	Aulomyrcia dichroma O. Berg	Myrcia guianensis (Aubl.) DC. var. guianensis	type	Moruca R.
1491			EBN	Diospyros paralea Steud.	Diospyros guianensis (Aubl.) Gürke		
1492	882		MLS	Rhynchanthera	Rhynchanthera dichotoma (Desr.) DC.		
1493	808		APO	Thevetia nerrifolia A. Juss. ex A. DC.	Thevetia peruviana (Pers.) K. Schum.		
1494			FAB	Machaerium leiophyllum (DC.) Benth.	Machaerium leiophyllum (DC.) Benth.		
1495	766		MLS	Miconia alternans Naudin	Miconia alternans Naudin		
1496	844		LOR				
1497	764		CMB	Combretum laxum Jacq.	Combretum laxum Jacq.		
1498	872		SAP	Monopteris guianensis Klotzsch [n.n.]	Matayba arborescens (Aubl.) Radlk.		
1499	814		MLP	Stigmaphyllon puberum (Rich.) A. Juss. var. schomburgkianum Benth.	Stigmaphyllon puberum (Rich.) A. Juss.	ht: K	
1500	819		VRB	Aegiphila macrantha Ducke	Aegiphila macrantha Ducke		
1501	873		MLP	Stigmaphyllon puberum (Rich.) A. Juss. var. puberum	Stigmaphyllon puberum (Rich.) A. Juss.	K	Waini R.
1502			MLS	Miconia	Miconia prasina (Sw.) DC.		
1503	920		LOR	Struthanthus flexistylis Miq.	Phthirusa retroflexa (Ruiz & Pav.) Kuijt		
1504	803		MLP	Banisteria lobulata E. Mey.	Banisteriopsis lucida (Rich.) Small	K	Waini R.
1505	874		CLU	Quapoya ligulata Klotzsch [n.n.]	Clusia myriandra (Benth.) Planch. & Triana		
1506			SPT	Chrysophyllum auratum Miq. var. majus Miq.	Chrysophyllum argenteum Jacq. subsp. auratum. (Miq.) T.D. Penn		
1507	813		LAU	Acrodiclidium oppositifolium Nees	Licaria oppositifolia (Nees) Kosterm.	tc: E	
1508	765	Oct 1843	SPT	Mimusops balata (Aubl.) Gaertn.	Manikara bidentata (A. DC.) A. Chev. subsp. bidentata		
1509	780	Oct 1843					Barama R.
1510	875	Oct 1843	BOR	Cordia guianensis Klotzsch [n.n.]	Cordia fallax I.M. Johnst.	tc: B G	Barama R.

No.	No.	Date	Fam.	Name	Current determination	Code	Locality
1510		Oct 1843	BOR	Cordia aubletii DC.	Cordia schomburgkii A. DC.	B	Barama R.
1511	761	Oct 1843	MLP	Tetrapterys crispa A. Juss.	Tetrapterys crispa A. Juss.	K	Barama R.
1512	806		RUB	Gonzalea spicata DC.	Gonzalagunia spicata (Lamb.) Gómez		
1513	911		ASC	Asclepias	Matelea delascioi Morillo		
1514			APO	Thyrsanthus schomburgkii Benth.	Forsteronia acouci (Aubl.) A. DC.		
1515	820		MIM	Calliandra portoricensis (Willd.) Benth.	Zapoteca portoricensis (Jacq.) H. Hern.		
1516	775		LOG	Strychnos smilacina Benth.	Strychnos mitscherlichii R.H. Schomb.		
1517		Oct 1843	MLP	Heteropterys platyptera DC.	Heteropterys multiflora (DC.) Hochr.	K	Barama R.
1518	772		SPT	Sideroxylon micranthum Klotzsch [n.n.]	Pouteria cuspidata (A. DC.) Baehni subsp. cuspidata		Roraima
1519	776	1843	EUP	Gaedawakka schomburgkiana Kuntze	Chaetocarpus schomburgkianus (Kuntze) Pax & K. Hoffm	B †	
1520	774		VIO	Alsodeia pubiflora Benth.	Rinorea pubiflora (Benth.) Sprague & Sandw. var. pubiflora		Roraima
1520*	774*		VIO	Alsodeia flavescens Spreng.	Rinorea pubiflora (Benth.) Sprague & Sandw. var. grandifolia (Eichl.) Hekking		
1521	879		EUP	Stilaginella laxiflora Tul.	Hyeronima alchorneoides Allemão var. alchorneoides		
1522	880		PLG	Triplaris vahliana Fisch. & Mey. ex C.A. Mey.	Triplaris weigeltiana (Rchb.) Kuntze		
1523	881		CSL	Cassia bacillaris L.f.	Senna sandwithiana H.S. Irwin & Barneby		Roraima
1524	919	Oct 1843	SAP	Serjania baramensis Klotzsch [n.n.]	Serjania membranacea Splitg.		Barama R.
1525			ANA	Spondias lutea L.	Spondias mombin L.		
1526	859		CSL	Cassia bacillaris L.f.	Senna sandwithiana H.S. Irwin & Barneby		
1527			SOL	Solanum			
1528	858		AMA	Iresine grandiflora Hook.	Pfaffia grandiflora (Hook.) Stützer	M P	
1529	856		FLC	Casearia parvifolia Willd.	Casearia decandra Jacq.		
1530	854		RUB	Gonzalea	Gonzalagunia		
1531	762		AST	Vernonia scorpioides (Lam.) Pers.	Cyrtocymura scorpioides (Lam.) H. Rob.		
1532	763	Oct 1843	CMM	Commelina platyphylla Klotzsch [n.n.] ('Commelyna')	Commelina platyphylla Klotzsch ex C.B. Clarke 'Commelyna'	st: B	
1533	777		CNV	Lysiostyles scandens Benth.	Lysiostyles scandens Benth.		
1534	781		CSL	Cynometra marginata Benth. var. guianensis Dwyer	Cynometra marginata Benth. var. guianensis Dwyer	U	
1535	804		STR	Sterculia ivira Sw.	Sterculia pruriens (Aubl.) K. Schum. var. pruriens		
1536		Oct 1843	CNN	Connarus punctatus Planch.	Connarus punctatus Planch.		
1537			MLP	Lophopterys splendens A. Juss.	Lophopterys euryptera Sandw.	ht: K	Barama R.
1538			VOC	Qualea muelleriana M.R. Schomb. [n.n.]	Ruizterania rigida (Stafleu) Marc.-Berti		Upper Cujang R.
1539							
1540			LAU	Acrodiclidium camara R.H. Schomb.	Licaria camara (R.H. Schomb.) Kosterm.	tc: G	Cuyuni R.

Index 5. Richard Schomburgk's collection series (1841–1844) – continued

No.	Coll.	Date	Fam.	Name on label	Current name	Type	Locality
1541	965		MRS	Cybianthus			
1542							
1543							
1544							
1545			MLS	Miconia rubiginosa (Bonpl.) DC.	Miconia rubiginosa (Bonpl.) DC.	ht: K	
1546			GEN	Tachia guianensis Aubl.	Tachia guianensis Aubl.		
1547							Cuyuni R.
1548		Dec 1843	ERI	Ledothamnus guyanensis Meisn.	Ledothamnus guyanensis Meisn.		
1549							
1550							
1551			APO	Dipladenia cordifolia Klotzsch [n.n.]	Galactophora crassifolia (Muell. Arg.) Woods		
1552			HUM	Vantanea emarginata Klotzsch [n.n.]	Vantanea minor Benth.	tc: B MO	
1553	969		OCH	Leitgebia guianensis Eichl.	Sauvagesia guianensis (Eichl.) Sastre	type	savanna
1554							
1555							
1556		1843	ASL	Asplenium salicifolium L.	Asplenium salicifolium L.	B UC	
1557							
1558							
1559							
1560							
1561			LOR				
1562							
1563			ERI	Vaccinium subcrenulatum Klotzsch [n.n.]	Vaccinium puberulum Klotzsch ex Meisn. var. subcrenulatum Maguire, Steyerm. & Luteyn	tc: NY	Upper Courantyne R.
1564			BML	Anoplophytum guianense Beer	Cottendorfia guianensis (Beer) Klotzsch ex Baker var. guianensis	ht: B	Upper Courantyne R.
1565							
1566		1843	TEA	Ternstroemia verticillata Klotzsch [n.n.]	Ternstroemia verticillata Klotzsch ex Wawra	F G	Courantyne R.
1567							
1568			RUB	Randia formosa (Jacq.) K. Schum.	Randia ruiziana DC.		
1569			BIG	Bignonia brachycalyx Klotzsch [n.n.]	Mussatia prieurei (DC.) Bureau ex K. Schum.	type	Courantyne R.
1570							
1571			FLC	Homalium racoubea Sw.	Homalium guianense (Aubl.) Oken	M	
1572							
1573		1843	SAP	Cupania affinis Klotzsch [n.n.]	Matayba macrostylis Radlk.		Upper Courantyne R.

No.	Coll. no.	Date	Code	Name	Accepted name	Type	Locality
1574			DSC	Dioscorea trifida L.f.	Dioscorea trifida L.f.	ht: B †	Courantyne R.
1575							
1576							
1577							
1578			LNT	Utricularia muscosa Benj.	Utricularia calycifida Benj.		
1579			ORC				
1580			LOG	Strychnos erichsonii M.R. Schomb. ex Progel	Strychnos erichsonii M.R. Schomb. ex Progel	tc: F	Courantyne R.
1581	982		HUM	Vantanea guianensis Aubl.	Vantanea guianensis Aubl.	BM NY US	Courantyne R.
1582	983		MEL	Moschoxylon propinquum Miq.	Trichilia quadrijuga Kunth subsp. quadrijuga		
1583		Oct 1843	AST	Gongylolepis benthamiana R.H. Schomb.	Gongylolepis benthamiana R.H. Schomb.	tc: B G	
1584		Oct 1843	MRT	Eugenia fasciculiflora O. Berg	Eugenia feijoi O. Berg	type	Upper Courantyne R.
1585							
1586							
1587			STR	Melochia graminifolia A. St.-Hil.	Melochia graminifolia A. St.-Hil.		
1588			MRT	Eugenia pyrroclada O. Berg	Eugenia pyrroclada O. Berg		
1589			CSL	Heterostemon mimosoides Desf.	Heterostemon mimosoides Desf. var. mimosoides		
1590			ELC	Sloanea laurifolia (Benth.) Benth.	Sloanea laurifolia (Benth.) Benth.		
1591			OCH	Elvasia calophyllea DC.	Elvasia calophyllea DC.		
1592						type ?	
1593							
1594			MRS	Conomorpha laxiflora (Mart.) A. DC.	Cybianthus spicatus (Kunth) G. Agostini		
1595							
1596							
1597			MLV	Malachra fasciata Jacq.	Malachra fasciata Jacq.		Upper Courantyne R.
1598							
1599	986	1842–43	CHB	Licania heteromorpha Benth.	Licania heteromorpha Benth. var. heteromorpha	B †	Courantyne R.
1600			PLG	Gussonia cuneata Klotzsch [n.n.]	Coccoloba acuminata Kunth	B	Courantyne R.
1601		1843	RPT	Rapatea paludosa Aubl.	Rapatea paludosa Aubl.	tc: B	Rupununi R.
1602		1843	LOR	Viscum guyanense Klotzsch [n.n.]	Phthirusa guyanensis Eichl.	ht: GOET; it: K NY	Upper Cuyuni R.
1603			MLP	Byrsonima coleostachya Griseb.	Byrsonima stipulacea A. Juss.	st: B †	
1604	987		CNN	Bernardia guianensis Schellenb.	Cnestidium guianense (Schellenb.) Schellenb.		
1605	979		MRT	Calyptranthes fasciculata O. Berg	Calyptranthes fasciculata O. Berg		
1606			fern				
1607			fern				
1608			fern				
1609			moss	Leucobryum martianum (Hornsch.) C. Müll.	Leucobryum martianum (Hornsch.) C. Müll.		
1610			fern				
1611			fern				

Index 5. Richard Schomburgk's collection series (1841–1844) – continued

1612	fern					
1613	fern					
1614	fern					
1615	fern					
1616	fern					
1617	fern					
1618	fern					
1619	fern					
1620	fern					
1621	fern					
1622	fern					
1623	fern					
1624	fern					
1625	fern					
1626	fern					
1627	fern					
1628	fern					
1629	fern					
1630	fern					
1631	fern					
1632	fern					
1633	fern					
1634	fern					
1635	fern					
1636	fern					
1637	fern					
1638	fern					
1639	fern					
1640	fern					
1641	fern					
1642	fern					
1643	fern					
1644	fern					
1645	fern					
1646	fern	1843	DST	Davallia imrayana Hook.	Ormoloma imrayanum (Hook.) Maxon st: B	Mt. Roraima

No.	Family	Name (as collected)	Year	Current name	Distribution	Locality
1647	LOM (fern)	Acrostichum decoratum Kunze		Elaphoglossum decoratum (Kunze) T. Moore	B BD	Mt. Roraima
1648	LOM	Acrostichum simplex Sw. ex Klotzsch		Elaphoglossum flaccidum (Fée) T. Moore	tc: B	
1649	GMM	Grammitis furcata Hook. & Grev.	1843	Cochlidium furcatum (Hook. & Grev.) C. Chr.	st: B	
1650	PLP	Polypodium richardii Klotzsch		Polypodium adnatum Kunze ex Klotzsch	ht: B; it: B NY frag	
1651	LOM	Acrostichum prieurianum (Fée) Klotzsch	1843	Lomariopsis prieuriana Fée	B CAY	
1652	PLP	Polypodium aureum L.		Phlebodium aureum (L.) J. Sm.	B	
1653	PLP	Polypodium decumanum Willd.	1843	Phlebodium decumanum (Willd.) J. Sm.	B	
1653*	PLP	Polypodium persicariifolium Schrad. ('persicariaefolium')	1843	Microgramma persicariifolium (Schrad.) C. Presl	B	
1654	ADI (fern)	Hemionitis palmata L.	1843	Hemionitis palmata L.	B	
1655	HMP	Hymenostachys diversifrons Bory		Trichomanes diversifrons (Bory) Mett. ex Sadeb.	B	
1656	CTH	Alsophila multiflora (Sw.) J. Sm.		Cyathea pungens (Willd.) Domin	B	
1657	CTH	Alsophila multiflora (Sw.) J. Sm.	1843	Cyathea multiflora J. Sm.	G	
1658	CTH	Alsophila multiflora (Sw.) J. Sm.	1843	Cyathea cyatheoides (Desv.) K.U. Kramer	B K	
1658*	DRY	Polybotrya caudata Kunze	1843	Polybotrya caudata Kunze	B	
1658**	LYC	Lycopodium cernuum L.	1843	Lycopodiella cernua (L.) Pic.-Ser.	B K	Mt. Roraima
1659	ASL	Asplenium formosum Willd.	1843	Asplenium formosum Willd.	B BR	
1660	HMP	Hymenophyllum asplenioides (Sw.) Sw.	1843	Hymenophyllum asplenioides (Sw.) Sw.	B BR	Mt. Roraima
1661	HMP	Sphaerocionium crispum (Kunth) Klotzsch var. amoenum Klotzsch		Hymenophyllum fendlerianum J.W. Sturm	ht: B; it: BR K P	
1664	ADI	Gymnogramma calomelanos (L.) Kaulf.	1843	Pityrogramma calomelanos (L.) Link	B	
1665	GMM	Polypodium cultratum Willd.	1843	Grammitis mollissima (Fée) Proctor	B BR	
1666	CTH	Alsophila pungens (Willd.) Kaulf.		Cyathea pungens (Willd.) Domin	B K LE UC	
1667	OLN	Nephrolepis ensifolia (Sw.) C. Presl	1843	Nephrolepis biserrata (Sw.) Schott	B	
1668	PLP	Polypodium percussum Cav.	1843	Pleopeltis percussa (Cav.) Hook. & Grev.	B	
1669	PLP	Polypodium attenuatum Humb. & Bonpl. ex Willd.	1843	Polypodium attenuatum Humb. & Bonpl. ex Willd.	B	
1670	PLP	Polypodium incanum Sw.	1843	Polypodium polypodioides (L.) Watt	B	
1671	GLC	Mertensia schomburgkiana J.W. Sturm	1843	Dicranopteris schomburgkiana (J.W. Sturm) C. Morton	ht: B; it: CAY NY UC US	Mt. Roraima
1672	PTR	Acrostichum aureum L.	1843	Acrostichum aureum L.	B	
1673	LOM	Acrostichum erythrodes (Fée) Kunze		Lomariopsis japurensis (Mart.) J. Sm.	B CAY	
1674	THL	Meniscium serratum Cav.		Thelypteris serrata (Cav.) Alston	B	
1674*	DRY	Aspidium hookerii Klotzsch		Cyclodium meniscioides (Willd.) C. Presl var. meniscioides	B UC	
1675	PLP	Polypodium fasciale Humb. & Bonpl. ex Willd.		Campyloneuron repens (Aubl.) C. Presl	B	
1676	TEC	Aspidium macrophyllum Sw.		Tectaria incisa Cav.		

Index 5. Richard Schomburgk's collection series (1841–1844) – continued

No.	No.	Date	Fam.	Determination	Current name	Note	Locality
1677							
1678			ORC	Pleurothallis ruscifolia (Jacq.) R. Br.	Pleurothallis ruscifolia (Jacq.) R. Br.		
1679*			ORC	Pseuderiopsis schomburgkii Rchb.f.	Eriopsis biloba Lindl.		
1680			ORC	Sobralia sessilis Lindl.	Sobralia sessilis Lindl.		
1681			CYP	Diplasia karataefolia Rich.	Diplasia karataefolia Rich.		
1682							
1683							
1684							
1685							
1686	439		CNN	Connarus incomptus Planch.	Connarus incomptus Planch.		
1687							
1688							
1689							
1690							
1691							
1692							
1693							
1694							
1695							
1696							
1697							
1698							
1699							
1700							
1701	989		RUB	Palicourea guianensis Aubl.	Palicourea guianensis Aubl.		
1702	990		CLU	Garcinia macrophylla Mart.	Rheedia benthamiana Planch. & Triana		
1703	1004	Mar 1844	LAU	Nectandra rodioei R.H. Schomb.	Chlorocardium rodiei (R.H. Schomb.) Rohwer, Richter & van der Werff		
1704	994		CUC	Anguira polyanthos Klotzsch [n.n.]	Gurania subumbellata (Miq.) Cogn.	it: B	Demerara R.
1705			MLP	Tetrapterys glaberrima Benth.	Tetrapterys mucronata Cav.		Demerara R.
1706	1010		ANA	Tapirira guianensis Aubl.	Tapirira guianensis Aubl.	K	Demerara R.
1707		Mar 1844	CLU	Quapoya colorans Klotzsch [n.n.]	Clusia panapanari (Aubl.) Choisy	tc: B	Demerara R.
1708							Demerara R.
1709			BIG	Pithecoctenium guianense Klotzsch [n.n.]	Distictella parkeri (DC.) Sprague & Sandw.		Demerara R.

				Name [as written]	Accepted name	Type/Herb.	Locality
1710	992	1844	SAP	Urvillea schomburgkii Klotzsch [n.n.]	Serjania paucidentata DC.	B	
1711		1844	ASL	Asplenium cuneatum Lam.	Asplenium cuneatum Lam.	B	
1712	996		CNN	Omphalobium opacum Klotsch [n.n.]	Connarus coriaceus Schellenb.		
1713	991	1844	MYS	Myristica sebifera Sw.	Virola sebifera Aubl.	G	Demerara R.
1714							
1715			CLU	Renggeria montana Klotzsch [n.n.]	Quapoya bracteolata Sandw.		
1716	993	Mar 1844	ANN	Guatteria vestita Klotzsch var. angustifolia Klotzsch [n.n.]	Guatteria schomburgkiana Mart.	B NY	Demerara R.
1717	1007	1844	PGL	Bredemeyera bracteata Klotzsch ex Hassk.	Bredemeyera bracteata Klotzsch ex Hassk.	ht: L	Demerara R.
1718	998		ACA	Trichanthera gigantea (Humb. & Bonpl.) Nees	Trichanthera gigantea (Humb. & Bonpl.) Nees var. guianensis Gleason		
1719	1006		CHB	Moquilea guianensis Aubl.	Licania guianensis (Aubl.) Griseb.	K	
1720	1003		LAU	Aiouea densiflora Nees	Aiouea guianensis Aubl.	ht: B †; it: US	Demerara R.
1721	995		MRS	Conomorpha magnoliifolia Mez	Cybianthus fulvopulverulentus (Mez) G. Agostini subsp. magnoliifolius (Mez) Pipoly		
1722	988		CSL	Parivoa	Eperua grandiflora (Aubl.) Benth. subsp. guyanensis R.S. Cowan		
1723	1000		FAB	Clitoria javitensis (Kunth) Benth.	Clitoria javitensis (Kunth) Benth.		
1724							
1725							
1726							
1727	1008		MLS	Acinodendron plukenetti Kuntze	Miconia plukenetti Naudin		
1728			HPC	Hippocratea floribunda Benth.	Elachyptera floribunda (Benth.) A.C. Sm.		
1729	1002		RUB	Malanea angustifolia Bartl. [n.n.]	Malanea obovata Hochr.	K	Demerara R.
1730	1009		LAU	Aydendron firmulum Nees	Beilschmiedia curviramea (Meisn.) Kosterm.		Demerara R.
1731							
1732	999	Apr 1844	CLU	Renggeria montana Klotzsch [n.n.]	Quapoya bracteolata Sandw.	tc: B	
1733	997		SCR	Alectra brasiliensis Benth.	Alectra aspera (Cham. & Schltdl.) L.O. Williams		
1734							
1735	1005		MEL	Trichilia roraimana C. DC.	Trichilia micrantha Benth.		
1736							
1737	1013		GNE	Gnetum nodiflorum Brongn.	Gnetum nodiflorum Brongn.		
1737*	1001		RUB	Commianthus schomburgkii Benth.	Retiniphyllum schomburgkii (Benth.) Muell. Arg.		
1738	1012	Apr 1844	LAU	Oreodaphne schomburgkiana Nees	Ocotea schomburgkiana (Nees) Mez	st: B US	Demerara R.
1739	1011		MLP	Banisteria	Mascagnia guianensis W.R. Anderson	K	Demerara R.
1740			LOR	Struthanthus squamulosus Klotzsch [n.n.]	Phthirusa rufa (Mart.) Eichl.		

INDEX 6. ROBERT AND RICHARD SCHOMBURGK'S SINE NUMERO COLLECTIONS.

Collector	Year	Fam.	Original name	Current name	Herbarium	Location
Rob. ser. 1	1836	ACA	Dipteracanthus humilis Pohl ex Nees var. diffusus Nees		st: K	Roraima
Rich.		ADI	Adiantum cajennense Willd. ex Klotzsch var. stenophyllum Hook.	Adiantum cajennense Willd. ex Klotzsch	syntype ?	
Rob. ser. 2	1842–43	ADI	Adiantum klotzschianum Hook.	Adiantum tomentosum Klotzsch	B	Roraima
Rob. ser. 1	1837	ADI	Adiantum leprieurii Hook.	Adiantum leprieurii Hook.	K	Berbice
Rob. ser. 1		ADI	Adiantum radiatum L.	Adiantopis radiata (L.) Fée	K	Carawaimi Mts.
Rich.		ADI	Gymnogramma calomelanos (L.) Kaulf.	Pityrogramma calomelanos (L.) Link.	NY	Roraima
Rob.		AMA	Iresine polymorpha Mart.	Iresine diffusa Humb. & Bonpl. ex Willd.		
Rob. or Rich.		ANA	Tapirira guianensis Aubl.	Tapirira guianensis Aubl.		
Rob.		ANN	Guatteria foliosa Benth.	Guatteria foliosa Benth.	LE	
Rob. or Rich.		ANN	Guatteria schomburgkiana Mart.	Guatteria schomburgkiana Mart.	tc: P	
Rob. or Rich.		ANN	Rollinia orthopetala A. DC.	Rollinia mucosa (Jacq.) Baill.		
Rob.		ANN	Xylopia grandiflora A. St.-Hil.	Xylopia aromatica (Lam.) Mart.	LE	
Rob. or Rich.		APO	Galactophora schomburgkiana Woodson	Galactophora schomburgkiana Woodson	ht: K MO	
Rob. ser. 1		APO	Plumeria bracteata A. DC.	Himatanthus bracteatus (A. DC.) Woodson	type	R. Padauiri
Rob. or Rich.		AQF	Ilex jenmanii Loes.	Ilex jenmanii Loes.	B K	
Rob. or Rich.		ARA	Dracontium dubium Kunth	Dracontium dubium Kunth	tc: B †	Mt. Curassawaka of the Kanuku Range, S of Nappi R.
Rich.		ARE	Attalea speciosa Mart.	Orbignya phalerata Mart.		
Rich.		ARE	Manicaria saccifera Gaertn.	Manicaria saccifera Gaertn.		
Rich.	1844	ASL	Asplenium alloeopteron Kunze ex Klotzsch	Asplenium radicans L. var. radicans	UC	
Rich.		ASL	Asplenium cuneatum Lam.	Asplenium cuneatum Lam.	tc: B CAY	
Rich.		ASL	Asplenium pteropus Kaulf.	Asplenium pteropus Kaulf.	W	Mt. Roraima
Rich.		ASL	Asplenium schomburgkianum Klotzsch	Asplenium serratum L.	BM	
Rob.		AST	Eupatorium schomburgkii Benth.	Baccharis brachylaenoides DC. var. brachylaenoides	tc: GH frag	
Rob. or Rich.		AST	Mikania racemulosa Benth.	Mikania psilostachya DC.	tc ?: GH	
Rob.		AST	Synedrella nodiflora (L.) Gaertn.	Synedrella nodiflora (L.) Gaertn.		
Rob. ser. 1		AST	Verbesina helianthoides Kunth			
Rob. or Rich.		AST	Vernonia dichocarpha Less. [n.n.]	Piptocoma vernonioides (Kunth) Pruski		Pakaraima Mts.
Rob. ser. 1		AST	Vernonia opaca Benth.	Piptocarpha opaca (Benth.) Baker		Roraima
Rob.		AST	Vernonia tricholepis DC. var. microphylla Benth.	Lepidaploa remotiflora (Rich.) H. Rob.		Serra Mey
Rich.		AST	Xiphochaeta aquatica Poepp. & Endl.	Stilpnopappus viridis Benth. ex Baker	U	

Collector	Date	Code	Name	Current name	Herb.	Locality
Rob. ser. 1	1838	BIG	Bignonia aequinoctialis L.	Cydista aequinoctialis (L.) Miers	BM	
Rob. or Rich.		BIG	Pyrostegia amabilis Miers	Pyrostegia dichotoma Miers ex K. Schum.		
Rich.		BIG	Spathodea consanguinea Klotzsch [n.n.]	Memora schomburgkii (DC.) Miers	tc: K	Takutu R.
Rob. or Rich.		BIG	Tabebuia insignis (Miq.) Sandw.	Tabebuia insignis (Miq.) Sandw. var. insignis		
Rob. or Rich.		BIG	Tanaecium albiflorum DC.	Tanaecium jaroba Sw.	P	
Rich.		BLE	Blechnum asplenioides Sw.	Blechnum asplenioides Sw.	BR	
Rob. ser. 1		BML	Brochinia reducta Baker	Brochinia reducta Baker	K	Essequibo R. and Mt. Warima
Rich.		BML	Puya macrostachya A. Dietr. [n.n.]	Puya altensteinii (Link, Klotzsch & Otto) Lem. var. gigantea (Hook.) Baker	B	
Rob. or Rich.		BML	Tillandsia	Pitcairnia maidifolia (C. Morren) Decne.	K	
Rob. or Rich.		BML	Tillandsia flexuosa Sw.	Tillandsia flexuosa Sw.		
Rob. or Rich.		BML	Tillandsia splendens Brongn.	Vriesea splendens (Brongn.) Lem.		
Rob. or Rich.		BML	Tillandsia usneoides (L.) L. var. filiformis (André) Mez	Tillandsia usneoides (L.) L.	type	
Rob.	1842	BMN	Dictyostega schomburgkii Miers	Dictyostega orobanchoides (Hook.) Miers subsp. orobanchioides	BM MICH U US	
Rob. or Rich.		BNN	Archytaea multiflora Benth.	Archytaea triflora Mart.		Roraima: near Brook Roné
Rob. or Rich.		BNP	Helosis guyanensis Rich.	Helosis cayenensis (Sw.) Spreng. var. cayemmensis		Berbice
Rob. or Rich.		BNP	Langsdorffia rubiginosa Wedd.	Langsdorffia hypogaea Mart.	BM	R. Orinoco
Rob. ser. 1	1836	BOR	Heliophytum passerinoides Klotzsch [n.n.]	Heliotropium ternatum Vahl	K	
Rob. or Rich.	1842	BRS	Hedwigia hostmannii Engl.	Tetragastris hostmannii (Engl.) Kuntze		
Rich.		CAB	Cabomba aquatica DC.	Cabomba aquatica DC.		
Rob. or Rich.		CAC	Cereus scandens Salm.-Dyck	Hylocereus scandens (Salm.-Dyck) Backeb.		
Rob. or Rich.		CEC	Cecropia palmata Willd.	Cecropia palmata Willd.		
Rich.		CEC	Coussapoa fagifolia Klotzsch	Coussapoa microcephala Trécul	F G	
Rob. or Rich.		CHB	Moquilea multiflora (Benth.) Walp.	Couepia multiflora Benth.	U	
Rob.		CLE	Clethra guianensis Klotzsch ex Meisn.	Clethra guianensis Klotzsch ex Meisn.	G	
Rob. ser. 1		CLU	Garcinia parviflora Benth.	Garcinia parviflora Benth.	type	Carawaimi Mts.
Rob. or Rich.		CMM	Commelina schomburgkiana Klotzsch ex Seub. var. angustifolia ('Commelyna')	Commelina schomburgkiana Klotzsch ex Seub. var.	B	
Rob. or Rich.		CNN	Rourea revoluta Planch.	Rourea revoluta Planch. var. revoluta	U	
Rob.		CNV	Aniseia ensifolia Choisy var. minor Choisy	Aniseia cernua Moric.	U	
Rob. or Rich.		CNV	Dicranostyles	Dicranostyles	BM	
Rob. or Rich.		CNV	Ipomoea evolvuloides Moric.	Jacquemontia evolvuloides (Moric.) Meisn.	U	
Rob. or Rich.	Apr 1842	CNV	Ipomoea potentilloides Meisn.	Merremia potentilloides (Meisn.) Hallier f.	tc: B	R. Branco
Rob. or Rich.		CNV	Pharbitis nil Choisy	Ipomoea nil (L.) Roth		
Rob. or Rich.		COT	Costus spicatus (Jacq.) Sw.	Costus scaber Ruiz & Pav.		
Rob. or Rich.		CSL	Cassia chrysocarpa Desv.	Senna chrysocarpa (Desv.) H.S. Irwin & Barneby		

Index 6. Robert and Richard Schomburgk's sine numero collections – continued

Collector	Year	Code	Name as cited	Accepted name	Type	Locality
Rob. ser. 1	1837	CSL	Cassia cultrifolia Kunth	Chamaecrista diphylla (L.) Greene	K	
Rob. ser. 1	1837	CSL	Cassia leiantha Benth.	Chamaecrista hispidula (Vahl) H.S. Irwin & Barneby	US	Berbice
Rich.		CSL	Cassia ramosa Vogel	Chamaecrista ramosa (Vogel) H.S. Irwin & Barneby var. ramosa	P U	
Rob. or Rich.		CSL	Cassia silvestris Vell.	Senna silvestris (Vell.) H.S. Irwin & Barneby var. silvestris	US	
Rob. or Rich.		CSL	Cassia uniflora Spreng.	Cassia uniflora Spreng.		dry savanna of the Rupununi
Rich.		CSL	Cynometra bauhiniifolia Benth. ('bauhiniaefolia')	Cynometra bauhiniifolia Benth. var. bauhiniifolia	G W	
Rich.		CSL	Cynometra marginata Benth. var. guianensis Dwyer	Cynometra marginata Benth. var. guianensis Dwyer	G U W	
Rob.		CSL	Elizabetha coccinea M.R. Schomb. ex Benth.	Elizabetha coccinea M.R. Schomb. ex Benth. var. coccinea	ht: K	Upper Essequibo R.
Rob.		CSL	Elizabetha macrostachya Benth.	Elizabetha macrostachya Benth.		
Rob.		CSL	Elizabetha princeps M.R. Schomb. ex Benth.	Elizabetha princeps M.R. Schomb. ex Benth.	ht: K	Roraima
Rich.		CSL	Outea acaciifolia Benth. ('acaciaefolia')	Macrolobium acaciifolium (Benth.) Benth.		Essequibo R. and Rupununi R. R. Branco
Rob. or Rich.		CSL	Outea multijuga DC.	Macrolobium multijugum (DC.) Benth. var. multijugum		
Rob. ser. 1		CSL	Schnella longipetala Benth.	Bauhinia longipetala (Benth.) Walp.	type	Pakaraima Mts. and Parima Mts.
Rob. ser. 1		CSL	Schnella splendens Benth.	Bauhinia guianensis Benth.		Barcelos on R. Negro
Rob.		CSL		Hymenaea oblongifolia Huber var. palustris (Ducke) Lee & Langenh.		
Rob. or Rich.		CTH	Alsophila gibbosa Klotzsch	Cyathea gibbosa (Klotzsch) Domin	tc ?: US	
Rob. or Rich.		CTH	Alsophila marginalis Klotzsch	Cyathea marginalis (Klotzsch) Domin	NY	
Rich.		CTH	Alsophila multiflora (Sw.) J. Sm.	Cyathea cyatheoides (Desv.) K.U. Kramer	B	
Rich.		CTH	Alsophila senilis Klotzsch	Sphaeropteris senilis (Klotzsch) R.M. Tryon	B	
Rich.		CTH	Hemitelia hostmanni Hook.	Cyathea cyatheoides (Desv.) K.U. Kramer	syntype ?	
Rich.		CTH	Pteris pungens Willd.	Cyathea procera (Willd.) Domin	ht: B NY frag from B	
Rich.		CTH	Sphaeropteris macrocarpa (C. Presl) R. Tryon	Cyathea macrocarpa (C. Presl) Domin	B	Mt. Roraima
Rob. ser. 1	1837	CUC	Anguria polyanthos Klotzsch [n.n.]	Gurania subumbellata (Miq.) Cogn.	BM	Roraima
Rob.		CYC	Carludovica kegeliana Lem.	Thoracocarpus bissectus (Vell.) Harling	K	Barima R.
Rob. ser. 1		CYC	Carludovica nana Gleason	Dicranopygium nanum (Gleason) Harling	tc: K CGE	Serra Mey
Rob. or Rich.		CYP	Fimbristylis dichotoma (L.) Vahl	Fimbristylis dichotoma (L.) Vahl		

Collector	Date	Name (as labelled)	Fam.	Accepted name	Herb.	Locality
Rob. or Rich.		Hypolyrum macrophyllum Boeck.	CYP	Mapania macrophylla (Boeck.) H. Pfeiff.	K	
Rob. or Rich.		Isolepis lantana Kunth	CYP	Bulbostylis lantana (Kunth) C.B. Clarke	U	
Rob. or Rich.		Lagenocarpus guianensis Nees	CYP	Lagenocarpus guianensis Lindl. & Nees ex Nees	type	
Rich.		Lagenocarpus tremulus Nees	CYP	Lagenocarpus rigidus (Kunth) Nees subsp. tremulus (Nees) T. Koyama & Maguire	ht: NY	
Rob. or Rich.		Scleria bracteata Cav.	CYP	Scleria bracteata Cav.		
Rob. or Rich.		Scleria scabra Willd.	CYP	Scleria scabra Willd.		
Rob. or Rich.		Aspidium denticulatum Sw.	DRY	Arachniodes denticulata (Sw.) Ching		Roraima
Rich.		Aspidium guianense Klotzsch	DRY	Cyclodium guianense (Klotzsch) van der Werff ex L.D. Gómez	BM	
Rich.		Aspidium hookerii Klotzsch	DRY	Cyclodium meniscioides (Willd.) C. Presl var. meniscioides	BM	
Rich.		Polybotrya caudata Kunze	DRY	Polybotrya caudata Kunze	B NY	Roraima
Rich.		Lindsaea dubia Spreng.	DST	Lindsaea dubia Spreng.	K	
Rich.	1842	Lindsaea falcata Dryand.	DST	Lindsaea lancea (L.) Bedd. var. falcata (Dryand.) Rosenst.	BM	Mt. Roraima
Rich.		Lindsaea falciformis Hook.	DST	Lindsaea falciformis Hook.	ht: K; it BM	Mt. Roraima
Rich.		Lindsaea reniformis Dryand.	DST	Lindsaea reniformis Dryand.	K	Mt. Roraima
Rob. or Rich.		Lindsaea rigescens Willd.	DST	Lindsaea stricta (Sw.) Dryand. var. stricta	BM	
Rich.		Lindsaea rufescens Kunze	DST	Lindsaea portoricensis Desv.	BM	
Rich.		Lindsaea schomburgkii Klotzsch	DST	Lindsaea schomburgkii Klotzsch	L	
Rob. or Rich.		Cavendishia duidae A.C. Sm.	ERI	Cavendishia callista J.D. Sm.	BM	
Rob. ser. 1	1839	Psammisia guianensis Klotzsch	ERI	Psammisia guianensis Klotzsch	U US frag	
Rob. or Rich.		Eriocaulon brevifolium Klotzsch ex Körn.	ERO	Eriocaulon klotzschii Moldenke		
Rich.		Eriocaulon humboldtii Kunth	ERO	Eriocaulon humboldtii Kunth		
Rich.		Paepalanthus eriocephala Klotzsch [n.n.]	ERO	Leiothrix flavesens (Bong.) Ruhland		
Rob. or Rich.		Paepalanthus guianensis Klotzsch [n.n.]	ERO	Paepalanthus dichotomus Klotzsch ex Körn.		
Rich.		Paepalanthus schomburgkii Klotzsch ex Körn.	ERO	Syngonanthus longipes Gleason		
Rob.		Paepalanthus umbellatus (Lam.) Kunth	ERO	Syngonanthus umbellatus (Lam.) Ruhland		
Rich.		Anisophyllum thymifolium Haw.	EUP	Euphorbia thymifolia L.	U	
Rob. or Rich.	Apr 1843	Dactylostemon schomburgkii Klotzsch	EUP	Actinostemon schomburgkii (Klotzsch) Hochr.	P K	
Rob. or Rich.		Discocarpus essequiboensis Klotzsch	EUP	Discocarpus essequiboensis Klotzsch	K L	
Rob.		Euphorbia erythrocarpa Klotzsch [n.n.]	EUP	Euphorbia dioeca Kunth	L U	
Rich.		Euphorbia graminea Jacq.	EUP	Euphorbia graminea Jacq.	L	Demerara R.
Rob. or Rich.		Euphorbia pilulifera L.	EUP	Euphorbia hirta L.	K	Rupununi R.
Rich.		Euphorbia thymifolia L.	EUP	Euphorbia thymifolia L.	U	
Rich.		Gymnanthes guianensis Klotzsch ex Muell. Arg.	EUP	Sebastiana guyanensis (Muell. Arg.) Muell. Arg.	K	Manakobi; Courantyne R.

Index 6. Robert and Richard Schomburgk's sine numero collections – continued

Collector	Year	Family	Name as on collection	Current name	Herb.	Locality
Rob. or Rich.		EUP	Jatropha urens L.	Cnidoscolus urens (L.) Arthur		
Rob. ser. 1		EUP	Macrocroton cuneatus Klotzsch [n.n.]	Croton cuneatus Klotzsch	st: B†	
Rob. or Rich.		EUP	Phyllanthus adianthoides Klotzsch	Phyllanthus adianthoides Klotzsch		Roraima
Rob. or Rich.		FAB	Alexandra imperatricis R.H. Schomb.	Alexa imperatricis (R.H. Schomb.) Baill.		Wanamu R.
Rob. ser. 1		FAB	Centrosema macrocarpum Benth.	Centrosema macrocarpum Benth.		
Rob. ser. 2		FAB	Clitoria leptostachya Benth.	Clitoria leptostachya Benth.		Upper Courantyne R.
Rob. or Rich.		FAB	Clitoria poitaei DC.	Clitoria arborescens R. Br.		
Rob. ser. 1		FAB	Crotalaria anagyroides Kunth	Crotalaria micans Link		R. Branco
Rob. ser. 1		FAB	Crotalaria genistella Kunth	Crotalaria pilosa P. Mill.		Upper Rupununi R.; moist savanna
Rob. ser. 1		FAB	Crotalaria leptophylla Benth.	Crotalaria maypurensis Kunth		Rupununi savanna
Rob. ser. 1		FAB	Desmodium elatum Kunth	Desmodium asperum (Poir.) Desv.		
Rob. or Rich.		FAB	Dipteryx crassifolia Spruce ex Benth.	Taralea crassifolia (Spruce ex Benth.) Ducke		
Rich.		FAB	Drepanocarpus inundatus Mart.	Macherium inundatum (Mart. ex Benth.) Ducke	BM G P Q	
Rob. or Rich.		FAB	Glycine crinita Kunth	Eriosema crinitum (Kunth) G. Don		
Rich.		FAB	Lonchocarpus densiflorus Benth.	Lonchocarpus densiflorus Benth.		
Rob. ser. 1		FAB	Machaerium nervosum Vogel	Macherium quinata (Aubl.) Sandw. var. parviflorum (Benth.) Rudd	W	
Rob. ser. 1		FAB	Neurocarpum flagellare Benth.	Clitoria flagellaris (Benth.) Benth.	ht: K	R. Branco
Rob. ser. 1		FAB	Neurocarpum longifolium Mart. ex Benth.	Clitoria guianensis (Aubl.) Benth.		
Rob. ser. 1		FAB	Ormosia aff. coccinea (Aubl.) Jackson	Ormosia aff. coccinea (Aubl.) Jackson		Pakaraima Mts.
Rob. or Rich.		FAB	Phaseolus lasiocarpus Mart. ex Benth.	Vigna juruana (Harms) Verdcourt		
Rob. ser. 2		FAB	Sabinea florida (Vahl) DC.			
Rob. or Rich.		FAB	Swartzia microstylis Benth.	Swartzia dipetala Willd. ex Vogel	ht: K	dry savanna
Rob. ser. 1		FLC	Carpotroche surinamensis Uittien	Carpotroche surinamensis Uittien	K	Kwitaro R.
Rob. ser. 1		FLC	Mayna laxiflora Benth.	Oncoba maynensis Poepp. & Endl. ex Eichl.	ht: K; it W	R. Orinoco; Esmeralda
Rob. ser. 1		GEN	Coutoubea reflexa Benth.	Coutoubea reflexa Benth.		
Rob. ser. 1		GEN	Lisianthus chelonoides L.f.	Irlbachia alata (Aubl.) Maas subsp. alata	type	moist savanna
Rob. ser. 1		GEN	Schultesia brachyptera Cham.	Schultesia brachyptera Cham.		
Rob. ser. 1		GEN	Schultesia stenophylla Mart.	Schultesia guianensis (Aubl.) Malme		moist savanna
Rob.	1839	GEN	Tachia gracilis Benth.	Tachia gracilis Benth.	type	
Rob. or Rich.		GEN	Tachia schomburgkiana Benth.	Tachia schomburgkiana Benth.	syntype	
Rob. ser. 1		GEN	Voyria acuminata Benth.	Voyria acuminata Benth.	ht: K	
Rob. or Rich.		GLC	Mertensia immersa Kaulf.	Gleichenia immersa (Kaulf.) Spreng.		Parima Mts.
Rich.		GLC	Mertensia pectinata Willd.	Dicranopteris pectinata (Willd.) Underw.	U	

Collector	No.	Code	Name on specimen	Accepted name	Herbarium	Locality
Rob. or Rich.		GLC	Mertensia pedalis Kaulf.			
Rich.		GLC	Sticherus bifidus (Willd.) Ching	Gleichenia bifida (Willd.) Spreng.	B	Roraima
Rob. or Rich.		GMM	Grammitis furcata Hook. & Grev.	Cochlidium furcatum (Hook. & Grev.) C. Chr.		
Rich.		GMM	Melpomene pilosissima (M. Martens & Galeotti) A.R. Sm. & R.C. Moran		st: B	
Rob. or Rich.		GMM	Polypodium subsessile Baker	Grammitis subsessilis (Baker) Morton	st: K	Mt. Roraima
Rob. or Rich.		GMM	Polypodium trichomanoides Sw.	Grammitis trichomanoides (Sw.) Ching	B	Roraima
Rob. ser. 1		GSN	Alloplectus patrisii DC.	Drymonia coccinea (Aubl.) Wiehler		
Rob. ser. 1		GSN	Besleria laxiflora Benth.	Besleria laxiflora Benth.		
Rob. ser. 1		GSN	Centrosolenia hirsuta Benth.	Nautilocalyx cordatus (Gleason) L.E. Skog		
Rob. ser. 1		GSN	Tapina	Tapina		
Rob. ser. 1		GSN	Tussacia rupestris Benth.	Chrysothemis rupestris (Benth.) Leeuwenb.		
Rob. ser. 1		GSN	Tussacia villosa Benth.	Chrysothemis villosa (Benth.) Leeuwenb.		Kanuku Mts.
Rob. or Rich.		HMP	Didymoglossum sphenoides C. Presl	Trichomanes punctatum Poir. subsp. labiatum (Jenman) W. Boer	B	Bartica
Rich.		HMP	Hymenophyllum polyanthos (Sw.) Sw.	Hymenophyllum polyanthos (Sw.) Sw.	U	
Rich.		HMP	Hymenophyllum trichmanoides Bosch	Hymenophyllum trichmanoides Bosch	st: BM CAY	
Rich.	1843	HMP	Trichomanes ankersii Parker ex Hook. & Grev.	Trichomanes ankersii Parker ex Hook. & Grev.	B	Demerara R.
Rich.		HMP	Trichomanes crispum L.	Trichomanes crispum L.	U	
Rich.		HMP	Trichomanes pedicellatum Desv.	Trichomanes subsessile Splitg.	P	Roraima
Rich.		HMP	Trichomanes prieurii Kunze	Trichomanes elegans Rich.	st: K L	
Rob. or Rich.		HMP	Trichomanes tenerum Spreng.	Trichomanes tenerum Spreng.		Parima Mts.
Rob. ser. 1		HMP	Trichomanes vittaria Poir.	Trichomanes vittaria Poir.		Carawaimi Mts.
Rob. or Rich.		HPC	Hippocratea micrantha Camb.	Elachyptera floribunda (Benth.) A.C. Sm.	ht: B; it: NY	Upper Rupununi R.
Rob. or Rich.		HPC	Tontelea longifolia Miers	Tontelea attenuata Miers		
Rob. or Rich.		HUM	Humiria floribunda Mart. ('Humirium floribundum')	Humiria balsamifera (Aubl.) J. St.-Hil. var. floribunda (Mart.) Cuatrec.		
Rob. or Rich.		HUM	Humiria laurina Klotzsch ex Urb. ('Humirium laurinum')		K	Upper Rupununi. Parinari Mts.
Rob. or Rich.		HUM	Humiria obovata Benth. ('Humirium obovatum')	Humiriastrum obovatum (Benth.) Cuatrec.		
Rob. ser. 1		LAM	Hyptis laciniata Benth.	Hyptis laciniata Benth.		Pakaraima Mts.
Rob. ser. 1		LAM	Hyptis simplex A. St.-Hil. ex Benth.	Hyptis simplex A. St.-Hil. ex Benth.		Pakaraima Mts.
Rich.		LAU	Aydendron firmulum Nees	Aniba firmula (Nees & Mart.) Mez		Demerara R.
Rob. or Rich.		LAU	Goeppertia reflectens Nees	Endlicheria reflectens (Nees) Mez		
Rich.		LAU	Nectandra pallida Nees	Nectandra amazonum Nees	P	
Rob. ser. 1	1838	LCY	Bertholletia excelsa Humb. & Bonpl.	Bertholletia excelsa Humb. & Bonpl.	ht: B; it: K BM CGE G K L P	

Index 6. Robert and Richard Schomburgk's sine numero collections – continued

Rob. or Rich.		LCY	Gustavia augusta L. var. guianensis O. Berg	Gustavia augusta L.	tc: K MEL	
Rob. or Rich.		LCY	Lecythis schomburgkii O. Berg	Lecythis schomburgkii O. Berg	B	
Rob. ser. 1	1839	LIL	Hippeastrum solandriflorum Herb. ('solandraeflorum')		BM	
Rob. or Rich.		LIL	Isidrogalvis guianensis Klotzsch [n.n.]	Isidrogalvis schomburgkiana (Oliver) Cruden		
Rich.		LNT	Utricularia cambelliana Oliver	Utricularia cambelliana Oliver	tc: K	
Rich.		LOG	Antonia ovata Pohl	Antonia ovata Pohl		Roraima
Rich.		LOG	Pagamea guianensis Aubl.	Strychnos guianensis (Aubl.) Mart.		Roraima
Rob.		LOG	Strychnos gardneri A. DC.	Strychnos gardneri A. DC.		
Rich.		LOM	Acrostichum lingua Raddi	Elaphoglossum lingua (Raddi) Brack.	K	Upper Courantyne R.
Rich.		LOM	Acrostichum cuspidatum Willd.	Elaphoglossum cuspidatum (Willd.) T. Moore		Berbice
Rich.		LOM	Acrostichum flaccidum Fée	Elaphoglossum rigidum (Aubl.) Urb.	BM	
Rich.		LOM	Acrostichum nigrescens Hook.	Elaphoglossum nigrescens (Hook.) T. Moore		
Rob. ser. 1	1837	LOM	Acrostichum plumosum Fée	Elaphoglossum plumosum (Fée) T. Moore	tc: K	Roraima
Rich.		LOM	Leptochilus nicotianifolia (Sw.) C. Chr.	Bolbitis nicotianifolia (Sw.) Alston	K	Berbice
Rich.		LOR	Struthanthus guianensis Klotzsch [n.n.]	Struthanthus syringifolius Mart. ('syringaefolius')	K	
Rob.		LOR	Struthanthus pyrifolius (Kunth) G. Don	Phthirusa pyrifolia (Kunth) Eichl.		
Rich.		LOR	Struthanthus terniflorus (Willd.) Klotzsch	Struthanthus dichotrianthus Eichl.		
Rich.		LOR	Struthanthus triceps Klotzsch [n.n.]	Struthanthus syringifolius Mart. ('syringaefolius')		
Rob.		LYT	Cuphea antisyphilitica Kunth	Cuphea antisyphilitica Kunth var. acutifolia Benth.	type	
Rob. or Rich.		MEL	Guarea aubletti A. Juss.	Guarea guidonia (L.) Sleumer		
Rob. or Rich.		MEL	Melia azedarach L.	Melia azedarach L.	BM	
Rob. or Rich.		MIM	Calliandra hookeriana R.H. Schomb.	Calliandra rigida Benth.		
Rob. or Rich.		MIM	Inga umbellifera (Vahl) Steud. ex DC.	Inga umbellifera (Vahl) Steud. ex DC.	ht: K	Kamarang R.
Rob. or Rich.		MIM	Mimosa microcephala Humb. & Bonpl. ex Willd.	Mimosa microcephala Humb. & Bonpl. ex Willd.		
Rob. ser. 1	1837	MLP	Tetrapterys discolor (G. Mey.) DC.	Tetrapterys discolor (G. Mey.) DC.	BM	Parima Mts.
Rob. ser. 1		MLS	Diplochita bracteata DC.	Miconia holosericea (L.) DC.		
Rob.		MLS	Macairea pachyphylla Benth.	Macairea pachyphylla Benth.	U	R. Negro
Rob. ser. 1		MLS	Macairea thyrsifolia DC.	Macairea thyrsifolia DC.		Mt. Roraima
Rob.		MLS	Meisneria cordifolia Benth.	Siphanthera cordifolia (Benth.) Gleason		R. Padauiri
Rob. ser. 1		MLS	Miconia nitens Benth.	Tococa nitens (Benth.) Triana		
Rob.		MLS	Miconia prasina (Sw.) DC.	Miconia prasina (Sw.) DC.	ht: K	sandstone regions
Rob. ser. 1		MLS	Miconia rubiginosa (Bonpl.) DC.	Miconia rubiginosa (Bonpl.) DC.		
Rob.		MLS	Microlicia recurva (Rich.) DC.	Acisanthera uniflora (Vahl) Gleason	U	
Rob.		MLS	Poteranthera pauciflora (Naudin) Wurdack	Acisanthera crassipes (Naudin) Wurdack	tc: P	R. Padauiri
Rich.		MLS	Rhynchanthera acuminata Benth. var. sublaevis Cogn.	Rhynchanthera grandiflora (Aubl.) DC.	ht: BR	

Collector	Year	Code	Original name	Current name	Type	Locality
Rich.		MLS	Rhynchanthera ambigua Naudin	Rhynchanthera grandiflora (Aubl.) DC.	ht: P	Mt. Roraima
Rob. ser. 1		MLS	Salpinga secunda Schranck	Salpinga secunda Schranck	type	Serra Mey
Rob. ser. 1		MLS	Spennera circaeoides Benth.	Aciotis purpurascens (Aubl.) Triana		R. Parimé or R. Parima
Rob.		MLS	Tibouchina aspera Aubl.	Tibouchina aspera Aubl. var. aspera		
Rob. ser. 1		MLS	Tococa roreimi Benth.	Tococa guianensis Aubl.	ht: K	Mt. Roraima
Rob. or Rich.		moss	Cryptangium schomburgkii Müll.	Hydropogonella gymnostoma (Schimp.) Card.		
Rob. or Rich.		moss	Octoblepharum cylindricum Mont.	Octoblepharum cylindricum Mont.		
Rob. ser. 1		MRC	Norantea guianensis Aubl.	Norantea guianensis Aubl.		Berbice R.
Rob. or Rich.		MRN	Ischnosiphon parkeri Körn.	Ischnosiphon parkeri Körn.		Aruka R.
Rob. or Rich.		MRN	Marantha obliqua Rudge	Ischnosiphon obliquus (Rudge) Körn.		savanna
Rob. or Rich.		MRS	Weigeltia grandiflora Mez	Cybianthus grandiflora (Mez) Agostini	tc: K	Kumut (Cumuti) Mts.
Rob. or Rich.		MRT	Eugenia sinamariensis Aubl.	Eugenia		Parima Mts.
Rob. or Rich.		MRT	Eugenia vismeaefolia Benth.	Myrciaria vismiifolia (Benth.) O. Berg		Kwitaro R.
Rich.		MRT	Marlierea schomburgkiana O. Berg	Marlierea schomburgkiana O. Berg	tc: B	Roraima
Rob. or Rich.		MRT	Myrcia subcordata DC.	Myrcia subcordata DC.		
Rob. or Rich.		MRT	Myrcia tomentosa (Aubl.) DC.	Myrcia tomentosa (Aubl.) DC.	BM	
Rob. or Rich.		MRT	Psidium aromaticum Aubl.	Campomanesia aromatica (Aubl.) Griseb.		
Rob. or Rich.		MRT	Psidium ciliatum Benth.	Psidium salutare (Kunth) O. Berg		
Rob. or Rich.		MRT	Psidium pyriferum L.	Psidium guajava L.		dry savanna
Rob. or Rich.		MTT	Danaea simplicifolia Rudge	Danaea simplicifolia Rudge		
Rob. ser. 1	1837	MTX	Amphidesmium blechnoides (Sw.) Klotzsch	Metaxya rostrata (Kunth) C. Presl	BM NY	
Rob. or Rich.	1841	MYS	Myristica sebifera Sw.	Virola sebifera Aubl.	G P	
Rich.		NYC	Pisonia eggersiana Heimerl	Guapira eggersiana (Heimerl) Lundell		
Rob. ser. 1	1837	NYM	Victoria regia Lindl.	Victoria amazonica (Poepp.) Sowerby	type	Berbice R.
Rob. ser. 1	1837	OCH	Gomphia schomburgkii Planch.	Ouretea schomburgkii (Planch.) Engl.		Berbice R. and Demerara R.
Rob. ser. 1		OCH	Olax macrophylla Benth.	Dulacia macrophylla (Benth.) Kuntze	ht: K	R. Negro; Mt. Padowan; R. Padauiri
Rich.		OLN	Aspidium nodosum Willd.	Oleandra articulata (Sw.) C. Presl	BM UC	
Rich.	1842	OLN	Nephrolepis biserrata (Sw.) Schott	Nephrolepis biserrata (Sw.) Schott	NY	Roraima
Rob. or Rich.		OLN	Nephrolepis pendula (Raddi) J. Sm.	Nephrolepis cordifolia (L.) C. Presl	UC	
Rob. or Rich.		ORC	Aganisia pulchella Lindl.	Aganisia pulchella Lindl.		Demerara R.
Rob. or Rich.		ORC	Anagraecum fasciola Lindl.	Campylocentrum fasciola (Lindl.) Cogn.		
Rob. or Rich.		ORC	Bifrenaria aurantiaca Lindl.	Rudolfiella aurantiaca (Lindl.) Hoehne		along Essequibo R. and Pomeroon R.
Rob. or Rich.		ORC	Bifrenaria longicornis Lindl.	Bifrenaria longicornis Lindl.		Demerara R.
Rob. or Rich.		ORC	Bulbophyllum exaltatum Lindl.	Bulbophyllum exaltatum Lindl.		
Rob.		ORC	Bulbophyllum quadrisetum Lindl.	Bulbophyllum quadrisetum Lindl.		
Rob. or Rich.		ORC	Catasetum deltoideum (Lindl.) Mutel	Catasetum deltoideum (Lindl.) Mutel	type	Berbice

Index 6. Robert and Richard Schomburgk's sine numero collections – continued

Collector	Year		Name cited	Current name	Type	Locality
Rob. or Rich.		ORC	Catasetum longifolium Lindl.	Catasetum longifolium Lindl.		
Rob. or Rich.		ORC	Catasetum macrocarpum Rich. ex Kunth	Catasetum macrocarpum Rich. ex Kunth		
Rob. ser. 1		ORC	Catasetum poriferum Lindl.	Catasetum poriferum Lindl.		
Rob. or Rich.		ORC	Catasetum saccatum Lindl.	Catasetum saccatum Lindl.		
Rob. ser. 1		ORC	Cleistes parviflora Lindl.	Cleistes parviflora Lindl.		
Rob. ser. 1		ORC	Cleistes rosea Lindl.	Cleistes rosea Lindl.		
Rob. or Rich.		ORC	Cleistes rosea Lindl.	Cleistes rosea Lindl.		savanna near lake Capoochy
Rich.		ORC	Cypripedium klotzscheanum Rchb.f.	Phragmipedium klotzscheanum (Rchb.f.) Rolfe	tc: BM L	
Rob. ser. 1		ORC	Cypripedium lindleyanum R.H. Schomb.	Phragmipedium lindleyanum (R.H. Schomb.) Rolfe		Roraima
Rob. ser. 1	1839	ORC	Cypripedium palmifolium Lindl.	Selenipedium palmifolium (Lindl.) Rchb.f.	BM P	sandy savanna
Rob. ser. 1		ORC	Diothonea imbricata Lindl.	Hexisea imbricata (Lindl.) Rchb. f.		Roraima Mts.
Rob. or Rich.		ORC	Elleanthus lilifolius C. Presl	Elleanthus graminifolius (Barb. Rodr.) Lojtnant		Roraima
Rob.		ORC	Epidendrum chloranthum Lindl.	Encyclia chloroleuca (Hook.) Neum.		Demerara R.
Rob.		ORC	Epidendrum clavatum Lindl.	Epidendrum purpurascens Focke		
Rob. ser. 1		ORC	Epidendrum decipiens Lindl.	Epidendrum ibaguense Kunth		
Rob.		ORC	Epidendrum durum Lindl.	Epidendrum dendrobioides Thunb.		
Rob. or Rich.		ORC	Epidendrum fuscatum Lindl.	Epidendrum anceps Jacq.		
Rob.		ORC	Epidendrum longicolle Lindl.	Epidendrum longicolle Lindl.		
Rob. ser. 1		ORC	Epidendrum microphyllum Lindl.	Epidendrum microphyllum Lindl.	type	Roraima
Rob. or Rich.		ORC	Epidendrum orchidiflorum Salzm. ex Lindl.	Epidendrum orchidiflorum Salzm. ex Lindl.		Berbice R.
Rob.		ORC	Epidendrum pachyanthum Lindl.	Encyclia pachyantha (Lindl.) Hoehne		
Rob. ser. 1		ORC	Epidendrum schomburgkii Lindl.	Epidendrum macrocarpum (Rich.) Garay	ht: K	
Rob.		ORC	Epidendrum viviparum Lindl.	Epidendrum viviparum Lindl.	type	
Rob. ser. 1		ORC	Epistephium parviflorum Lindl.	Epistephium parviflorum Lindl.	type	Lake Tapakuma and along Berbice R.
Rob. or Rich.		ORC	Fernandezia elegans Lodd.	Lockhartia imbricata (Lam.) Hoehne		
Rob. or Rich.		ORC	Galeandra devoniana R.H. Schomb. ex Lindl.	Galeandra devoniana R.H. Schomb. ex Lindl.		Barcelos on R. Negro
Rob. ser. 1		ORC	Habenaria heptadactyla Rchb.f.	Habenaria heptadactyla Rchb.f.		Orinoco R.
Rob. or Rich.		ORC	Habenaria seticauda Lindl.	Habenaria seticauda Lindl.		Pirara
Rob. or Rich.		ORC	Houlletia vittata Lindl.	Braemia vittata (Lindl.) Jenny	type	Acarai Mts.
Rob. or Rich.		ORC	Huntleya sessiliflora Batem. ex Lindl.	Huntleya sessiliflora Batem. ex Lindl.	type	
Rob. or Rich.		ORC	Huntleya violacea Lindl.	Bollea violacea (Lindl.) Rchb.f.		Essequibo R.
Rob. or Rich.		ORC	Maxillaria alba Lindl.	Maxillaria alba Lindl.		
Rob. ser. 1		ORC	Maxillaria eburnea Lindl.	Maxillaria eburnea Lindl.	MO	Mt. Maravaca

Collector	Year	Family	Name	Current name	Herb.	Locality
Rob. or Rich.		ORC	Maxillaria monacensis Kraenzl.	Maxillaria monacensis Kraenzl.	MO	Essequibo R.
Rob. or Rich.		ORC	Maxillaria steelii Hook.	Scuticaria steelii (Hook.) Lindl.	type	Mt. Roraima
Rob. or Rich.		ORC	Oncidium iridifolium Kunth	Psygmorchis pusilla (L.) Dodson & Dressler	type	Pirara
Rob. or Rich.		ORC	Oncidium nigratum Lindl. & Paxt.	Oncidium nigratum Lindl. & Paxt.	type	Mt. Roraima
Rich.	1842	ORC	Oncidium pirarense Rchb.f.	Oncidium pirarense Rchb.f.		Barima R.
Rob. or Rich.		ORC	Oncidium pulchellum Hook.	Oncidium pulchellum Hook.		
Rob. or Rich.		ORC	Paphinia cristata (Lindl.) Lindl.	Paphinia cristata (Lindl.) Lindl.		
Rob. or Rich.		ORC	Peristeria pendula Hook.	Peristeria pendula Hook.		
Rob. ser. 1		ORC	Physurifolia physurifolia Rchb.f.	Psilochilus physurifolius (Rchb. f.) Løjtnant	tc: K	
Rob.		ORC	Pleurothallis stenopetala Lodd. ex Lindl.	Pleurothallis sclerophylla Lindl.	type	Christmas cataracts on the Berbice R.
Rob. ser. 1		ORC	Pogonia surinamensis Lindl.	Triphora surinamensis (Lindl. ex Benth.) Britton		
Rob.		ORC	Schomburgkia crispa Lindl.	Schomburgkia marginata Lindl.	type	Demerara R.
Rob.		ORC	Schomburgkia gloriosa Rchb.f.	Schomburgkia gloriosa Rchb.f.	US	Essequibo R.
Rob.		ORC	Schomburgkia marginata Lindl.	Schomburgkia marginata Lindl.		
Rob.		ORC	Sobralia liliastrum Lindl.	Sobralia liliastrum Lindl.	type	
Rob. or Rich.		ORC	Sobralia sessilis Lindl.	Sobralia sessilis Lindl.		
Rob. or Rich.		ORC	Sobralia stenophylla Lindl.	Sobralia stenophylla Lindl.	type	
Rob. or Rich.		ORC	Spiranthes elata (Sw.) Rich.	Cyclopogon elatus (Sw.) Schltr.		
Rob. or Rich.		ORC	Spiranthes picta Lindl.	Sarcoglottis acaulis (J.E. Sm.) Schltr.	type	
Rob. ser. 1		ORC	Vanilla bicolor Lindl.	Vanilla bicolor Lindl.		
Rob. or Rich.		ORC	Zygopetalum rostratum Hook.	Zygosepalum labiosum (Rich.) C. Schweinf.	L	
Rob. or Rich.		PGL	Catocoma cuneata Klotzsch	Bredemeyera cuneata Klotzsch ex Hassk.	L P	
Rob. or Rich.		PGL	Catocoma lucida Benth.	Bredemeyera lucida (Benth.) Klotzsch ex Hassk.	G U	
Rob. or Rich.		PGL	Polygala appressa Benth.	Polygala appressa Benth.	P	Pirara
Rob. or Rich.		PGL	Polygala hebeclada DC.	Polygala hebeclada DC.	G	
Rob. ser. 1	1838	PGL	Polygala hygrophila Kunth	Polygala hygrophila Kunth	L	Roraima
Rob. or Rich.		PGL	Polygala timoutou Aubl.	Polygala timoutou Aubl.		
Rob. or Rich.		PGL	Polygala variabilis Kunth	Polygala variabilis Kunth	G	
Rob.		PGL	Polygala violacea Aubl.	Polygala violacea Aubl. emend. Marques	G	Pirara
Rob. or Rich.		PIP	Artanthe oblongifolia Klotzsch	Piper oblongifolium (Klotzsch) C. DC.	BM	Parima Mts.
Rob. or Rich.		PIP	Artanthe olfersianum Kunth	Piper hispidum Sw.	NY	
Rob.		PLG	Coccoloba moritzii Klotzsch ex Meisn.	Coccoloba ovata Benth.	type	R. Parimé or R. Parima
Rob. or Rich.		PLG	Coccoloba parimensis Benth.	Coccoloba parimensis Benth.		Upper Rupununi R.
Rob. or Rich.		PLG	Coccoloba pubescens L.	Coccoloba pubescens L.	tc ?: NY	
Rob. or Rich.		PLG	Ruprechtia tenuifolia Benth.	Ruprechtia tenuifolia Benth.	B	
Rich.	1843	PLG	Symmeria paniculata Benth.	Symmeria paniculata Benth.		
Rob. ser. 1		PLG	Triplaris schomburgkiana Benth.	Triplaris americana L.	type	Kwitaro R.

Index 6. Robert and Richard Schomburgk's sine numero collections – continued

Collector	Code	Year	Name	Determination	Herbaria	Locality
Rob. or Rich.	PLP		Phymatodes schomburgkiana J. Sm.	Microgramma megalophylla (Desv.) de la Sota	tc: U	
Rich.	PLP		Polypodium angustissimum Fée	Polypodium angustissimum Fée	B	
Rob. or Rich.	PLP		Polypodium bombycinum Maxon	Polypodium bombycinum Maxon	CAY	
Rob. or Rich.	PLP	1842–43	Polypodium lepidopteris (Langsd.) Kunze	Polypodium lepidopteris (Langsd.) Kunze	BM U	
Rob. ser. 1	POA	1839	Arundinaria schomburgkii Benn.	Arthrostylidium schomburgkii (Benn.) Munro	st: GH US	Parima Mts.
Rob. ser. 1	POA		Arundinaria schomburgkii Benn.	Arthrostylidium schomburgkii (Benn.) Munro	ht: BM NY; it: GH K	Mt. Maravaca
Rob. or Rich.	POA		Bambusa surinamensis Rupr.	Bambusa vulgaris Schrad. ex J.C. Wendl.	US	
Rob. ser. 1	PON	1837	Eichhornia heterosperma Alexander	Eichhornia heterosperma Alexander	G K P	Berbice R.
Rob. or Rich.	PRT		Andripetalum sessilifolium Klotzsch	Panopsis sessilifolia (Rich.) Sandw.		Roraima
Rich.	PSL		Psilotum nudum (L.) P. Beauv.	Psilotum nudum (L.) P. Beauv.	K	
Rich.	PTR		Acrostichum aureum L.	Acrostichum aureum L.	G P TCD US	
Rob. or Rich.	PTR		Pteris propinqua Agardh	Pteris polita Link		
Rob. or Rich.	QII	1842	Quiina guianensis Aubl.	Quiina guianensis Aubl.	K	
Rob. or Rich.	RHM		Rhamnus ulei Pilg.	Rhamnus ulei Pilg.	P	
Rob.	RPT		Spathanthus unilateralis (Rudge) Desv.	Spathanthus unilateralis (Rudge) Desv.		Roraima
Rich.	RUB		Amaioua guianensis Aubl.	Amaioua guianensis Aubl.		Demerara R. and Pomeroon R.
Rob. ser. 1	RUB		Amaioua saccifera Mart.	Amaioua saccifera Mart.		R. Padauiri
Rob.	RUB		Calycophyllum stanleyanum R.H. Schomb.	Calycophyllum stanleyanum R.H. Schomb.	tc: NY	near Rupununi R. and Takutu R.
Rob. or Rich.	RUB		Coffea calycina Benth.	Morinda calycina (Benth.) Steyerm.		Curassawaka
Rob. ser. 1	RUB	1836	Coussarea schomburgkiana (Benth.) Benth. & J.D. Hook		BM	
Rich.	RUB		Malanea macrophylla Bartl. ex Griseb.	Malanea macrophylla Bartl. ex Griseb.		
Rob. or Rich.	RUB		Perama humilis Benth.	Perama humilis Benth.	syntype	Roraima
Rob. ser. 1	RUB	1837	Posoqueria longiflora Aubl.	Posoqueria longiflora Aubl.	BM	
Rob. or Rich.	RUB		Psychotria crassa Benth.	Psychotria crassa Benth.	type	Mt. Maravaca
Rob. ser. 1	RUB		Psychotria hyptoides Benth.	Psychotria hyptoides Benth.		Parima Mts.
Rob. ser. 1	RUB		Psychotria spicata Benth.	Psychotria phaneroloma Standley & Steyerm. var. phaneroloma	type	Pakaraima Mts.
Rob. Ser. 1	RUB		Remijia densiflora Benth.	Remijia densiflora Benth.	tc: K	Parima Mts. and R. Negro
Rob. or Rich.	RUB		Rondeletia capitata Benth.	Coccocypselum aureum (Spreng.) Cham. & Schltdl. var. capitatum (Benth.) Steyerm.	type	Mt. Roraima
Rob. ser. 1	RUB		Sabicea guianensis (Aubl.) Baill.	Patima guianensis Benth.	BM K	Mt. Canaupang

Author	Year	Name on specimen	Fam.	Current name	Herb./type	Locality
Rob. or Rich.		Ophiocaryon paradoxum R.H. Schomb.	SAB	Ophiocaryon paradoxum R.H. Schomb.		Essequibo R. near junction Mazaruni R. and Cuyuni R. moist savanna
Rob. or Rich.		Angelonia salicariifolia Humb. & Bonpl. ('salicariaefolia')	SCR			Parima Mts.
Rob. ser. 1		Escobedia scabrifolia Ruiz & Pav.	SCR	Escobedia scabrifolia Ruiz & Pav.	syntype	
Rob. ser. 1		Stemodia foliosa Benth.	SCR	Stemodia pratensis (Aubl.) C.P. Cowan		Serra Mey
Rob. ser. 1		Actinostachys trilateralis J. Sm.	SCZ	Schizaea penicillata Humb.		
Rob. or Rich.		Anemia villosa Humb. & Bonpl. ex Willd.	SCZ	Anemia villosa Humb. & Bonpl. ex Willd.	B	
Rob. or Rich.		Schizaea elegans (Vahl) Sw.	SCZ	Schizaea elegans (Vahl) Sw.		
Rob. or Rich.		Solanum schomburgkii Sendtn.	SOL	Solanum schomburgkii Sendtn.		
Rob. or Rich.		Solanum sempervirens Dunal	SOL	Solanum pensile Sendtn.	B	banks of the Barima R.
Rob. ser. 1		Herrania mariae Goudot	STR	Herrania lemniscata (R.H. Schomb.) R.E. Schult.	type	savanna Upper Rupununi R.
Rob. or Rich.		Melochia oblonga Benth.	STR	Melochia simplex A. St.-Hil.		
Rich.		Waltheria viscosissima A. St.-Hil.	STR	Waltheria viscosissima A. St.-Hil.	UC	Roraima
Rich.		Dryopteris funesta (Kunze) C. Chr.	TEC	Triplophyllum funestum (Kunze) Holttum var. funestum	UC	
Rich.		Aspidium gongylodes Schkuhr	THL	Thelypteris interrupta (Willd.) Iwatsuki		Berbice
Rob. or Rich.		Nephrodium subcuneatum Baker	THL	Thelypteris abrupta (Desv.) Proctor	K	Annai
Rob.		Luehea rufescens A. St.-Hil.	TIL	Luehea speciosa Willd.		Rupununi savanna
Rob. or Rich.		Turnera parviflora Benth.	TNR	Turnera odorata Rich.	type	
Rob. ser. 2		Sponia micrantha (L.) Decne.	ULM	Trema micrantha (L.) Blume		
Rob. or Rich.	1841	Alsodeia laxiflora Benth.	VIO	Rinorea brevipes (Benth.) S.F. Blake	Fl LE	
Rob. or Rich.		Alsodeia pubiflora Benth.	VIO	Rinorea pubiflora (Benth.) Sprague & Sandw. var. pubiflora	M	Roraima
Rob. or Rich.		Barbacenia alexandrinae R.H. Schomb.	VLL	Vellozia tubiflora (A. Rich.) Kunth		
Rob. or Rich.		Aegiphila salutaris Kunth	VRB	Aegiphila mollis Kunth var. mollis		
Rob. or Rich.		Clerodendron fragrans Vent.	VRB	Clerodendron fragrans Vent.		
Rob. ser. 1		Lippia schomburgkiana Schauer	VRB	Lippia origanioides Kunth		Upper Rupununi R.
Rob. or Rich.		Stachytarpheta elatior Schrad. ex Schult. & Roem. ('Stachytarpha')	VRB			
Rob. ser. 1	1837	Vitex capitata Vahl	VRB	Vitex capitata Vahl	U	Essequibo R.
Rob. or Rich.		Vitex umbrosa Sw.	VRB	Vitex compressa Turcz.		
Rich.		Hecistopteris pumila (Spreng.) J. Sm.	VTT	Hecistopteris pumila (Spreng.) J. Sm.	B K	
Rich.	1843	Taeniopsis lineata J. Sm.	VTT	Vittaria lineata (L.) J. Sm.		
Rich.		Vittaria angustifolia (Sw.) Baker	VTT	Vittaria costata Kunze	US	Mt. Roraima

202

8. REFERENCES

Agostini, G. 1980. Una nueva clasificación del género *Cybianthus* (Myrsinaceae). Acta Biol.Ven. 10(2): 129–185.

Andersson, L. 1981. Revision of the *Thalia geniculata* complex (Marantaceae). Nord. J. Bot.1: 48–56.

Anonymus, 1860. Verzeichnis der Mitglieder und Beamten der Kaiserlich Leopoldino-Carolinische Deutschen. Akademie der Naturforscher. Leopoldina 1: 57–72.

Arbo, M.M. 1995. Flora Neotropica Monograph 67: 1–156. Turneraceae Part I. *Piriqueta.* The New York Botanical Garden Press.

Austin, D.F. & P.B. Cavalcante. 1982. Convolvuláceas da Amazônia. Publicações Avulsas do Museu Goeldi, Brazil.

Barneby, R.C. & J.W. Grimes. 1996. Silk Tree, Guanacaste, Monkey's Earring. A generic system for the synandrous Mimosaceae of the Americas. Part 1. *Aberema, Albizia,* and allies. Mem. New York Bot. Gard. 74 (1): 1–292.

Barneby, R.C. & J.W. Grimes. 1997. Silk Tree, Guanacaste, Monkey's Earring. A generic system for the synandrous Mimosaceae of the Americas. Part 2. *Pithecellobium, Cojoba,* and *Zygia.* Mem. New York Bot. Gard. 74 (2): 1–149.

Bentham, G. 1839a. Enumeration of plants collected by Mr. Schomburgk, British Guiana. Ann. Nat. Hist. 2: 105–111, 441–451.

Bentham, G. 1839b. Enumeration of plants collected by Mr. Schomburgk, British Guiana. Ann. Nat. Hist. 3: 427–438.

Bentham, G. 1840. Contributions towards the Flora of South America. Enumerations of plants collected by Mr. Schomburgk in British Guiana. J. Bot. (Hooker) 2: 38–103, 127–146, 210–223, 286–324.

Bentham, G. 1841. Contributions towards the Flora of South America. Enumerations of plants collected by Mr. Schomburgk in British Guiana. J. Bot. (Hooker) 3: 212–250.

Bentham, G. 1842a. Contributions towards the Flora of South America. Enumerations of plants collected by Mr. Schomburgk in British Guiana. J. Bot. (Hooker) 4: 99–133, 321–323.

Bentham, G. 1842b. Contributions towards the Flora of South America. Enumerations of plants collected by Mr. Schomburgk in British Guiana. London J. Bot. 1: 193–203.

Bentham, G. 1842c. Notes on Mimoseae, with a synopsis of species. J. Bot. (Hooker) 4: 323–418.

Bentham, G. 1843. Contributions towards the Flora of South America. Enumerations of plants collected by Mr. Schomburgk in British Guiana. London J. Bot. 2: 42–52, 359–378, 670–674.

Bentham, G. 1844. Notes on Mimoseae, with a synopsis of species. London J. Bot. 3: 195–226.

Bentham, G. 1845a. Notes on Mimoseae, with a synopsis of species. London J. Bot. 4: 577–622.

Bentham, G. 1845b. Contributions towards the Flora of South America. Enumerations of plants collected by Mr. Schomburgk in British Guiana. London J. Bot. 4: 622–637.

Bentham, G. 1846a. Contributions towards the Flora of South America. Enumerations of plants collected by Mr. Schomburgk in British Guiana. London J. Bot. 5: 351–365.

Bentham, G. 1846b. Notes on Mimoseae, with a synopsis of species. London J. Bot. 5: 75–108.

Bentham, G. 1848. Contributions towards the Flora of Guiana. Enumerations of plants collected in British, Dutch, and French Guiana, by Sir Robert and Richard Schomburgk, Dr. Hostmann, M. Leprieur, and others. London J. Bot. 7: 116–137.

Berg, C.C. 1992. Flora of the Guianas. Series A, fascicle 11: 20. Ulmaceae. 21. Moraceae. 22. Cecropiaceae. 23. Urticaceae. 26. Casuarinaceae. Koeltz Scientific Books, Koenigstein.

Boggan, J., V. Funk, C. Kelloff, M. Hoff, G. Cremers & C. Feuillet. 1997. Checklist of the plants of the Guianas (Guyana, Surinam, French Guiana). 2nd Edition. Biological diversity of the Guianas Program. Washington D.C.

Britten, J. 1891. Obituary. London J. Bot. 29: 224.

Cowan, R.S. 1968. Flora Neotropica Monograph 1: 1–228. *Swartzia* (Leguminosae, Caesalpinioideae (Swartzieae). Hafner Press New York.

Cowan, R.S. & J.C. Lindeman, 1989. Flora of the Guianas. Series A, fascicle 7: 88. Caesalpiniaceae p.p. Koeltz Scientific Books, Koenigstein.

Cremers, G. & K.U. Kramer, 1991. Flora of the Guianas. Series B, fascicle 4. Dennstaedtiaceae. Koeltz Scientific Books, Koenigstein.

Cremers, G. & K.U. Kramer, 1993. Flora of the Guianas Series B, fascicle 6. Nephrolepidaceae. Koeltz Scientific Books, Koenigstein.

Cremers, G. & K.U. Kramer, 1993. Flora of the Guianas Series B, fascicle 6. Oleandraceae. Koeltz Scientific Books, Koenigstein.

Cremers, G., K.U. Kramer, R.C. Moran & A.R. Smith, 1993. Flora of the Guianas Series B, fascicle 6. Dryopteridaceae. Koeltz Scientific Books, Koenigstein.

Cuatrecasas, J. 1961. A taxonomic revision of the Humiriaceae. Contr. U.S. Natl. Herb. 35 (2): 1–217.

Dwyer, J.D. 1954. The Tropical American genus *Tachigalia* Aubl. (Caesalpiniaceae). Ann. Missouri Bot. Gard. 41: 223–260.

Dwyer, J.D. 1958. The New World species of *Cynometra*. Ann. Missouri Bot. Gard. 45(4): 313–345.

Ek, R.C. 1990. Flora of the Guianas. Supplementary Series, fascicle 1. Index of Guyana plant collectors. Koeltz Scientific Books, Koenigstein.

Ewan, J. 1962. Synopsis of the South American species of *Vismia*. Contr. U.S. Natl. Herb. 35(5): 293–377.

Feuillet, C. & O. Poncy, 1998. Flora of the Guianas. Series A, fascicle 20: 10. Aristolochiaceae. Royal Botanic Gardens, Kew.

Forero, E. 1983. Flora Neotropica Monograph 36: 1–208. Connaraceae. The New York Botanical Garden Press.

Fryxell, P.A. 1999. Flora Neotropica Monograph 76: 1–234. *Pavonia* (Malvaceae). The New York Botanical Garden Press.

Gates, B. 1982. Flora Neotropica Monograph 30: 1–237. *Banisteriopsis, Diplopterys* (Malpighiaceae). The New York Botanical Garden Press.

Gentry, A.H. 1992. Flora Neotropica Monograph 25 (II): 1–370. Bignoniaceae-Part II (Tribe Tecomeae). The New York Botanical Garden Press.

Gillespie, L.J. 1993. Euphorbiaceae of the Guianas: Annotated species checklist and key to the genera. Brittonia 45(1): 56–94.

Goldberg, A. 1967. The genus *Melochia* L. Contr. U.S. Natl. Herb. 34(5): 191–363.

Goodall, E.A. 1977. Sketches of Amerindian tribes, 1841–1843. British Museum publications, London.

Görts-van Rijn, A.R.A. & A.M.W. Mennega, 1994. Flora of the Guianas. Series A, fascicle 16: 110. Hippocrateaceae. Koeltz Scientific Books, Koenigstein.

Gouda, E.J. 1987. Flora of the Guianas. Series A, fascicle 3: 189. Bromeliaceae, subfamily Tillandsioideae. Koeltz Scientific Books, Koenigstein.

Grear, J.W. 1970. A Revision of the American species of *Eriosema* (Leguminosae-Lotoideae). Mem. New York Bot. Gard. 20(3): 1–98.

Hansen, B. 1993. Flora of the Guianas. Series A, fascicle 14: 107. Balanophoraceae. Koeltz Scientific Books, Koenigstein.

Harling, G. 1958. Monograph of the Cyclanthaceae. Acta Horti Berg. 18: 1–428.

Haynes, R.R. & L.B. Holm-Nielsen. 1992. Flora Neotropica Monograph 56: 1–34. The Limnocharitaceae. The New York Botanical Garden Press.

Haynes, R.R. & L.B. Holm-Nielsen. 1994. Flora Neotropica Monograph 64: 1–112. The Alismataceae. The New York Botanical Garden Press.

Hekking, W.H.A. 1988. Flora Neotropica Monograph 46: 1–207. Violaceae Part I. *Rinorea* and *Rinoreocarpus*. The New York Botanical Garden Press.

Hensold, N. 1991. Revisionary studies in the Eriocaulaceae of Venezuela. Ann. Missouri Bot. Gard. 78(2): 460–464.

Hensold, N. & A.M. Giulietti. 1991. Revision and redefinition of the genus *Rondonanthus* Herzog (Eriocaulaceae). Ann. Missouri Bot. Gard. 78(2): 441–459.

Hiepko, P. 1987. The collections of the Botanical Museum Berlin-Dahlem (B) and their history. Englera 7: 219–252.

Hiepko, P. 1993. Flora of the Guianas. Series A, fascicle 14: 102. Olacaceae. 103. Opiliaceae. Koeltz Scientific Books, Koenigstein.

Holmgren, P.K., N.H. Holmgren, and L.C. Barnett. 1990. Index Herbariorum (ed.8). Part I: The herbaria of the world. New York Botanical Garden Press.

Hopkins, H.C.F. 1986. Flora Neotropica Monograph 43: 1–124. *Parkia* (Leguminosae: Mimosoideae). The New York Botanical Garden Press.

Horn, C.N. 1994. Flora of the Guianas. Series A, fascicle 15: 197. Pontederiaceae. Koeltz Scientific Books, Koenigstein.

Irwin, H.S. 1964. Monographic studies in *Cassia* (Leguminosae-Caesalpinioideae) I. Section *Xerocalyx*. Mem. New York Bot. Gard. 12(1): 1–114.

Irwin, H.S. & R.C. Barneby. 1982. The American Cassiinae. A synoptical revision of Leguminosae tribe Cassieae subtribe Cassiinae in the New World. Mem. New York Bot. Gard. 35: 1–918.

Jansen-Jacobs, M.J. 1988. Flora of the Guianas. Series A, fascicle 4: 148. Verbenaceae. Koeltz Scientific Books, Koenigstein.

Jansen-Jacobs, M.J. & W. Meijer. 1995. Flora of the Guianas. Series A, fascicle 17: 49. Tiliaceae. Koeltz Scientific Books, Koenigstein.

Jeffrey, C. 1978. Further notes on Cucurbitaceae: IV. Some New World taxa. Kew Bull. 33: 347–380.

Johnston, I.M. 1935. Studies in Boraginaceae X. The Boraginaceae of northeastern South America. J. Arnold Arbor. 16: 1–64.

Johnston, M.C. & L.A Johnston. 1978. Flora Neotropica Monograph 20: 1–96. *Rhamnus*. The New York Botanical Garden Press.

Jones, H.G. 1973. The genus *Schomburgkia*, a study in the history and bibliography of plant taxonomy. Taxon 22: 229–239.

Judziewicz, E.J. 1990. Flora of the Guianas. Series A, fascicle 8: 187. Poaceae. Koeltz Scientific Books, Koenigstein.

Kaastra, R.C. 1982. Flora Neotropica Monograph 56: 1–198. Pilocarpinae (Rutaceae). The New York Botanical Garden Press.

Killip, E.P. 1938. The American species of Passifloraceae. Publ. Field. Mus. Nat. Hist., Bot. Ser. 19: 1–613.

Koeppen, R. & H.H. Iltis. 1962. Revision of *Martiodendron* (Cassieae, Caesalpiniaceae). Brittonia 14: 191–209.

Knuth, R. 1924. Dioscoreaceae. In A. Engler, Das Pflanzenreich IV. 43: 1–387.

206

Kobuski, C.E. 1942. Studies in the Theaceae, XII notes on the South American species of *Ternstroemia*. J. Arnold Arb. 23: 298–343.

Kopp, L.E. 1966. A taxonomic revision of the Genus *Persea* in the Western Hemisphere (*Persea*- Lauraceae). Mem. New York Bot. Gard. 14(1): 1–120.

Kral, R. 1994. Flora of the Guianas. Series A, fascicle 15: 182. Xyridaceae. Koeltz Scientific Books, Koenigstein.

Krukoff, B.A. & J. Monachino 1942. The American species of *Strychnos*. Brittonia 4: 248–322.

Krukoff, B.A. 1965. Supplemantary notes on the American species of *Strychnos*. Mem. New York Bot. Gard. 12(2): 1–94.

Kubitzki, K. 1971. *Doliocarpus, Davilla* und verwandte Gattungen. Mitt. Bot. Staatssaml. München 9: 1–105.

Kubitzki, K. & S. Renner. 1982. Flora Neotropica Monograph 31: 1–125. Lauraceae I (*Aniba* and *Aiouea*). The New York Botanical Garden Press.

Lamshed, M. 1955. The people's Garden. A centenary history of the Adelaide Botanic Garden, 1855–1955. The Botanic Garden Adelaide, South Australia.

Landrum, L.E. 1986. Flora Neotropica Monograph 45: 1–178. *Campomanesia, Pimenta, Blepharocalyx, Legrandia, Acca, Myrrhinium*, and *Luma* (Myrtaceae). The New York Botanical Garden Press.

Lasègue, A. 1970. Musée Botanique de M. Benjamin Delessert. J. Cramer, Lehre.

Lellinger, D.B. 1994. Flora of the Guianas. Series B, fascicle 3. Hymenophyllaceae. Koeltz Scientific Books, Koenigstein.

Leeuwenberg, A.J.M. 1994. A revision of *Tabernaemontana*. Two. The New World Species and *Stemnadenia*. Royal Botanic Gardens, Kew, Richmond, Surrey, England.

Lleras, E. 1978. Flora Neotropica Monograph 28: 1–73. Trigoniaceae. The New York Botanical Garden Press.

Lleras, E. 1998. Flora of the Guianas. Series A, fascicle 21: 124. Trigoniaceae. Royal Botanic Gardens, Kew.

Luteyn, J.L. 1983. Flora Neotropica Monograph 35: 1–290. Ericaceae Part I. *Cavendishia*.The New York Botanical Garden Press.

Luteyn, J.L. 1995. Flora Neotropica Monograph 66: 1–560. Ericaceae Part II. The superior-ovaried genera (Monotropoideae, Pyroloideae, Rhododendroideae, and Vaccinioideae p.p.). The New York Botanical Garden Press.

Maas, P.J.M. 1985. Flora of the Guianas. Series A, fascicle 1: 192. Musaceae. 193. Zingiberaceae. 195. Cannaceae. Koeltz Scientific Books, Koenigstein.

Maas, P.J.M., H. Maas-van de Kamer, J.van Benthem, H.C.M. Snelders, and T. Rübsamen, 1986. Flora Neotropica Monograph 42: 1–189. Burmanniaceae. The New York Botanical Garden Press.

Maas, P.J.M. & H. Maas-van de Kamer, 1989. Flora of the Guianas. Series A, fascicle 5: 174. Triuridaceae. Koeltz Scientific Books, Koenigstein.

Maas, P.J.M. & H. Maas-van de Kamer, 1994. Flora of the Guianas. Series A, fascicle 15: 198. Haemodoraceae. Koeltz Scientific Books, Koenigstein.

Maas, P.J.M. & P. Ruyters, 1986. Flora Neotropica Monograph 41: 1–93. *Voyria* and *Voyriella* (Saprophytic Gentianaceae). The New York Botanical Garden Press.

Maas, P.J.M, L.Y. Th. Westra and Collaborators. 1992. Flora Neotropica Monograph 57: 1–188. *Rollinia*. The New York Botanical Garden Press.

Maguire, B., J.J. Wurdack, & Collaborators. 1953. The Botany of the Guayana Highland, Part I. Mem. New York Bot. Gard. 8(2): 87–160.

Maguire, B., J.J. Wurdack, & Collaborators. 1957. The Botany of the Guayana Highland, Part II. Mem. New York Bot. Gard. 9(3): 1–392.

Maguire, B., J.J. Wurdack, & Collaborators. 1958. The Botany of the Guayana Highland, Part III. Mem. New York Bot. Gard. 10(1): 1–156.

Maguire, B., J.J. Wurdack, & Collaborators. 1961. The Botany of the Guayana Highland, Part IV. Mem. New York Bot. Gard. 10(2): 1–37.

Maguire, B., J.J. Wurdack, & Collaborators. 1961. The Botany of the Guayana Highland, Part IV (2). Mem. New York Bot Gard. 10(4): 1–87.

Maguire, B., J.J. Wurdack, & Collaborators. 1964. The Botany of the Guayana Highland, Part V. Mem. New York Bot. Gard. 10(5): 1–278.

Maguire, B., J.J. Wurdack, & Collaborators. 1965. The Botany of the Guayana Highland, Part VI. Mem. New York Bot. Gard. 12(3): 1–285.

Maguire, B., J.J. Wurdack, & Collaborators. 1967. The Botany of the Guayana Highland, Part VII. Mem. New York Bot. Gard. 17(1): 1–439.

Maguire, B., J.J. Wurdack, & Collaborators. 1969. The Botany of the Guayana Highland, Part VIII. Mem. New York Bot. Gard. 18(2): 1–290.

Maguire, B., J.J. Wurdack, & Collaborators. 1972. The Botany of the Guayana Highland, Part IX. Mem. New York Bot. Gard. 23: 1–832.

Maguire, B., J.J. Wurdack, & Collaborators. 1978. The Botany of the Guayana Highland, Part X. Mem. New York Bot. Gard. 29: 1–288.

Maguire, B., J.J. Wurdack, & Collaborators. 1981. The Botany of the Guayana Highland, Part XI. Mem. New York Bot. Gard. 32: 1–391.

Maguire, B., J.J. Wurdack, & Collaborators. 1984. The Botany of the Guayana Highland, Part XII. Mem. New York Bot. Gard. 38: 1–84.

Marcano-Berti, L. 1998. Flora of the Guianas. Series A, fascicle 21: 123. Vochysiaceae. 123a. Euphroniaceae. Royal Botanic Gardens, Kew.

Mitchell, J.D. 1997. Flora of the Guianas. Series A, fascicle 19: 129. Anacardiaceae. Royal Botanic Gardens, Kew.

Mitchell, J.D. & S. Mori. 1987. The cashew and its relatives (*Anacardium*: Anacardiaceae). Mem. New York Bot. Gard. 42: 1–76.

Moran, R.C. 2000. Monograph of the neotropical species of *Lomariopsis* (Lomariopsidaceae). Brittonia: 52(1): 55–111.

208

Mori, S.A. & G.T. Prance, 1992. Flora of the Guianas. Series A, fascicle 12: 53. Lecythidaceae. Koeltz Scientific Books, Koenigstein.

Morton, C.V. & D.B. Lellinger, 1966. The Polypodiaceae subfamily Asplenioideae in Venezuela. Mem. New York Bot. Gard. 15: 1–49.

Nees von Esenbeck, C.G.D. 1840. Cyperaceae a Schomburgkio in Guiana Anglica collectae, ex Herbario Lindleyano. J. Bot. (Hooker) 2: 393–399.

Payne, P. 1992. Dr Richard Schomburgk and Adelaide Botanic Garden 1865–1891. Ph.D. thesis, University of Adelaide.

Pennington, T.D. 1981. Flora Neotropica Monograph 28: 1–470. Meliaceae. The New York Botanical Garden Press.

Pennington, T.D. 1990. Flora Neotropica Monograph 52: 1–770. Sapotaceae. The New York Botanical Garden Press.

Philcox, D. 1965. Revision of the New World species of *Buchnera* L. Kew Bull. 18: 275–315.

Pipoly, J.J. 1987. A Systematic Revision of the Genus *Cybianthus* Subgenus *Grammadenia* (Myrsinaceae). Mem. New York Bot. Gard. 43: 1–76.

Prance, G.T. 1971. An index of plant collectors in Brazilian Amazonia. Acta Amazonica 1(1): 25–65.

Prance, G.T. 1972. Flora Neotropica Monograph 9: 1–410. Chrysobalanaceae. Hafner Press New York.

Prance, G.T. 1972. Flora Neotropica Monograph 10: 1–83. Dichapetalaceae. Hafner Press New York.

Prance, G.T. 1974. *Victoria amazonica* ou *Victoria regia* ? Acta Amazonica 4(1): 5–8.

Prance, G.T. 1989. Flora of the Guianas. Series A, fascicle 2: 85. Chrysobalanaceae. Koeltz Scientific Books, Koenigstein.

Prance, G.T. & M. Freitas da Silva. 1973. Flora Neotropica Monograph 12: 1–75. Caryocaraceae. Hafner Press New York. Radlkofer, L. 1921. Sapindaceae americanae novae vel emendatae. Repert. Spec. Nov. Regni Veg. 17: 355–365.

Radlkofer, L. 1931–1934. Sapindaceae. In A. Engler, Das Pflanzenreich IV. 165: 1–1539.

Rivière, P.G. 1995. Absent minded imperialism: Brittain and the Expansion of Empire in Nineteenth-century Brazil. Tauris Academic Studies. London.

Rivière, P.G. 1998. From science to imperialism: Robert Schomburgk's humanitarianism. Arch. Nat. Hist. 25(1): 1–8.

Rodway, J. 1889. The Schomburgks in Guiana. Timehri, ser. 2, 3(1): 1–29.

Robinson, H. 1990. Studies in the *Lepidaploa* complex (Vernonieae: Asteraceae) VII. The genus *Lepidaploa*. Proc. Biol. Soc. Wash. 103(2): 464–498.

Rogers, D.J. & S.G. Appan. 1973. Flora Neotropica Monograph 13: 1–272. *Manihot, Manihotoides* (Euphorbiaceae). Hafner Press New York.

Rohwer, J.G. 1986. Prodromus einer Monographie der Gattung *Ocotea* Aubl., sensu lato. Mitt. Inst. Allg. Bot. Hamburg 20: 3–278.

Rohwer, J.G. 1993. Flora Neotropica Monograph 60: 1–332. Lauraceae: *Nectandra*. The New York Botanical Garden Press.

Roon, A.C. de, 1994. Flora of the Guianas. Series A, fascicle 16: 112. Icacinaceae. Koeltz Scientific Books, Koenigstein.

Sandwith, N.Y. 1931. Contributions to the Flora of Tropical America VIII. Bull. Misc. Inform. 1931: 467–492.

Sandwith, N.Y. 1936. Contributions to the Flora of Tropical America XXVI. Bull. Misc. Inform. 1936: 210–221.

Schomburgk, M.R. 1848. Reisen in British-Guiana in den Jahren 1840–1844. 3 vols. J.J. Weber, Leipzig. Translated and edited by W.E. Roth in 1922.

Schomburgk, M.R. 1876. The Flora of British Guiana. Botanical Reminiscences. pp. 81–90.

Schomburgk, M.R. 1879. On the Urari: The deadly arrow-poison of the Macusis, Indian tribe in British Guiana. Spiller, Adelaide.

Schomburgk, R.H. 1832. Remarks on Anegada. J. Roy. Geogr. Soc. 2: 152–170.

Schomburgk, R.H. 1836. Report of an expedition into the interior of British Guayana, in 1835–1836. J. Roy. Geogr. Soc. 6: 224–284.

Schomburgk, R.H. 1837a. Diary of an ascent of the River Corentyn in British Guayana, on October 1836. J. Roy. Geogr. Soc. 7: 285–301.

Schomburgk, R.H. 1837b. Diary of an ascent of the River Berbice in British Guayana, in 1836–1837. J. Roy. Geogr. Soc. 7: 302–350.

Schomburgk, R.H. 1838. On the Ant Tree of Guiana (*Triplaris americana*). Ann. Nat. Hist. 1(4): 264–267.

Schomburgk, R.H. 1840a. A description of British Guiana geographical and statistical: Exhibiting its resources and capabilities together with the present and future condition and prospects of the colony. Simpkin, Marshall, and Co., London.

Schomburgk, R.H. 1840b. Description of the Snake-nut Tree of Guiana. Ann. Nat. Hist. 5(30): 202–204.

Schomburgk, R.H. 1841a. On the Urari, the arrow poison of the Indians of Guiana; with a description of the plant from which it is extracted. Ann. Mag. Nat. Hist. 7: 407–427.

Schomburgk, R.H. 1841b. Report of the third expedition to the interior of Guayana, comprising the journey to the sources of the Essequibo, to the Carumé Mountains, and to Fort São Joaquim, on the Rio Branco, in 1837–8. J. Roy. Geogr. Soc. 10: 159–190.

Schomburgk, R.H. 1841c. Journey from Fort São Joaquim, on the Rio Branco, to Roraima, and thence by rivers Parima and Merewari to Esmeralda, on the Orinoco, in 1838–9. J. Roy. Geogr. Soc. 10: 191–247.

210

Schomburgk, R.H. 1841d. Journey from Esmeralda, on the Orinoco, to San Carlos and Moura on the Rio Negro and thence by Fort São Joaquim to Demerara, in the spring of 1839. J. Roy. Geogr. Soc. 10: 248–267.

Schomburgk, R.H. 1842a. Expedition to the lower parts of the Barima and the Guiana Rivers in British Guiana. J. Roy. Geogr. Soc. 12: 169–178.

Schomburgk, R.H. 1842b. Excursion up the Barima and Cuyuni Rivers, in British Guiana in 1841. J. Roy. Geogr. Soc. 12: 178–196.

Schomburgk, R.H. 1843a. Information respecting Scientific Travellers. Ann. Nat. Hist. 12:190–202.

Schomburgk, R.H. 1843b. Visit to the sources of the Takutu, in British Guiana, in the year 1842. J. Roy. Geogr. Soc. 3: 18–74.

Schomburgk, R.H. 1844a. Description of a new species of *Calycophyllum* from British Guiana. London J. Bot. 3: 621–623.

Schomburgk, R.H. 1844b. Two new species of the family Laurineae from the forest of Guiana. London J. Bot. 3: 624–631.

Schomburgk, R.H. 1845a. Journal of an expedition from Pirara to the Upper Corentyne, and from thence to Demerara. J. Roy. Geogr. Soc. 15: 1–103.

Schomburgk, R.H. 1845b. A description of *Ophiocaryon paradoxum*, on the Snake Nut tree of Guiana. London J. Bot. 4: 375–378.

Schomburgk, R.H. 1847. Beschreibung dreier neuen Pflanzen aus dem Flussgebiete des Carimani oder Carimang, eines Zuflusses des Mazaruni. Linnaea 20: 751–760.

Schomburgk, R.H. 1859. Autobiographie von Robert Hermann Schomburgk. Leopoldina 1: 34–39.

Schomburgk, R.H. 1860. Fishes of British Guiana. Naturalist's Library, Edinburgh.

Schweinfurth, C. 1967. Orchidaceae of the Guyana Highland. Mem. New York Bot. Gard. 14(3): 69–214.

Silva, M.F. da, 1986. Flora Neotropica Monograph 44: 1–128. *Dimorphandra* (Caesalpiniaceae). The New York Botanical Garden Press.

Simpson, B.B. 1998. Flora of the Guianas. Series A, fascicle 21: 126. Krameriaceae. Royal Botanic Gardens, Kew.

Sleumer, H.O. 1954. Proteaceae americanae. Bot. Jahrb. Syst. 76: 139–211.

Sleumer, H.O. 1980. Flora Neotropica Monograph 22: 1–499. Flacourtiaceae. The New York Botanical Garden Press.

Smith, A.R. & R. C. Moran. 1992. *Melpomene*, a new genus of Grammitidaceae (Pteridophyta). Novon 2(4): 426–432.

Smith, A.R. 1993. Flora of the Guianas. Series B, fascicle 6. Thelypteridaceae. Koeltz Scientific Books,Koenigstein.

Smith, L.B. 1967. Bromeliaceae of the Guyana Highland. Mem. New York Bot. Gard. 14(3): 15–68.

Smith, L.B. 1974. Flora Neotropica Monograph 14 (1): 1–658. Bromeliaceae. Hafner Press New York.

Snow, N. & N. Holton, 2000. Additions to Weber's Three-Letter Family Acronyms based on results of the Angiosperm Phylogeny Group. Taxon 29: 77–78.

Stafleu, F.A. & R.S. Cowan, 1967–1988. Taxonomic literature (ed.2), 7 vols. Bohn, Scheltema & Holkema, Utrecht.

Steinberg, C.H. 1977. The collectors and collections in the Herbarium Webb. Webbia 32(1): 1–49.

Stergios, B. 1996. Contributions to South American Caesalpiniaceae. II. A Taxonomic update of *Campsiandra* (Caesalpinieae). Novon 6: 434–459.

Steyermark, J.A. 1981. Erroneous citations of Venezuelan localities. Taxon: 30: 816–817.

Steyermark, J.A. & J.L. Luteyn. 1980. Revision of the genus *Ochtocosmus*. Brittonia 32: 128–143.

Steyermark, J.A., P.E. Berry & B. K. Holst. 1995. Flora of the Venezuelan Guayana. Volume 2. Pteridophytes.

Spermatophytes Acanthaceae-Araceae. Missouri Botanical Garden. St. Louis, USA.

Steyermark, J.A., P.E. Berry & B. K. Holst. 1997. Flora of the Venezuelan Guayana. Volume 3. Araliaceae-Cactaceae. Missouri Botanical Garden. St. Louis, USA.

Steyermark, J.A., P.E. Berry & B. K. Holst. 1998. Flora of the Venezuelan Guayana. Volume 4. Caesalpiniaceae-Ericaceae. Missouri Botanical Garden. St. Louis, USA.

Steyermark, J.A., P.E. Berry & B. K. Holst. 1999. Flora of the Venezuelan Guayana. Volume 5. Eriocaulaceae-Lentibulariaceae. Missouri Botanical Garden. St. Louis, USA.

Stevenson, D. 1991. Flora of the Guianas. Series A, fascicle 9: 208. Cycadaceae. 208.1. Zamiaceae. 211. Podocarpaceae. Koeltz Scientific Books, Koenigstein.

Stevenson, D. & T. Zanoni, 1991. Flora of the Guianas. Series A, fascicle 9: 209. Gnetaceae. 210. Pinaceae. Koeltz Scientific Books, Koenigstein.

Todzia, C.A. 1988. Flora Neotropica Monograph 48: 1–138. Chloranthaceae: *Hedyosmum*. The New York Botanical Garden Press.

Tryon, A.F. 1970. A monograph of the fern genus *Eriosorus*. Contr. Gray Herb. 200: 54–174.

Weber, W.A. 1982. Mnemonic three-letter acronyms for the families of vascular plants: a device for more effective herbarium curation. Taxon 31: 74–88.

Woodson, R.E., R.W. Schery and Collaborators. Flora of Panama Part IX. Family 184. Compositae. Ann. Missouri Bot. Gard. 62: 835–1322.

Wurdack, J.J., T. Morley & S. Renner, 1993. Flora of the Guianas. Series A, fascicle 13: 99. Melastomataceae. Koeltz Scientific Books, Koenigstein.